The Cambridge Cockpit and the
Paradoxes of Fatigue, 1940–1977

The Cambridge Cockpit and the Paradoxes of Fatigue, 1940–1977

DAVID BLOOR

The University of Chicago Press
Chicago and London

The University of Chicago Press, Chicago 60637
The University of Chicago Press, Ltd., London
© 2025 by The University of Chicago
All rights reserved. No part of this book may be used or reproduced in any manner
whatsoever without written permission, except in the case of brief quotations in
critical articles and reviews. For more information, contact the University of Chicago
Press, 1427 E. 60th St., Chicago, IL 60637.
Published 2025

34 33 32 31 30 29 28 27 26 25   1 2 3 4 5

ISBN-13: 978-0-226-84232-5 (cloth)
ISBN-13: 978-0-226-84234-9 (paper)
ISBN-13: 978-0-226-84233-2 (e-book)
DOI: https://doi.org/10.7208/chicago/9780226842332.001.0001

Library of Congress Cataloging-in-Publication Data

Names: Bloor, David, author.
Title: The Cambridge cockpit and the paradoxes of fatigue, 1940–1977 / David Bloor.
Description: Chicago : The University of Chicago Press, 2025. |
    Includes bibliographical references and index.
Identifiers: LCCN 2024058209 (print) | LCCN 2024058210 (ebook) |
    ISBN 9780226842325 (cloth) | ISBN 9780226842349 (paperback) |
    ISBN 9780226842332 (ebook)
Subjects: LCSH: Aviation psychology. | Air pilots—Psychological testing. |
    Mental fatigue. | Psychological tests.
Classification: LCC TL555 .B56 2025 (print) | LCC TL555 (ebook) |
    DDC 155.9/65—dc23/eng/20250131
LC record available at https://lccn.loc.gov/2024058209
LC ebook record available at https://lccn.loc.gov/2024058210

For Max and Konrad

At the same time problems of pilot fatigue became pressing and urgent. Nobody could doubt either that they were real problems, or that existing laboratory studies could throw very little light on them.

F. C. BARTLETT

Contents

Illustrations

Figures

Tables

Introduction

Anthropologists call them "war stories." These are the tales that older members of a group tell younger members, tales the youngsters then embroider and circulate amongst themselves. They tell of victories and defeats, of bravery and cunning, of wisdom and impetuosity, and sometimes of betrayal or loss. All enduring and close-knit groups have their war stories, in one form or another. Such stories both express and create a sense of belonging, and they contribute to the cohesion of the group. To use an elegant, if dated, formulation from the anthropologist Franz Boas (1858–1942), the stories present "an autobiography of the tribe."[1] Like all autobiographical accounts, they need to be read with a critical eye.

Even those who work in scientific laboratories have their war stories. I first heard mention of the Cambridge Cockpit in just such a story when I was a student in the Cambridge Psychological Laboratory, well over fifty years ago. Because the events that were recounted were set in 1940, it was a war story in both the broad, anthropological sense and the narrower, literal sense. The story told of an eminent professor, a resourceful man of subtle intellect, and his brilliant young colleague. The country was at war and in danger of invasion. The air force was fighting a desperate, defensive battle for air superiority. The pilots were exhausted, and losses were mounting. New recruits had to be found. The right men had to be located and the pressure they were under had to be understood in order to be resisted. The worldly professor knew his way around Whitehall and had friends in high places. He persuaded the authorities to let him bring the remains of a Spitfire aircraft into his laboratory in Cambridge. His ingenious colleague turned it into a machine for measuring flying skill, endurance, and fatigue. This machine was the Cambridge Cockpit. The experimental subjects were trainee pilots of the Royal Air Force

(RAF). "Oh yes," said my informant, a fellow undergraduate who had heard the story from a research student, "they could tell who would be shot down."

I was skeptical about this final embellishment, but the image invoked—that of a Spitfire aircraft inside the Psychological Laboratory, and the interaction between the pilots and the psychologists that it must have involved—has always stayed with me. The professor in the story was Frederic Charles Bartlett (1886–1969), the first professor of experimental psychology to be appointed in Cambridge (see fig. 1.1).[2] The young colleague was Kenneth Craik (1914–1945), an Edinburgh-trained psychologist who began as a new research student in Cambridge in October 1936 (see fig. 1.2).[3] Craik arrived with the reputation of being a genius, and Bartlett soon became convinced that the attribution was well-deserved. Not only did Craik have a flair for building apparatus and a ready willingness to help others with apparatus problems, he pursued

FIGURE I.1. Frederic Bartlett (1886–1969). Bartlett was the first professor of experimental psychology at the University of Cambridge. Although primarily a social psychologist, he suggested to Kenneth Craik that pilot fatigue should be studied by using the cockpit of a real aircraft as if it were a piece of laboratory apparatus. By permission of the Royal Society of London.

FIGURE I.2. Kenneth Craik (1914–1945). Craik was an Edinburgh-trained psychologist who became a research student in Bartlett's laboratory in 1936. Craik had impressive engineering skills and, through Bartlett's influence as a member of the Flying Personnel Research Committee, was commissioned to construct what became known as the "Cambridge Cockpit." From Bartlett 1946, facing 109.

his doctoral research on sensory physiology with vigor and success. Behind Craik's breezy and enthusiastic manner, Bartlett detected a darker, less happy side. In the interests of science Craik showed an almost reckless disregard for personal safety. To investigate the origin of afterimages, he deliberately cut off the blood supply to the retina of one eye. Another experiment involved staring at the sun to create temporary blindness. His reputation as an investigator and experimentalist grew rapidly, and he was soon publishing papers with eminent Cambridge colleagues such as the Nobel Laureate physiologist E. D. Adrian (1899–1977).[4] When, in 1939, Bartlett put the Psychological Laboratory at the disposal of the war effort, Craik was given free rein.

In 1945, the war finally came to an end and the value of Craik's work had become apparent. He had continued his significant experimentation on the practical problem of the dark adaptation of the human eye.[5] He had also contributed reports to the important Flying Personnel Research Committee and to the corresponding committees in other branches of the armed forces. For the army, he had studied the problems of effective tank gunnery and anti-aircraft gunnery.[6] For the Navy, he had studied radar watchkeeping. Craik was on the Servo-Panel of the Ministry of Supply and was at work on a book in which he applied the ideas of servo-engineering to the explanation of human behavior. He embraced the technical novelty of the cybernetic approach to psychology and argued that the human operator could be understood as an intermittent correction servo.[7]

Despite the wartime pressure, in 1943 Craik had already published a small but intriguing book on scientific methodology. This work, mostly written at night in Craik's desperately limited spare time, appeared under the title *The Nature of Explanation*. The title was a gentle play on words. The explanatory process in science, said Craik, was not merely directed at natural objects, it was itself a natural object. Craik wanted an experimental and inductive approach to questions that had hitherto been the prerogative of the philosophers. In Craik's opinion, the "logical" and "analytical" methods of the philosopher were little more than a disguised form of the method of introspection that psychologists had now largely discarded. In the place of pedantic intuitions about the meanings of words, the paradoxes often generated by empirical data were to be resolved by the advance of empirical knowledge. Craik's thesis was that human thinking, like all animal thinking, involved the construction, in the brain, of an internal model of external reality. The brain was essentially a combination of analogue computers. The nervous system, he said, was to be "viewed as a calculating machine capable of modelling or paralleling external events." He gave the example of Kelvin's celebrated tide predictor. This device was a complex arrangement of wheels, strings, and pulleys that, given a body of data, could be used to calculate the Fourier coefficients needed to model the cyclical rise and fall of the tides.[8]

Bartlett made sure that, in 1944, the industrious and active Craik was appointed to the directorship of the newly established Applied Psychology Research Unit in Cambridge. The unit was financed by the Medical Research Council and was administratively distinct from the university, but Bartlett housed it in his own departmental laboratory. Craik published a short note in the *British Medical Bulletin* that was a manifesto for the new unit.[9] In future, argued Craik, the aim of the psychologist of work would go beyond finding ways of suiting the man for the job. Selection techniques, especially when bad mechanical design has made jobs unnecessarily difficult, result in high rates of rejection and unemployment. The postwar world would be better. The new priority would be to suit the job to the man. And jobs would soon be different as robots came to play an increasing role in the workplace. Here Craik cited the possibility of "robot pilots" replacing human pilots. Such robots would guide machines through the air without any trace of fatigue. Given that the editor of the *Bulletin* felt the need to add a footnote to Craik's article to explain to the reader the meaning of the word "radar," talk of "robots" must have seemed visionary indeed.

The immediate future of Cambridge psychology, however, was not to be as Bartlett envisaged it. Craik died in 1945 at the very end of the war, in fact, just one day before the officially designated "Victory in Europe Day." He was killed in a road accident when cycling along King's Parade in Cambridge. A second book, on which he was still working, and which was meant to push forward the cybernetic approach to psychological research, was never finished.[10] It is said that in the personal papers Craik left behind there were premonitions of the fatal accident.[11]

On a practical level, Craik brought to the construction of the Cambridge Cockpit a personal fascination and talent for engineering. It is less clear what Bartlett brought to the project. Bartlett once declared, "If I am to say what sort of psychologist I am, I think I can say only that I am a Cambridge psychologist." To explain what he meant, he invoked the names of his Cambridge teachers and friends William Rivers (1864–1922) and Charles Myers (1873–1946), who, he said, "put a social and ethnological stamp upon Cambridge psychology."[12] This social and ethnological orientation is evident in the title of Bartlett's early book *Psychology and Primitive Culture*, published in 1923. It contained an extensive discussion of folk stories. One of Bartlett's insights in that book is worth keeping in mind in the forthcoming discussion. It has a bearing on the story that circulated in his own laboratory—the story of the Cambridge Cockpit. Bartlett insisted that "it is not the institution that is derived from the story, but the story from the institution." He meant: A story about a society, told by itself to itself, is always part of that society, and hence, for the student of society, it is part of what has to be analyzed and explained.[13]

Some of Bartlett's admirers think that *Psychology and Primitive Culture* was his best book, but his international reputation was based on another work, his 1932 *Remembering: A Study in Experimental and Social Psychology*.[14] The centerpiece of *Remembering* was an experimental study in which subjects were required to remember a folk story that had the title *The War of the Ghosts*. What, asked Bartlett, was the relation between the social function of such stories and the manner and accuracy with which they might be recounted? This somewhat esoteric concern raises the question: What did Bartlett bring to the Cambridge Cockpit work? Expertise in folk stories is not an obvious qualification for working with fighter pilots. What use were investigations of the kind described in *Remembering* to fighting a real war? Indeed, why was Bartlett involved at all?

The answer to these questions involves filling out the limited picture so far glimpsed of the scope of Bartlett's own experimentation. When he stressed the social and ethnographic side of the Cambridge tradition he was speaking truly, but he was not being wholly candid. There were other dimensions of commitment and institutional affiliation that were at work, including his own experiments in the interwar years on aircraft-detection systems.[15] But, at a deeper level, it will emerge that Bartlett's sociological and anthropological

concerns *were* as important as Craik's constructive, apparatus-building abili-
ties. That link, however, is far from obvious. Even Bartlett struggled when he
tried to articulate it, so it seems that it was barely evident to Bartlett himself.
Nevertheless, I shall show that the link was real and important. When properly understood, the argument in *Remembering* can be used to explain (i) why
Bartlett chose to build the Cambridge Cockpit in the first place, and (ii) why
Bartlett and Craik planned the particular experiments that they did. Only
when these connections and contingencies are teased out will we understand
the actual events that were refracted through the medium of the laboratory
war story.

The existence of the Cambridge Cockpit is well known to psychologists
and historians of science, but the experimental program of which the Cockpit experiments were a part has never, to my knowledge, been subject to a
detailed study or assessment. Nor, I suspect, has the long-term influence of
this work been fully appreciated. In 1948, the Air Ministry published *Pilot Error: Some Laboratory Experiments*—by the Cambridge psychologist D. Russell Davis.[16] This document provided a useful but limited and, I shall argue, a
one-sided, account of the Cambridge work. At that time, it was not possible
for other psychologists to put this report in context because the papers of
the Flying Personnel Research Committee had not been fully declassified.
They were closed until the 1970s. Soon after their declassification, however,
the collection *Aircrew Stress in Wartime Operations*, edited by E. J. Dearnaley
and P. B. Warr, gave easy access to some of these reports, though the material
in their book mainly concerned the official response of the military authorities to the general problem of stress.[17] Nevertheless, *Aircrew Stress* included
the report of the first experiments on the Cockpit and an important study
of landing accidents, neither of which had been available before. The vital
minutes of the Flying Personnel Research Committee, however, were still lying, largely unexploited, on the shelves of the Public Record Office, and these
documents were my starting point.[18]

Before beginning my analysis of the Cockpit experiments, I want to offer
some orienting remarks. The label "Cambridge Cockpit" was used by Bartlett
and his colleagues, though originally Craik had simply called the machine
he had built a "fatigue apparatus." Obviously, the word "cockpit" can refer,
generally, to any theatre of conflict.[19] My title thus carries the hint that the
Cambridge experiments were controversial as well as being, themselves, embedded in a context of war. The hint is correct. Disagreements soon arose
about the meaning of the results generated by use of the Cockpit, and, in the
background, there rumbled long-standing theoretical and ideological conflicts. These background conflicts must be examined because they informed

the theoretical analysis that Bartlett gave to the Cockpit results. Ideas that had long found favor in Cambridge, or parts of Cambridge, were viewed with acute disfavor elsewhere. Even amongst Cambridge psychologists themselves, there were tensions.

What of the word "paradox" in the subtitle? I first noticed the tendency to describe fatigue as "paradoxical" when reading a paper by a non-Cambridge author—Helmut von Bracken's 1956 "Paradoxien der Ermüdung."[20] Bracken identified three problematic phenomena confronting any theory of fatigue. He grouped them under the headings (i) *Frisch nach Arbeit*, being refreshed rather than tired by work; (ii) *Müde ohne Arbeit*, being fatigued without having performed work; and (iii) *Weniger müde während der Arbeit als hinterher*, being less tired during work than afterward. How, asked Bracken, can these divergent effects all result from one underlying cause? He drew the conclusion that fatigue might be diagnosed but never measured. Bracken's stance was not unusual. Everybody is familiar with fatigue from everyday life, but for many years, and despite great effort, it has proven extraordinarily difficult to pin down in the laboratory: it is both omnipresent and elusive. At the beginning of the period discussed in this book, consensus regarding the nature of fatigue still eluded the scientific community. Once alerted to the word, I realized that many others, including Cambridge psychologists, had used the words "paradox" and "paradoxical" to describe the frustrating results of experiments on fatigue.

The temptation of researchers in the field of fatigue was, and still is, to try to avoid these paradoxes by a flight into convoluted, definitional precision. Craik and Bartlett, by contrast, sternly rejected this tendency as mere pedantry. "Whenever a paradox *does* arise," said Craik in his 1943 book, philosophers attribute it to inexactness, and many psychologists think they should follow this example. They seek more precise and elaborate definitions, but they

> fail to see that their remedy of exact definition may be impossible and unattainable by the very nature of the physical world and of human perception, and that their definition should be corrected in the way of greater extensiveness and denotative power, rather than greater analytical, intensive, or connotative exactitude.[21]

Empirical advance, rather than logical retreat, and experimental analysis rather than conceptual analysis, were the order of the day for Craik.

Craik's references to "the physical world" and to the "world of human perceptions" point to another endemic source of trouble for the study of fatigue. Psychologists operate in the realms of both mind and brain, and it seems necessary to assume that fatigue plagues both realms. How was this dual-ism to be handled? During the course of the twentieth century, psychologists had increasingly distanced themselves from the speculative preoccupations of the philosophers and had moved closer to the methods of thinking of the physiologist. Bartlett embraced this strategy but appreciated the disciplinary diplomacy and conceptual finesse required to make it work. Psychology, he said, was to be a biological science, but psychologists had their own agenda. They had much to learn from physiologists but also much to offer. While the focus for the physiologist was on the interior, material structure of the organism, the focus for the psychologist had to encompass the situation and surroundings of the organism.[22] For Cambridge psychologists, the question was how the organism responded to the tasks presented by the material and social environment.

Certainly, by the onset of WWII, Cambridge psychologists focused pri-marily on their subjects' active behavior. They were interested in how people and animals performed certain tasks. They sometimes solicited the opinions of their human subjects—and we shall see that this was true of the relation between Cambridge psychologists and the pilots they studied. But, unlike previous generations of psychologists, they had little interest in the subjects' more refined introspections, that is, the passive contemplation and descrip-tion of the qualitative character of their inner, psychic states. They no longer pressed their subjects to make fine distinctions or discriminations between subjective states such as images, sensations, and thoughts. The questions and answers that now interested the psychologists were designed to elicit the ca-pacity to make discriminations between external stimuli or between finely tuned responses to experimental tasks. "Behavior," therefore, encompassed not only limb movements but also various classes of verbal signals. Verbal signals were also accepted as signs and symptoms of emotionality and the propensities that are called character traits.

The Cambridge psychologists who studied fatigue took for granted many, though not all, of the commonsense relationships between the material and the mental. Just as that everyday vocabulary becomes refined outside the laboratory by a smattering of medical and anatomical knowledge, so, within the laboratory, it was refined by physiological knowledge and systematic ex-perimentation. It was considered prudent for the psychologist to learn how to gloss behavior using the currently favored physiological concepts. Psy-chologists readily employed certain broad hypotheses about the structure and function of the nervous system but learned not to commit themselves too deeply, or to become too entangled in the details of the physiological processes involved.

The traditional ways that early Cambridge psychologists studied fatigue illustrates this emerging mixture of professional caution and eclecticism. The standard textbook written for the first generation of Cambridge students of experimental psychology was Myers's *Textbook of Psychology*, first published in 1911. Although Myers saw a place for introspection, the primary orientation was toward laboratory work and was heavily influenced by the publications of German psychologists. Thus, Myers's account of fatigue was accompanied by a revealing picture of Kraepelin's formidable ergograph (see fig. 1.3). The experimental subject had to keep lifting a weight attached to the middle finger of the hand while the rest of the hand was clamped firmly in place. The finger movement was registered on the smoked surface of a rotating drum (known as a kymograph), which gave a trace that could be observed and measured. A typical trace, or "ergogram," is shown in figure 1.4. There is a progressive diminution in the size of the response, until finally the subjects declare that they can make their finger muscles do no more work. Physiologically, the muscle tissue was thought to fatigue because of the accumulation of toxins such as lactic acid generated by the exertion.

The cessation of measured movement suggested that, eventually, the muscle tissue becomes completely fatigued. But is this inference correct? Myers explained to his student readers why the inference cannot be correct; the phenomenon is central rather than peripheral:

> In the intact living body, however, the apparent fatigue produced by volitional muscular exercise seems to have a more central origin. For when a series of contractions have been volitionally obtained, and when ultimately no further contractions can be volitionally produced, they may nevertheless be evoked by electrical stimulation of the motor nerve.[23]

Myers's language in this passage moves back and forth, albeit in a somewhat stilted fashion, from a psychological vocabulary to a physiological vocabulary. The transition is aided by the fact that terms such as "stimulation" and "response" belong to both realms. The discrimination between "peripheral" and "central" processes is derived from commonsense anatomical categories or their simple extrapolation. On a practical level, the language at the interface of the two disciplines of psychology and physiology can often be employed with relatively little friction. In the quotation from Myers, we can also see how it was used to express some striking and non-commonsensical findings about muscles, muscular control, and fatigue. Central, or psychological, fatigue, it seems, might indicate that a biological mechanism has evolved that protects the peripheral muscles of the body from a more radical insult and a more destructive form of physiological fatigue. Subjective fatigue seems to be a warning signal.

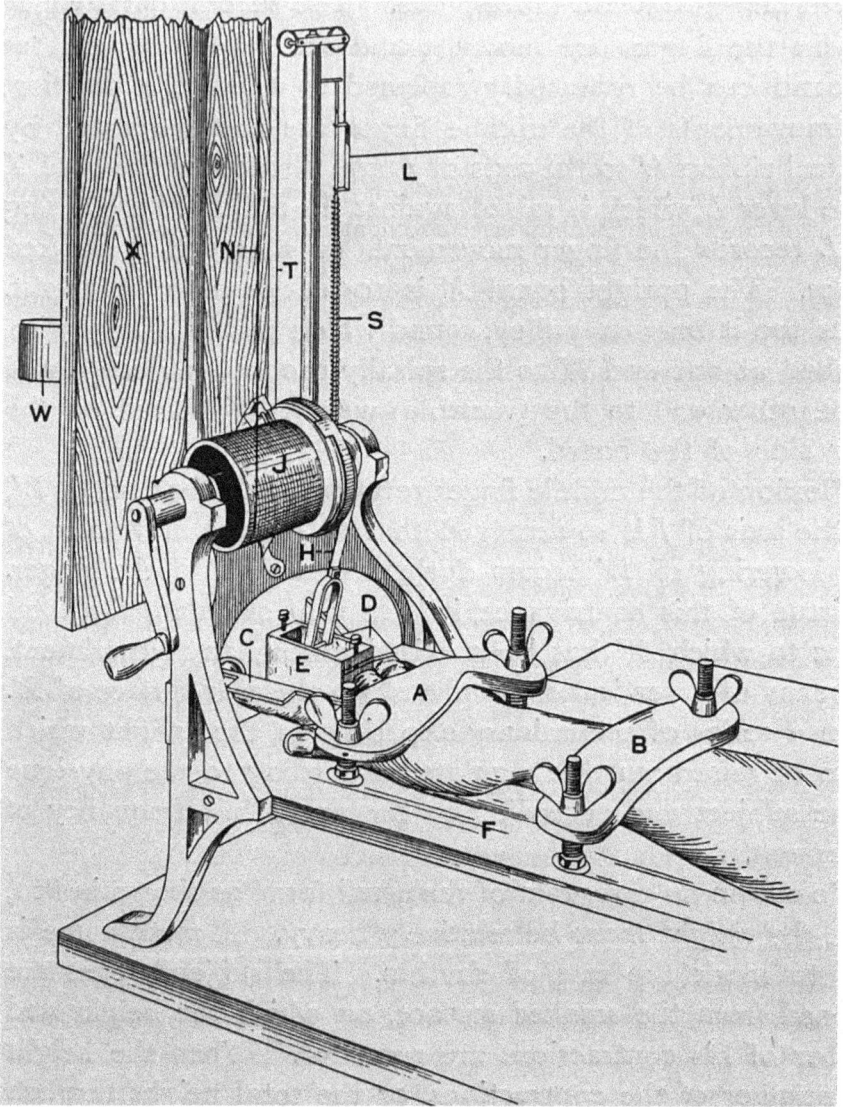

FIGURE I.3. Kraepelin's Ergonometer. This device was an apparatus for testing fatigue in the response of individual muscle groups in the fingers of one hand. The source of the fatigue was the work involved in the task of repeatedly lifting and lowering a measurable load attached to the finger. From Myers 1911, 59.

The Cambridge Cockpit clearly represented a break—perhaps even a revolutionary break—from the ergographic tradition, though the thought-provoking result about the warning-signal character of fatigue, mentioned by Myers, retains its plausibility despite the change in experimental practice.[24] The material differences, at the level of apparatus, between the ergograph and the Cockpit are manifest. Less easy to discern are the thought processes that carried Bartlett and his colleagues from the one mode of experiment to the other. My first task, in chapter 1, will be to describe the Cockpit experiments themselves, and then, in chapter 2, I will examine the theoretical interpretation of the results and explore the intellectual resources informing that interpretation. In chapter 3, I go back to the checkered historical beginnings of the theoretical analysis that Bartlett and Craik employed so creatively in their Cockpit work. That analysis will be referred to throughout this book as the "Jacksonian" understanding of the structure of the nervous system, after its development in the late nineteenth century by the renowned neurologist John Hughlings Jackson (1835–1911). In chapter 4, I shall describe a major empirical and theoretical challenge to the Cambridge Cockpit work on fatigue. This challenge derived from a study of landing accidents using statistical data from the Royal Air Force Bomber Command. Following Kuhn's terminology, I refer to the challenge as the Landing-Accident Anomaly. In subsequent chapters I explore the consequences of this challenge, and the Cambridge reaction to it, in the postwar years. Echoes of the dispute can still be detected in today's thinking about fatigue.

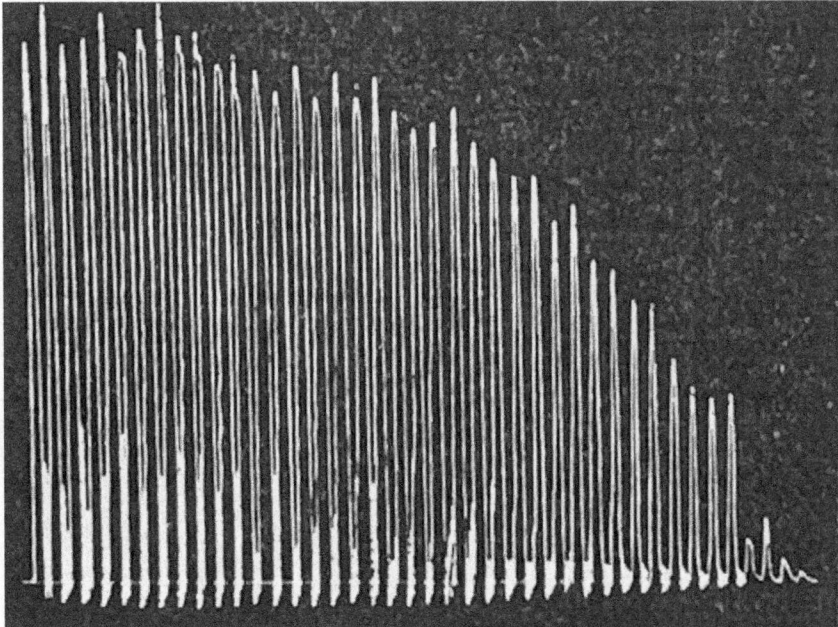

FIGURE I.4. An ergograph indicating fatigue. The Ergonometer, when used in conjunction with a kymograph, could be used to give an objective record of the work done during the course of an experimental session. That record, drawn on the surface of a rotating drum, was known as an ergogram. It could provide seemingly clear indications of the onset of fatigue, though in fact the precise interpretation of the phenomenon was an enduring scientific problem. From Myers 1911, 174.

In the course of my analysis (specifically in chapters 7 and 8), I make the case for the novelty of the work carried out by Bartlett, Craik, and their colleagues. I compare and contrast the British wartime studies of pilot fatigue with the nearest corresponding German and American wartime studies. I pose the question: Was the Cambridge Cockpit unique? My purpose in asking this question is not to press a priority claim or celebrate uniqueness or lament its lack. I treat the answer to this question to be a matter of fact rather than an implied evaluation. For me, the interest of the answer is that it provides a litmus test to determine how far these differently positioned researchers shared an understanding of the problems raised by fatigue. Were their assumptions and practices the same as those of the Cambridge psychologists? If so, then why? If not, then why not?

I should add that in addressing and answering these questions I am not attempting to provide a general history of the study of fatigue. I make comparisons across national boundaries, but I am not aiming to produce an overall survey. Nor am I trying to emulate the impressive scope of important studies such as Rabinbach's *The Human Motor: Energy, Fatigue, and the Origins of Modernity*, or Kehrt's *Moderne Krieger: Die Technikerfahrungen deutscher Militärpiloten 1910–1945*.[25] I am, however, wholly at one with these authorities in believing that science cannot be understood without an understanding of modernity, and modernity cannot be understood without understanding the nature of science. Science and technology are social phenomena that are central to our lives, and those complex facts themselves need to be understood in a scientific way.

In *The Cambridge Cockpit*, I am reporting and analyzing a limited case study in the history and sociology of science. A case study approach provides the opportunity to examine the details of scientific argumentation and the fine structure of the choices that scientists have to make at the level of apparatus and experimental design. Such an approach is well-fitted to cast light on the contingencies of day-to-day experimental science and its practices—and it is a precondition of any sociological analysis of the content of scientific knowledge. The focus of my attention in this book is, therefore, a small group of British scientists, where the scientists under study happen to be psychologists, and psychologists at war.

1

The Cambridge Cockpit

The Cambridge Cockpit research was commissioned by the Medical Research Council's Flying Personnel Research Committee. The committee consisted of around ten representatives of the upper echelons of science, medicine, the military, and the civil service. It was set up, hurriedly, in January 1939, by the secretary of state for air, shortly before the outbreak of war on September 3. The remit of the committee was to identify scientific problems associated with the performance and efficiency of pilots, navigators, radio operators, and other aircrew, and to plan the research designed to illuminate and solve these problems. As an indication of the importance of the committee, it was resolved that it should report directly to the secretary of state.[1]

The Flying Personnel Research Committee

The impetus for the creation of this body seems to have come after an official visit to Germany in May 1937 by a medical officer of the Royal Air Force, Wing Commander Philip Livingston. Livingston was shown around the lavishly equipped aviation-physiology facilities in Berlin by one of Germany's leading specialists, Professor Hubertus Strughold (1898–1986).[2] Strughold demonstrated their low-pressure chamber for simulating high-altitude flight. After Berlin, Livingston went to Hamburg, where yet more facilities were on display. Livingston was alarmed at the disparity between the level of British and German research. On his return, and to the irritation of more senior officers, he sought to put pressure on the British authorities to upgrade the research facilities at the disposal of the Royal Air Force. Eventually, the Medical Advisory Committee of the Air Ministry set in motion the process that created the Flying Personnel Research Committee or, for brevity, the FPRC.[3]

Both Bartlett, as professor of psychology at Cambridge, and Sir Joseph Barcroft (1872–1947), professor of physiology at Cambridge, were on this older Air Ministry committee. Both were invited to join the new Flying Personnel Research Committee. Barcroft declined, while Bartlett became a founding member.[4] Why was Bartlett even in the frame for consideration for this role? The immediate answer is that Bartlett had previously served on a number of other Medical Research Council committees, for example, the Physiology of Hearing Committee. For the moment it is enough to say that Bartlett was something of a committee man and, according to the obituary written by his colleague A. T. Welford, he was good at it:

> He had something of a flair for committee meetings which he treated as though they were cricket matches, disposing his forces to outwit the other side and win the day, although always strictly by fair means—he could be formidable but never devious.[5]

The chairman of the Flying Personnel Research Committee was the influential Sir Edward Mellanby (1884–1955), who was also the secretary of the Medical Research Council. Mellanby, himself a distinguished medical researcher, was adept at working behind the scenes and had contacts at the highest level of government, for example with the Committee of Imperial Defence, the Privy Council, and the War Cabinet.[6] Air Vice Marshall H. H. Whittingham was to be the chief executive officer of the FPRC.[7]

The committee held its first meeting on February 6, 1939, and drew up a list of research priorities. Fatigue was on the list.[8] Later, in FPRC report no. 58 (September 1939), Whittingham summed up the situation:

> In view of the fact that the government is preparing for a three-year war, it is strongly felt that the problem of fatigue will seriously occupy our attention in the near future, more so as time passes. It is, therefore, of great urgency that we begin these researches into fatigue at once, in the hope that the answer will be available at a time when it is really needed.[9]

The three-year timetable turned out to be optimistic, and what an "answer" to the problem of fatigue would look like was not specified. But a start had to be made somewhere. How, then, was the problem of fatigue to be investigated?

Conception and Construction

Bartlett told the story of how, as part of their new war duties, he and Craik had been sent to inspect some anti-aircraft guns and equipment. On the return journey to Cambridge that night, speeding through dark country lanes in a chauffeur-driven military car, Bartlett had broached the subject of pilot fatigue. He asked Craik whether some better ways could be found to study the subject of fatigue—better, that is, than the old-fashioned laboratory methods presented in Myers's textbook. Surely they could improve on Kraepelin's ergograph. Could a real cockpit be used, asked Bartlett. Craik immediately pulled out his notebook, bulging with loose notes on the experiments he was planning, and began to generate ideas. Quick sketches were made of possible mechanisms, recalled Bartlett, and indecipherable jottings were added. Then the car skidded on the greasy road and collided with the embankment, narrowly missing some trees in the hedgerow. Fortunately no one was hurt. Bartlett insisted that Craik hardly noticed the incident. He rubbed his arm and continued to jot down ideas and draw diagrams.[10]

In the telling of this story, Bartlett was clearly trading on the stereotype of the absent-minded, otherworldly scientist. Not that Craik would have minded: despite his engineering practicality, it was a pose he tended to assume.[11] Perhaps the most important part of this story, however, was something else that happened on the journey. Bartlett says he raised with Craik a question about the relation between fatigue and work. It was widely assumed that, given equal commitment to a task, a fatigued person performed less work than an unfatigued person. Fatigue was often *defined* as a diminution in the capacity to perform work. Bartlett thought this point should be treated as a matter of fact, not of definition. Even if it were true for muscular work, was there really a diminution in work capacity when mental fatigue was at issue? And what about skill fatigue? Bartlett wanted to challenge the assumption that fatigue implied a reduction in the rate of work and, with Craik's cooperation assured, things moved quickly.[12]

At the sixth meeting of the Flying Personnel Research Committee, on October 4, 1939, the Minute Book records that

> Professor Bartlett said it was intended to fit up in his laboratory, at Cambridge, a cockpit of a Spitfire aircraft for the purpose of assessing what factors contribute to fatigue while flying, with special reference to instruments and controls. It was hoped by this means to determine and measure the criteria of fatigue set up by flying.[13]

Whittingham assured Bartlett that a Spitfire would be delivered at the end of the month—it was to be brought from Farnborough. The airfield and hangers at Farnborough, in Hampshire, housed the British center for aeronautical research and was the headquarters of the Royal Aeronautical Establishment and the Royal Air Force Physiological Laboratory.[14] When the aircraft finally arrived at Cambridge its controls were intact, but it was devoid of instruments.

Craik immediately got to work to replace the instruments and fit them out for use in laboratory tests.

The Flying Personnel Research Committee dealt with a wide range of problems apart from fatigue, and Bartlett, despite his committee prowess, did not always get his own way. One problem was that of improving communication with bomber crews. Should the current radio-telephone equipment be replaced? At the committee's sixth meeting, on October 4, 1939, Bartlett proposed that his Cambridge colleague, A. F. Rawdon-Smith, an expert on audition, should work on this problem. The committee agreed. At the end of January, however, the radio set proposed by Rawdon-Smith was rejected by Farnborough experts. They said that the old sets were better, once they had been fitted with new earpieces. Then there were misunderstandings with the manufacturer, Pye, and, further to this misunderstanding, Bartlett felt compelled to ask the committee to let Rawdon-Smith see some of the relevant reports because of doubts about their accuracy. The matter rumbled on. It was finally agreed that Rawdon-Smith's apparatus was indeed superior, but, to Bartlett's irritation, it was deemed only slightly better than the equipment currently issued.[15] For the most part, however, Bartlett's preferences were met. In the case of the Cockpit, Bartlett and Craik were readily given everything they asked for—at least initially.

Craik's Fatigue Apparatus

To appreciate the problems that Craik confronted when building his fatigue apparatus, it is useful to be aware of some basic features of aircraft control and instrumentation. When the pilot inclines the control stick and rudder bar, the control surfaces (the ailerons, elevator, and rudder) likewise assume a new angle. This change modifies the motion of the aircraft, and the change in motion continues at a rate proportional to the angle. The change in motion comes to an end when the controls are centralized. The situation registered on the aircraft's instruments will be the time integral of the rate of change of the motion that has been initiated. If Craik was to build a cockpit whose instruments responded in a realistic way, he needed a mechanism that would perform the required integration. Fortunately, Craik, with his fascination for engineering, knew about these things.

For the engineer, any continuous gear acts as an integrator. This principle had been exploited mathematically in the integrating machines built by Lord Kelvin and his brother in the 1870s.[16] The basis of these machines was a device known as an integrating disc (see fig. 1.1). The integrating disc consisted of a horizontal disc that drives a vertical wheel by friction. The distance from

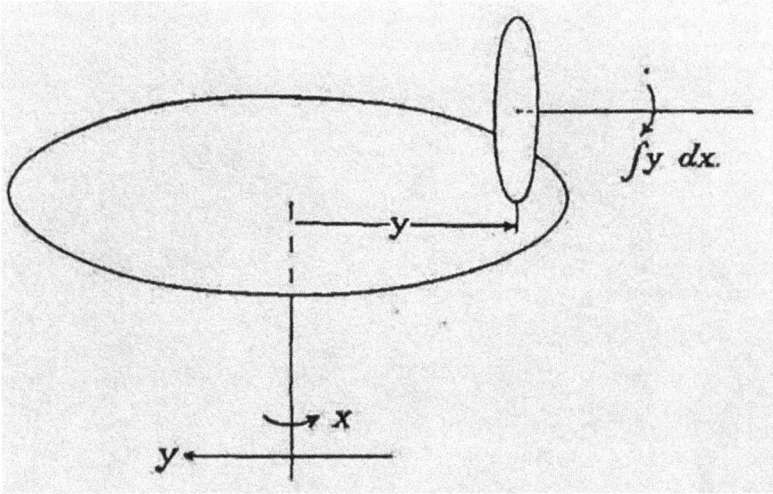

FIGURE 1.1. Kelvin's integrating disc. The physicist and engineer Lord Kelvin (1824–1907) developed a mechanical device that performed the mathematical operation of integration. In its simplest form, it involved a horizontal, rotating disc that turned a vertical wheel. The relative position of the wheel to the center of the disc could be altered. The mathematical meaning of the relationship between the components is indicated in the diagram. From Hartree 1935, 940.

the center of the disc to the point of contact with the wheel can be varied by displacing the disc. This displacement represents the value of the function to be integrated, while the rotation of the disc represents the variable of integration. The rotation of the wheel then represents the result of the integration.[17] Craik exploited these principles when he described his proposal for a "Fatigue Apparatus" to the Flying Personnel Research Committee.

In FPRC report number 119 of March 1940, Craik explained that he had refitted the damaged Spitfire cockpit with a set of new instruments.[18] These were (i) an airspeed indicator, (ii) an artificial horizon, (iii) a vertical-speed indicator, (iv) an altimeter, (v) a directional gyro, and (vi) a turn-and-bank indicator. To the left of the instrument panel, he fitted a clock, an undercarriage indicator, and signal lights for communication between the pilot and the experimenter. To the right on the panel, there was a petrol gauge and various temperature and pressure gauges. The end result, as it appeared to the pilot-subject, can be seen in figure 1.2.

The difficult question was how to make the instruments respond realistically to the cockpit controls. The point, Craik explained,

is that all the instrument readings, except the rate of turn indicator, represent integrals of the movements of the controls. Thus, a certain deflection

FIGURE 1.2. Instrument display panel of the Cambridge Cockpit. The central panel has six instruments. From left to right, the top row shows (i) airspeed indicator, (ii) artificial horizon, and (iii) vertical-speed indicator. The bottom row shows (iv) altimeter, (v) directional gyro, and (vi) turn-and-bank indicator. From Russell Davis 1948, 5.

of the rudder bar, lasting for a certain time, produces a progressive alteration in course. A given movement of the control column in the aileron plane causes a steady increase in angle of bank; and a certain movement of the stick produces an increasing angle of climb, and finally a loop, whose diameter is proportional to the amount by which the stick is held back. The integration is obtained mechanically by Kelvin integrating discs.[19]

Craik then moved from general principles to mechanical details.[20] He explained how he had constructed the crucial component from a brass disc in contact with a rubber friction wheel. Thus:

A wire cable from the control column pulls a friction wheel a . . . with a hard rubber tyre across a brass disc b, four inches in diameter, which is mounted on ball bearings and driven by an electric motor. If the friction wheel is centrally placed on the rotating disc it will remain stationary; otherwise it will turn

with a velocity proportional to its radial distance from the centre of the disc,
an[d] in a direction determined by the direction in which it has been moved
from the centre of the disc. A cord fastened through a hole in this shaft and
wound several times round it will consequently wind or unwind according to
the direction of rotation of the friction wheel, which is determined in turn by
the position of the control column.[21]

Craik went on to describe the rest of the mechanism and its connection to
the spindles that actually move the needles of the instruments on the panel in
front of the pilot. Thus:

> The effect of throttle on air-speed is obtained by a pulley "looped" in a cord
> passing to the air-speed meter and fixed on the end of a movable lever . . . as
> in Kelvin's harmonic integrator and tide predictor.[22]

The responses of the pilot and machine are recorded by four pencil lines
drawn on a paper tape activated by the recording cams and moving at about
4 feet per hour. Provided that a precise schedule of the required maneuvers
was specified in advance, the recording mechanism could be ingeniously arranged so that a perfectly executed and timed maneuver resulted in a straight
line on the recording paper. Errors would generate measurable deviations
from the straight line.

The complexity of Craik's "Fatigue Apparatus," with its levers, cams, cords,
pulleys, spindles, springs, pawls, and gears, can be appreciated from figure 1.3.
The four Kelvin integrating discs can be discerned on the right-hand side of
the figure, while the recording cams and the moving paper strips can be seen
on the left side. Despite all its complexity, Bartlett proudly announced to the
committee that Craik had managed to build the Cambridge Cockpit for the
sum of twelve pounds sterling.[23]

One point not mentioned in Craik's report, though it must have come up
in the discussions of the committee, was why the investigators did not modify
a Link Trainer, rather than rebuild a real cockpit. Link Trainers were simple
and widely used cockpit simulators. Like the Cockpit, they were anchored to
the ground, but unlike the Cockpit, they tilted to some degree in response to
the controls.[24] The reason the Link Trainer was not used was that it proved
too difficult to modify it in a way that allowed the experimenters to make the
multiple, simultaneous recordings of performance that they wanted.

Little was said in Craik's report of the thinking that lay behind the proposed experiment or the design of the experiment itself. These considerations, along with the response concerning the Link Trainer, only emerged
when the final report of the experiment was submitted to the committee in
December 1940. Report number 227 had the title *An Experimental Study of*

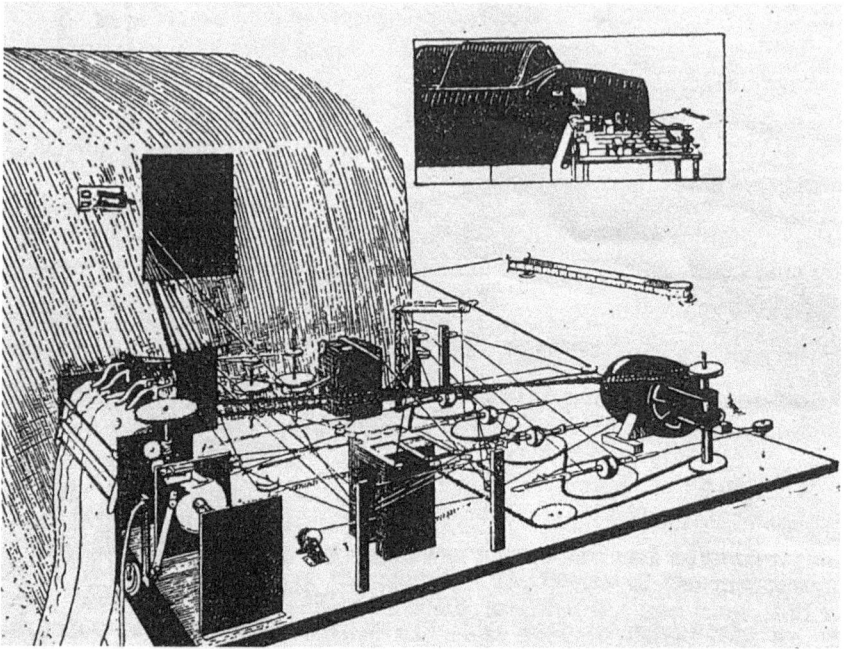

FIGURE 1.3. The Cambridge Cockpit apparatus. The cockpit (detached from the rest of the fuselage) is enclosed and covered to simulate night flying. The pilot depends on the instruments in the cockpit in order to perform the schedule of required maneuvers. The four Kelvin integrating discs can be seen on the right-hand side of the mechanism outside the cockpit, while the recorded traces of the control movements can be seen on the left. From Russell Davis 1948, 5.

Mental Fatigue.[25] Working under Bartlett's guidance, the experiment was performed by George Drew, who had helped Craik with the construction of the Cockpit. Drew was already established in the Cambridge laboratory as a versatile experimenter.[26] He was now charged with the responsibility of using the Cockpit as a novel tool of laboratory research.

The Experimental Design

The most important feature of the experimental design lay not in something that was done but in something that was not done. The Cambridge experimenters did not begin with an a priori definition of fatigue and then seek the conditions that would yield effects that satisfied the definition. In particular, they did not assume that fatigue diminished the capacity to work. In fact, said Drew, "we have reversed the normal procedure."[27] They started from the broad assumption that long hours of instrument flying would be "fatiguing" in some sense, and they then defined fatigue empirically. Fatigue was whatever happened under these and similar conditions. The "signs and symptoms" of fatigue might be a progressive incapacity to perform the task, or a fluctuating incapacity, or they might take the form of an increasing distractibility from the task, or some other, quite different behavioral manifestation. And what applied to the manifestations of fatigue also applied to its explanations. The task had to be engaging and complex, in order to prevent boredom. But whether fatigue was central or peripheral (that is, whether it involved brain or muscle), perceptual or motor, excitatory or inhibitory, or the role, if any, played by the uniformity or novelty of the cockpit environment, were all treated as open questions.

The subjects in Drew's experiment were pilots drawn from a Royal Air Force Initial Training Wing stationed near Cambridge. (The flat fenlands around Cambridge made excellent airfields.) The number of subjects was cited as 140. The idea was to require the trainee pilots to go through a nonstop series of repeated "blind-flying" maneuvers lasting for an hour. That is to say: their "flying" depended entirely on the instruments. To avoid distraction and to simulate the conditions of flying at night, the cockpit cover was blacked out. The subjects (in full flying gear) were first introduced to the machine, and the dials on the instrument panel were carefully explained to them. These explanations were important because the instruments had a limited excursion as well as other idiosyncrasies. If the aircraft were put into a simulated high-speed dive, the needle on the airspeed indicator would go right around the dial and begin a second circuit. It was important for the pilot to be aware of this fact in order to interpret the reading correctly.

The subjects were then given a period of practice. When they seemed to have mastered the instruments and controls, they were given their instructions. Their flight plan, describing the required sequence of maneuvers, was in printed form. In pilot fashion, the page of instructions was mounted on a small board, strapped to one leg, so that it could be seen at a glance by the seated pilot. After removing some teething troubles (instruments jammed, and the recording cams generated spurious error readings), pilots used the following flight plan.[28] The experiment began with a period of straight and level flying (on a course of 10° at 8500 feet). The pilots were then required to fly seven "course units" sometimes, but not always, separated by a period of straight and level flying. A course unit was specified as:

(i)   Dive to 5000 feet at a rate of 3500 feet per minute;
(ii) Rate 1 turn to right on a course of 220°;
(iii) Climb to 8500 feet at a rate of 3500 feet per minute;
(iv) Rate 1 turn to left on a course of 10°.

A Rate 1 turn is a turn that would take the aircraft through 180° in one minute. Each course unit lasted about ten minutes. Each of these maneuvers had to be initiated at a specified time. During the straight and level flying, the pilot was instructed to adopt an airspeed of 100 mph. Overall, the sequence was: straight and level, course unit one, straight and level, course unit two, course unit three, straight and level, course unit four, course unit five, straight and level, course unit six, course unit seven. Throughout the flight, the pilot was expected to check the fuel gauges at ten-minute intervals and reset the fuel indicator. After completing the seven course units, a simulated landing procedure took place that involved activating the control that would have lowered the undercarriage. The overall procedure, involving the initial explanation, practice, and flying, lasted about two hours. After the simulated landing, the experimenter elicited the pilot's own opinion of how the flight had gone.

The Question of Stability

Drew's report revealed something about the design of the Cambridge Cockpit that had not been made explicit in Craik's FPRC report no. 119. The Cambridge Cockpit was unstable. "It should be mentioned" said Drew,

> that the machine was purposely designed to be unstable, so that these level flying periods do not represent a complete rest, but rather a reduction in the degree of concentration demanded, and in the number of things that have to be considered at any given time.[29]

The question of aerodynamic stability has always been a matter of importance for the design and piloting of aircraft.[30] Unstable machines are difficult and dangerous to fly, although the instability might permit rapid maneuverability. Stability is necessary in a training aircraft, while a degree of instability is desirable in a fighter aircraft flown by an experienced pilot. The Cambridge Cockpit, it appears, behaved like a machine of the unstable kind. Presumably, Craik's idea was that an unstable aircraft would be fatiguing to fly, and thereby suited the purpose for which the Cockpit was constructed. Craik's design decision may also have had other, unintended consequences. Later, more will be said on the subject.[31]

Craik's apparatus allowed the experimenter to form a far more detailed picture of a skilled motor performance than had ever before been available in a laboratory. It permitted four main flying characteristics to be recorded simultaneously, namely:

(i)   sideslip,
(ii) airspeed,

(iii) altitude, and

(iv) compass course.

The significance of accurately maintaining a specified altitude and compass course is evident, but something should be said about sideslip and airspeed. Sideslip occurs when the aircraft is in a state of aerodynamic imbalance. It can be exploited by a skilled pilot in order to lose altitude rapidly, but it also occurs inadvertently when a novice, or a distracted pilot, initiates a turn. Drew and Bartlett wanted to identify and measure this sort of error, and Craik's apparatus made it possible to do so.

The second measure, airspeed, is sensitive to the nose-up or nose-down attitude of the aircraft. It is vital to maintain airspeed because below a certain speed (which varies depending on the attitude of the aircraft), a stall occurs and the wings fail to provide lift. An unstable machine will tend to have stall characteristics that are difficult to anticipate and control. Drew was looking out for errors in all four of these dimensions of flying skill because he wanted to know whether the frequency and magnitude of errors was greater in the later course units compared to the earlier course units. If the flying got worse, it would be prima facie evidence for fatigue. An analysis of the performance records on the paper strips might then shed light on the causes and nature of the fatigue.

Bartlett's Hints

The experiments started in January 1940, and throughout the next few months, Bartlett would drop hints to the committee about what he and Drew were finding. A report submitted in April 1940 to the secretary of state for air on the activities of the committee said:

> Such observations as have been made to date suggest that a very important part of high-level fatigue is due to an increasing relaxation of inhibition, as a result of which the fatigued person is more liable to be influenced by extra stimuli which, when he is fresh, he is able to keep himself from bothering about: the fatigue machine offers a good way of testing this possibility.[32]

The emphasis here is on perceptual processes rather than motor processes: that is, on the intake of information rather than the output of responses. Some feature of this intake, Bartlett was hinting, seemed to be the locus of the fatigue process. Bartlett's comments provide a good example of the characteristic inductivist style in which Cambridge psychologists handled experimental data: they began by inferring some of the general characteristics of the machinery of the nervous system from the general characteristics of an organism's behavior. Details were to come later, but fatigue, it seemed, was emerging as a process in which there was a failure to inhibit sources of distraction.

On March 29, 1940, in the committee's ninth meeting, Bartlett floated the idea that the pilot's performance in the Cockpit might reveal who would make a good fighter pilot and who would be more suited to pilot a bomber. Some subjects seemed to have markedly more endurance than others, and these individuals would be best suited as bomber pilots who had to fly long missions.[33] Then at the next meeting, the tenth, on May 15, 1940, Bartlett reported that fatigued subjects interpreted their bodily sensations differently to fresh subjects.[34] Again, this analysis was perceptual. At the eleventh meeting, on June 28, 1940, Bartlett reported that a steady flow of seventy subjects from Training Command, each with a minimum of twenty hours solo flight experience, was expected in August.[35] The Cockpit experiments were now taking place at the precise period that the war in the air over the south coast was at its most intense. Finally, in December 1940, Drew was able to submit his report.[36] The committee was impressed.[37]

The Experimental Results

The evidence for the importance of fatigue and the dangerous nature of its effects was dramatic. When the paper recordings issuing from Craik's apparatus were examined to identify flying errors, and when graphs were plotted to show their distribution over the seven course units, it was clear that by the end of the experiment the pilots were making some 40 percent to 50 percent more errors in sideslip and airspeed than they had been at the beginning. The results were statistically significant at the 0.01 level. A graph showing the increase in errors in responding to the airspeed indicator is shown in figure 1.4.

The evidence from altitude and compass-course errors was equally revealing but more complicated to interpret. The error curves for altitude and compass were a different shape than those for sideslip and airspeed. For the purpose of analyzing the altitude and compass results, Drew broke down the overall pattern of errors into two subgroups. First, there were errors that resulted from making the wrong maneuver or badly executing the required maneuver. Secondly, there were errors that resulted from making the right maneuver but making it at the wrong time. Once all the errors had been separated out into categories, a clear pattern was detectable. Most of the errors were of the second kind: they were due to bad timing. After an initial improvement in performance (presumably a practice effect), the timing deteriorated. As

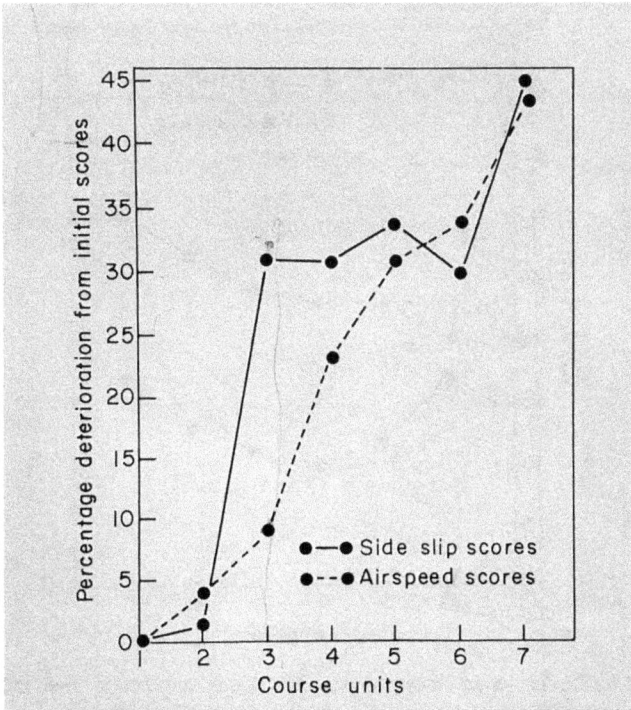

FIGURE 1.4. Errors in airspeed and sideslip. Sideslip and airspeed together provide a measure of the control exercised by the pilot. The data points show the percentage deterioration from the initial score. By the end of the test the pilots were making 40 percent to 50 percent more errors. From Drew 1940, 8.

the experiment progressed, the pilots increasingly did the right thing but at the wrong time. Scrutiny of the pencil traces showed that the quality of the maneuvers themselves did not deteriorate but even tended to improve. The improvement was not statistically significant, but the deterioration in timing, once it had set in, was statistically significant at the 0.01 level for both altitude and compass scores. Drew concluded that fatigue had a differential impact on the timing and the detailed execution of the pilot's actions. Fatigue left the details of the action more or less untouched, but it was the enemy of tim-ing. This important, empirically based, differentiation between the organiza-tion of the local motor response and the central administration of its timing is clearly shown in Drew's graphs. For example, figure 1.5 shows the overall curve for altitude errors and then the breakdown into the two different sorts of error: the maneuver itself and the timing. Bartlett's theoretical interpreta-tion of this distinction will be discussed in the next chapter.

Two further results reinforced the picture of fatigue as a potent source of danger. First, after initially attending to the fuel gauge at the required intervals,

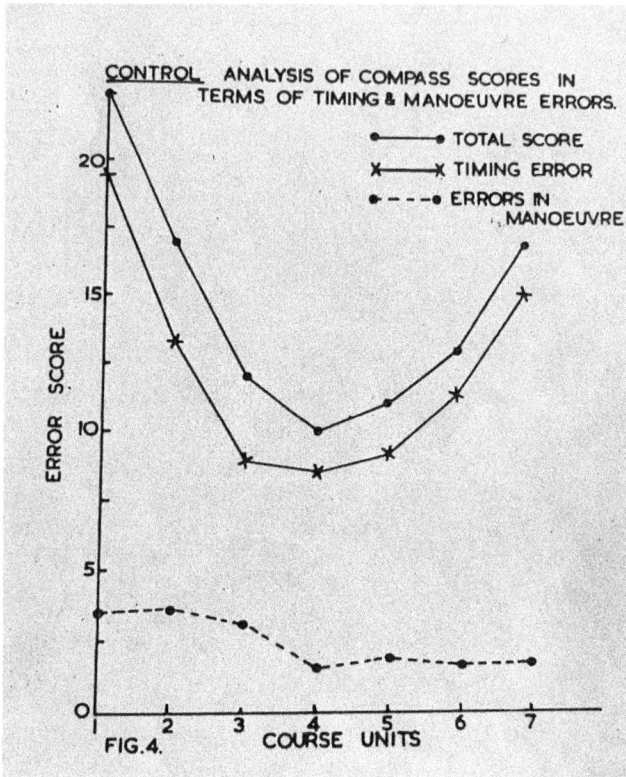

FIGURE 1.5. Analysis of altitude scores. Drew decomposed the overall error scores in altitude control into two components. One component consisted of errors in timing the achievement of the required altitude. The other component measured the accuracy of the maneuver. The result indicated that fatigue was the enemy of good timing rather than good maneuvering. From Drew 1940, 9.

the pilots increasingly tended to forget it. There were comparatively few lapses in checking the gauge and resetting the indicators during the first hour of the test, said Drew, but "after that the curve of forgetting rises very steeply, until 60% have forgotten about them at the end."[38] Secondly, a similar amnesia was at work when the time came to go through the motions of lowering the undercarriage during the simulated landing at the end of the experiment. In fact, reported Drew, "more than 80 of the 140 pilots tested landed with the undercarriage up."[39]

Fatigue and Bodily Sensations

Drew selected some of the 140 pilots and used them as subgroups to explore a number of specific hypotheses. He did not report on the numbers of subjects involved in these more specialized studies. The results must therefore be counted as impressionistic rather than well-established, but they helped to fill out the emerging picture. They also reveal something of the thinking behind the experiment.

Normally, military pilots sat on their parachute pack, but in the Cockpit, the parachute was replaced by a cushion. For a subgroup of subjects, Drew inserted a football bladder beneath the cushion and arranged it so that it could be inflated or deflated at will by the experimenter but without the knowledge of the subject. He then carried out the standard form of the experiment on these subjects but inflated the cushion during a climb and deflated it during a dive. (Subjects were simply told that they might feel some movement of the apparatus but were given no further explanation.) The question was: Did fresh pilots respond differently to those who were fatigued?

Drew found a clear difference. Seventy-five percent of the fresh pilots were convinced that the Cockpit was moving in response to the movement of the controls and that their bodily sensations were caused by this movement. None of the fatigued pilots experienced the changes in this way. "The fatigued pilot" said Drew, "sees no relationship between his panel readings and any sensations he may get from the seat."[40] Fatigued pilots thought they were getting cramp. They located the source of the sensations they were feeling within their own body rather than externally in the behavior of the aircraft. Nevertheless, both groups showed some temporary improvement in performance even though they interpreted the stimulation differently.[41]

A similar temporary improvement was found when pilots in another subgroup were given new flight instructions during the test itself. Some of the pilots had to follow the new instructions during course unit 2, when they were fresh, and some during course unit 6, when they were presumed to be fatigued. The instructions in fact required no objective change in the flight plan but merely represented the old instructions in disguised terms that were more difficult to follow. The pilots were told that they would receive a buzzer signal that required them to consult the new instructions, which were in a map pocket in the Cockpit. They were given five minutes to study the new instructions, during which time the experimenter would take control of the Cockpit. Everyone confronted by this rather complex procedure was baffled by the new instructions and failed to spot that they really amounted to no-change. Despite this bafflement, the flying technique of both fresh and fatigued alike seemed to derive some benefit from the challenge, although the benefit wore off more quickly for the more fatigued group.

The Causes of Gross Deterioration

From the evidence available, said Drew, "it has been possible to assign some causes to the gross deterioration in behaviour noted."[42] Three important processes could be discerned. The first of these was identified as a lowering of standards, the second was described as a splitting up of the task, and the third was the changing emotional state of the pilot. I shall describe what Drew said about each of these processes.

As the pilots become tired, the performance become less efficient, but did this drop in efficiency mean that they were no longer physically capable of doing the task well? Drew and Bartlett rejected this idea. Rather, they argued, the subjects set themselves progressively easier tasks as they got tired. They lowered their standards. On this basis, the researchers made two predictions:

> (i) that the subjects should be unaware of the process, since, in setting themselves a progressively easier task, they should feel that they are tackling it at least equally well throughout, and (ii) that errors in side-slip, for example, show a progressive shift from the small to the large type.[43]

Both predictions, said Drew, were corroborated by the evidence. Indeed, the prediction about the progressive shift from small to large errors proved true for errors in airspeed scores as well as sideslips.

The prediction about the unwitting character of the shift in standards was supported by the postflight discussion between the pilot and experimenter. A passage from Drew's report in which he dealt with this theme is worth quoting in full:

> The subjects, almost without exception, finish the test perfectly satisfied that they have improved steadily throughout. They are, even at the end, still aware of their dissatisfaction with their attempts at control in the early stages, and they are also aware of a gradual change in their attitude to satisfaction in the later stages. They are completely unaware, however, that this change represents a satisfaction with an increasingly lower standard of performance.[44]

In general, the subjective self-reporting of the pilots regarding what had happened in the course of the experiment was highly unreliable. There was, said Drew, "a subjective sense of doing well which seemed to develop in the pilots when they were fatigued."[45] An even more striking finding was that events that had not taken place were confidently reported, and events that had taken place were confidently denied. Although Drew did not say so, this finding had sobering implications for the accuracy of the material given to the RAF intelligence officers who routinely interviewed ("debriefed") pilots after operational flights.[46]

Drew described the second causal process as the "splitting up of the task." Instrument flying depends on systematic and repeated scanning of the instrument panel. This scanning must be accompanied by a correct interpretation

of how the information from the various instruments is to be brought together to provide a true, overall picture of the position and attitude of the
aircraft. In synthesizing this picture, the instrument readings cannot be taken
in isolation because the meaning of a given instrument reading may depend
on the reading of another instrument. Drew was able to show how this process of synthesis and updating broke down when a pilot was fatigued. Thus:

> When the pilots are fresh, the 6 central flying instruments are regarded by him
> as closely interconnected, and a movement in any one instrument is associ-
> ated with the corresponding movements of other instruments.[47]

Drew went on to say that when the pilot is fatigued, he "regards the task in
more simple terms as that of keeping six instruments to certain definite readings." He had split the task into its component parts and was responding to
these parts on what Drew called a "stimulus-response" basis.[48]

The third causal mechanism concerned the emotions. Fatigue manifested
itself in irritability and frustration, which in turn led to an increasingly violent handling of the controls. The handles for changing fuel tanks would be
wrenched so hard that the pilots frequently broke a thick copper-wire cable.
In fact, said Drew, "all the cables had to be strengthened considerably in view
of the violence of the behaviour of the fatigued pilots."[49] Given that fatigued
pilots were reported to have a complacent sense that they were doing well
in the test, the question arose as to how that dimension of satisfaction could
coexist with violent expressions of frustration, presumably derived from the
pilot's acute sense that he was doing badly.

This behavior was described and explained when Bartlett used the experimental results from the Cockpit as the basis for the prestigious Ferrier Lecture that he gave to the Royal Society of London in 1941. The lecture proved
to be a somewhat strange performance, and many years later, Donald Broadbent, one of Bartlett's most outstanding pupils, lamented that the "lecture became little known outside his own circle; perhaps because it was so cryptic."[50]
Cryptic or not, the lecture holds the key to the theoretical understanding that
informed the Cambridge analysis of the Cockpit results.

2

A Cryptic Lecture and a Scientific Gamble

Members of the Flying Personnel Research Committee were expressly told that their reports were secret.[1] Despite this restriction, on May 29, 1941, Bartlett gave the Ferrier Lecture to the Royal Society and based it on Craik's FPRC report no. 119 and Drew's FPRC report no. 227. To respect the demands of military secrecy, however, Bartlett expunged all reference to cockpits or aircraft or pilots from his lecture. His bland title was "Fatigue Following Highly Skilled Work."[2] There was no mention of the fact that the work in question was that of flying an aircraft in combat. Instead, Bartlett spoke in abstract terms of a "machine" which presented interconnected sources of information to an "operator." There were dials to be read, levers to be pulled, and switches to be thrown. The machine itself was likened to "a rather complex calculating machine"[3]—and that was all the audience was told. Bartlett apologized for being so unspecific but indicated that his reticence was necessary because of the war. It is difficult to know how seriously Bartlett was taking the requirement of secrecy. Perhaps his hints that there were things of which he could not speak were little more than theatre, and everyone knew or guessed that he was talking about pilot fatigue. Whatever the truth of the matter, the lack of a clear picture of the skill under discussion must have made parts of his argument difficult to follow.

Bartlett's Ferrier Lecture

Bartlett drew a contrast between two very different approaches to the study of fatigue and associated each approach with the name of an eminent scientific authority. One authority was Emil Kraepelin (1856–1926) (see fig. 2.1). The other was John Hughlings Jackson (1835–1911) (see fig. 2.2). Kraepelin's name was introduced at the beginning of the lecture; Jackson's was mentioned toward the end, in the course of summarizing the experimental data. Despite, or perhaps because of, the opacity generated by the wartime restrictions, the contrast drawn between Kraepelin and Jackson stood out as simple and far from cryptic. Bartlett's message was that Kraepelin was wrong and Jackson was right.[4]

Bartlett admitted that excellent physiological and psychological work had been done on the fatigue of single muscle groups presented with repeated stimulation. Nevertheless, he argued, it was a downright failure of imagination to think that analogous techniques could be used to study the special sort of mental fatigue associated with complex, skilled behavior. Anyone who cared to look could see that the skilled tasks of everyday life do not involve the repetition of simple responses. Bartlett declared that asking subjects to spend boring hours in the laboratory performing simple arithmetical calculations, or crossing out letters on a printed page, was wholly misguided. This, he said, was the fundamental error committed by Kraepelin. Simple

FIGURE 2.1. Emil Kraepelin (1856–1926). Kraepelin carried out pioneering and hugely influential work on fatigue in Germany. He also became a specialist in psychiatry. He used the Ergonometer to study motor fatigue, and he employed specially printed workbooks of repetitive, arithmetical tasks to study mental fatigue. Source: Wellcome Collection, London.

FIGURE 2.2. John Hughlings Jackson (1835–1911). Jackson was a renowned clinical neurologist and a mentor of Henry Head and W. R. Rivers during their London medical studies. Jackson was strongly influenced by the evolutionary thinking of Herbert Spencer and postulated a hierarchy of levels in the nervous system, the higher levels inhibiting the lower. Source: Wellcome Collection, London.

repetition was the essence of the Kraepelin Ergonometer, and the use of Kraepelin's *Rechenhefte* had transferred the same assumptions to the experimental approach to mental fatigue. Kraepelin, said Bartlett, "did far more harm than good, I think, in this field."[5] The direction Kraepelin gave to fatigue research was wrong because

> Through it all runs one great, unverified guess: fatigue must consist of diminished efficiency of specific performance due to repetition of that performance.[6]

The right approach, Bartlett concluded, was to examine complex, non-repetitive, and realistic tasks. The commitment to complexity and realism was the rationale for the experiments whose true nature, on that occasion, could not be mentioned. Bartlett conceded that there must also be a great unverified guess running through his own preferred, alternative approach.

His guess, he said, was that, in skilled work, "abstentions are as important as performances."[7] The word "abstention," in this context, can mean two things. It can mean the failure to make a response when one should be made, but it also means not responding to stimuli that are irrelevant to the task at hand. It means controlling impetuosity and excitement. Bartlett explained:

> When work begins, and the central nervous system is alert, keen and high in vigilance, its inhibitory activities are in perfect trim. Additional, irrelevant, unwanted and distracting stimuli are not within the effective field at all. But as work continues and fatigue grows, the inhibitions perhaps relax, until the skill-tired man is doing, not less work, but more, much more.[8]

Kraepelin's approach, and apparatus like the Ergonometer, could hardly reveal anything but a diminution in work. By contrast, inhibition and the release of activity as inhibition broke down were the two key concepts that Bartlett gave his Royal Society audience. The interest of the subsequent theoretical analysis then lay in the way that the nervous system was presumed to incorporate and utilize these processes of inhibition and release. It was in this arena that the name John Hughlings Jackson came into the picture.

Jackson was a London-based neurologist who specialized in the treatment of epilepsy and aphasia. He was renowned for his clinical and diagnostic subtlety. Later, more will be said about his links to Cambridge psychology. For the moment, the important point is that Jackson envisaged the nervous system as a structure composed of a number of levels. The levels formed a hierarchy, with the lower levels having emerged early in the stages of evolutionary development and the higher levels at later stages. The lower levels, he supposed, subserved rigid and crude responses, while the higher levels subserved more subtle and flexible responses. The higher levels held the lower levels in check: they inhibited their functions, but those functions remained, and under the right circumstances they could emerge into view again. The higher levels were also said to be more vulnerable than the lower to disease, injury, or other forms of insult. Alcoholic intoxication finds a simple explanation on this basis. Physiologically, alcohol is a depressant although it is typically thought to "stimulate" and "excite" bad behavior. From a Jacksonian perspective, what really happens is that the alcohol progressively disables the higher levels of the personality, thus releasing the characteristic crudities of the intoxicated.

In his lecture, Bartlett argued that a version of this schema allowed psychologists to rationalize the rich body of data that had been gathered from the Cockpit experiment. Fatigue undermines the higher levels of motor control. If a skill is an organized sequence of movements, then the activation of each individual movement, or each routine sequence of movements, can be thought of as "low level," while the initiation of the sequence as a whole, and the process by which the order of the elements of the sequence is determined, can be thought of as "high level." The lowering of standards of performance and the breakdown of timing were the predicted retreats from the higher levels to the lower levels. A dial on the control panel would signal that the machine was off course, but increasingly the signal was ignored, and the response came later and later. Then the correction was made hurriedly, and too strongly, and a further correction was needed so the later maneuvers in the schedule were disrupted. Each of the individual control movements in the sequence of attempted corrections and counter-corrections was routine and instinctive, and in that sense the "right response" had been made. The problem was that there were too many or too few responses, and they were made at the wrong time.

So much for the objective motor performance: the Jacksonian analysis of the subjective dimensions of the test was even more striking. Bartlett warmed to this theme, though he could only speak of a "machine" rather than a cockpit, and of "operators" rather than exuberant or fatigued young pilots:

> When the operator began, absorbed in his task, he was usually silent. As he went on, sighs and shufflings emerged from the machine. Then mild expletives took the place of sighs. By the end of the experiment, which lasted two hours or more, most operators kept up a flow of the most violent language they knew. And all the time their handling of the controls became more and more rough, so that they were doing more work not less, as the task went on.[9]

As fatigue increases, said Bartlett, "the general drift of behaviour is towards a less closely organized and effective central control." The result was that the power of inhibition was lost, and "the operator's temperament surged up and took charge of his behaviour."[10] This surging up corresponded to the release from inhibition of the lower Jacksonian levels.

At first sight, said Bartlett, "the subjective symptoms of fatigue characteristic of long continued skill all seem highly paradoxical."[11] He was referring to the fact that the subjects' intellectual sense of their own performance and the causes of that performance became increasingly contradictory. The experimental subject

> is at once more optimistic about his performance and pessimistic about his state. He asserts that he is doing well, and at the same time he blames something or somebody else for making him do badly.[12]

The paradoxical character of these responses makes Jacksonian sense: the conflicting messages are emerging from, or prompted by, the different levels of the nervous system, depending on their state of inhibition and fatigue. Paradoxical or not, Bartlett was sufficiently confident of the results overall to make a modest claim to scientific priority. He declared that he had just offered his audience "the first reasonably complete representation that has been drawn of the fatigue following highly skilled work."[13]

"Muscio's Paradox"

Those in the audience familiar with the state of psychology as a discipline might well have been surprised both by Bartlett's choice of topic and his claim to have made some manner of breakthrough. Fatigue was no longer as fashionable or as popular as a subject of research as it had been in earlier decades. Of course, during the previous Great War, there had been extensive research on industrial fatigue in munitions factories.[14] Bartlett had served on the Committee on Industrial Fatigue (later renamed the Industrial Fatigue Research Board), which had overseen these studies. Valuable conclusions had been reached about the need to avoid long, counterproductive, and accident-ridden working hours.[15] Nevertheless, fundamental questions about the nature of fatigue remained unanswered, and in the interwar years there had been a growing feeling that these questions might even be unanswerable. Bartlett argued that, for a number of years, there had been stagnation in the field of fatigue research. In his opinion, experimentation on fatigue had become dull and routinized:

> Probably everybody who is interested in the course of scientific development has noticed how scientific ideas, and especially scientific practice and technique, may linger on, setting people to work, long after they have ceased to have promise in the way of new developments. This happened in the experimental investigation of fatigue.[16]

The main culprit was the Kraepelin paradigm of experimentation. While the experimental techniques hardly changed, said Bartlett, all the time there was "an increasing number of interpretations and theories, milling about in a rather disorderly scientific playground."[17] In 1922, the industrial psychologist Bernard Muscio (1887–1926), who had worked as a demonstrator in the Psychological Laboratory in Cambridge, published a report for the Fatigue Research Board in which he reflected on the state of the field. Muscio posed the question: Is a fatigue test possible? He concluded that it was not possible.

Muscio's argument was that a fatigue test presupposed consensus about the meaning of the concept of fatigue, but there could be no such consensus because there is no independent criterion of the state of fatigue. The implication Muscio drew was that if psychologists thought they could ever know what they meant by the word "fatigue," they were deceiving themselves. He duly recommended that "the term *fatigue* be absolutely banished from precise scientific discussion."[18] Muscio's argument is highly questionable, but it has resonated down the years.[19] For example, some ninety years after this radical recommendation, some of the contributors to the 2012 *Handbook of Operator Fatigue* still felt trapped by these alleged circularities and definitional problems. As one of the contributors to the handbook observed, "This remains Muscio's paradox with respect to fatigue."[20]

Bartlett agreed with his friend Muscio about the absence of a simple test, but he had no intention of banishing the word "fatigue" from his vocabulary.[21] Nor did he have any intention of becoming entangled with Muscio's arguments. He adopted the stance of moving forward empirically rather than retreating into definitional considerations. At first glance, the minuted discussions of the Flying Personnel Research Committee might suggest that the Cambridge Cockpit was itself meant to be the long-sought test of fatigue. In the stilted language of the minute-taker, in the committee's fifth meeting, Whittingham asserted that it was "desired that suitable tests of fatigue should be devised." Bartlett, however, explained that "no satisfactory test for fatigue existed, but he thought that a new approach to this problem should be tried."[22] Bartlett was as good as his word and offered the committee a new, Jacksonian approach to the problem of fatigue. That new approach did not promise any quick and easy "test" of fatigue but took the form of a research program centered on the Cambridge Cockpit. This sequence of events raises two historical questions: Why did Bartlett think that, in this way, he could reverse the trend in fatigue research and make progress in a field that others were abandoning? And why did his new approach to fatigue take the particular and adventurous form that it did? The Cambridge Cockpit was a gamble. Why did Bartlett take this gamble?

The Gamble

The answer lies in Bartlett's 1932 book *Remembering*. If one knows where to look, it is possible to see that the book contains all the essential resources that Bartlett used in his fatigue research. The argument in *Remembering* provided the methodological template that he used to shape the new approach he had promised the committee. Bartlett had spotted an analogy between the problems that he had solved in *Remembering*, almost a decade earlier, and the problems regarding fatigue that he was now asked to address. He saw that, to exploit the analogy, he needed to perform a new sort of experiment on fatigue. The new experiment had to embody the same sort of novelty as the experiments that he had previously performed on memory. As every reader of *Remembering* knows, the memory experiments were prefaced and justified by a systematic criticism of the experimental techniques developed by Hermann Ebbinghaus (1850–1909). The insight and courage of these criticisms deserves to be appreciated because Ebbinghaus's methods seemed to be supported by cast-iron arguments. They had already become established as standard in the field of memory research. To many psychologists, they pointed the way to a truly scientific psychology. My argument is that Bartlett's criticism of Ebbinghaus was cogent and provided the model for his criticism of Kraepelin. Bartlett's gamble was that the methodological moves he had made in reacting against Ebbinghaus could be repeated successfully against Kraepelin.[23]

In order to justify these claims, I now need to put the theme of fatigue aside for a moment and explore the context of Bartlett's 1932 book on memory. Some of this background will be familiar to psychologists and historians of psychology, but, for the purposes of my argument, there are aspects of his 1932 position that still call for investigation. Only when such investigation has been carried out will it be possible to return to the theme of fatigue and see how Bartlett came to make the important and innovative step that he did.

Anthropology in the Laboratory

In a brief autobiographical publication, Bartlett revealed that he had originally wanted to be an anthropologist. Unfortunately, he had been too young to participate in the famous Cambridge Torres Strait Expedition of 1898.[24] Bartlett's *Remembering* can be seen as his vicarious and belated contribution to the expedition. An object of lively debate amongst anthropologists before the Great War was whether societies change as a result of an internal, predetermined developmental process or whether they change as a result of contact and interaction with one another. The discussion was framed as a choice: evolution or diffusion?[25] Bartlett's mentor William Rivers had shifted from the evolution camp to the diffusion camp in 1911. But if diffusionism is the true account, or part of the true account, how exactly does the process work? *Remembering* was designed to shed light on the psychological mechanics of diffusion, cultural contact, and cultural interaction.[26] As his representative element of culture, Bartlett chose the folk story. But if he was going to use folk stories as his material when he studied remembering and forgetting, he had to overcome a powerful methodological objection. This is where Hermann Ebbinghaus enters the story.

The Strength and Weakness of Ebbinghaus

In his 1885 classic *Über das Gedächtnis* (On Memory), Ebbinghaus pioneered a method of studying memory that required experimental subjects to memorize long lists of nonsense syllables.[27] The experimenter could then analyze the subject's attempts to recall the list after various intervals of time and could plot graphs showing the decline in the number of syllables that were correctly recalled. All manner of other variables could be manipulated—for example, the number of syllables to be learned, the practice schedules, the time interval before attempted recall, and the interference between different lists of syllables. Ebbinghaus exploited the fact that the time for relearning material that had once been learned but then forgotten was shorter than the time taken by the original learning. He developed a quantitative saving score to provide a measure of this fact. The subjectively obscure benefits of prior attempts to learn could now be measured and manipulated. In this way, Ebbinghaus sought to develop an objective science of memory and of the elementary, associative processes that link ideas together and sustain the image of past events in the mind.

Über das Gedächtnis was an impressive empirical achievement. Methodologically, there was a powerful justification for operating with nonsense syllables. Their use was meant to rule out interfering factors that might contaminate the experimental results. If meaningful material had been used, argued Ebbinghaus, then it would mean different things to different people. If a passage from Goethe or Shakespeare had to be memorized, then some subjects might already be familiar with it. Some subjects might find the material impressive and interesting (which would help them remember it), while others might find it boring or beyond their intellectual grasp (which would slow their learning and speed up their forgetting). Psychological processes and contingencies other than those involved in the pure exercise of the faculty of memory would then pollute the results. The use of nonsense syllables was meant to remove these uncontrolled factors from the experiment. But what ruled out Goethe and Shakespeare also ruled out the folk stories Bartlett wanted to use as his instrument of investigation.[28]

Bartlett argued that Ebbinghaus had committed a fallacy. He had assumed that a simple stimulus would evoke a simple response and expose the elements out of which all responses are constructed and compounded. While Ebbinghaus assumed that the complexity of a response depends on the complexity of the stimulus, Bartlett said that the opposite is true. The complexity of a response depends on the complexity of the responding organism. The use of nonsense syllables may strip the stimulus of meaning, but it invites the subject to engage in a spontaneous search for meaning and to make arbitrary and subjective associations. The complexity of the response mechanisms available to the organism would be engaged regardless of how seemingly simple the stimulus was made to be. Ebbinghaus had not rigorously controlled all the relevant variables; he had unwittingly introduced new and uncontrolled variables into the experiment.

Bartlett pointed out that all experiments in a psychological laboratory were, in and of themselves, specialized patterns of social behavior. They involved a specific mental stance and the exercise of considerable discipline on behalf of the subject. Participation required and promoted certain *habits*. This was evident in *Über das Gedächtnis* because, heroically, Ebbinghaus had experimented on himself, but it was equally true when subject and experimenter were distinct. By imposing the regime of meaningless nonsense syllables on the subject, whoever they were, Ebbinghaus had, as Bartlett put it, "forced them into the mould of habit." The subjects were confined to modes of behavior that had then been rendered automatic. The responses, said Bartlett, would then lose just that special character which initially made them objects of interest.[29]

Ebbinghaus's methods seemingly enhanced the objectivity of memory experiments because they made it easy to evaluate the accuracy of the subject's performance. The experimenter simply counted the number of items on the list of nonsense syllables that had been correctly recalled. Bartlett was unimpressed because he refused to equate "remembering" with correct recall. Bartlett's stance here may seem counterintuitive. Surely, one has only truly *remembered* something when one has remembered it *correctly*.[30] Even some of Bartlett's own students, for example Oliver Zangwill and Carolus Oldfield, were reluctant to follow Bartlett and take this step.[31] But they were missing the methodological point that Bartlett was making. Bartlett's aim was to offer an analysis that embraced *both* the making *and* the challenging of memory claims. The psychologist must study such interactions empirically and as a whole. Of course, said Bartlett, what we call memories "claim the confirmation of the past," but "psychologically speaking" that claim should not be taken too seriously, "whatever may be the logic of the matter."[32] What is called "remembering" is a function of daily life and has developed in ways that meet those demands. Scientifically, said Bartlett, "remembering" should be defined, simply and generally, as the capacity of bringing the past to bear on the present. How that capacity operates then becomes an empirical matter for the investigator. These are open questions, not to be foreclosed by an a priori definition of "remembering."

The War of the Ghosts

How did Bartlett put these anti-Ebbinghaus principles to work in his laboratory studies on recall? In one of his early and best-known experiments, he
asked some twenty Cambridge students to read and then recall a folk story.
The subjects had to read the story twice, to themselves, and then, after about
fifteen minutes, write down what they could remember of the story. "Other reproductions" he reported, "were effected at intervals as opportunity offered."[33]
Bartlett collected and compared these different drafts and carefully analyzed
the changes. It was characteristic of his methods that he did not try to secure
uniformity of the time intervals involved. As he put it: "Equalising intervals of
any length in no way equalises the effective conditions of reproduction in the
case of different subjects."[34] A rigid schedule would have been an empty ritual.
The psychological processes at work were, however, rendered visible in the sequence of qualitative changes in the reproduced material, and this is what interested Bartlett.[35] The anthropological purpose behind the experiment is evident
from the fact that the folk story he used was drawn from a different culture to
that of the experimental subjects. Bartlett used a North American Indian story,
"The War of the Ghosts," that had been recorded by Franz Boas in 1891. It was
indeed a "war story," but it was a story whose significance and social function
would be opaque to the subjects trying to recall it. This initial strangeness is
what made Bartlett's experiment an analogue of the psychological processes
and perplexities so often generated by cultural contact.[36] The story given to
Bartlett's Cambridge subjects told of two young men, of rather different temperaments, who set out on a hunt.[37] Here is the story in full:

The War of the Ghosts

One night two young men from Egulac went down to the river to hunt seals,
and while they were there it became foggy and calm. Then they heard war-
cries and they thought: "Maybe this is a war-party." They escaped to the shore,
and hid behind a log. Now canoes came up, and they heard the noise of paddles, and saw one canoe coming up to them. There were five men in the canoe,
and they said:

"What do you think? We wish to take you along. We are going up the river
to make war on the people."

One of the young men said: "I have no arrows."

"Arrows are in the canoe," they said.

"I will not go along. I might be killed. My relatives do not know where I
have gone. But you," he said, turning to the other, "may go with them."

So one of the young men went, but the other returned home.

And the warriors went up the river to a town on the other side of Kalama.

The people came down to the water, and they began to fight, and many were killed. But presently the young man heard one of the warriors say: "Quick, let us go home: that Indian has been hit." Now he thought; "Oh, they are ghosts." He did not feel sick, but they said he had been shot.

So the canoes went back to Egulac, and the young man went ashore to his house, and made a fire. And he told everybody and said: "Behold, I accompanied the ghosts, and we went to fight. Many of our fellows were killed, and many of those who attacked us were killed. They said I was hit and I did not feel sick."

He told it all, and then he became quiet. When the sun rose he fell down. Something black came out of his mouth. His face became contorted. The people jumped up and cried.

He was dead.

Bartlett's question was: What would happen when his subjects recalled this story?

Convention and Creativity

The sequential recollections of the story by the different subjects typically followed a distinctive pattern. As one might expect, the story, as it was recalled, became shorter overall, and details such as place names were rapidly lost. Canoes became boats, and paddles became oars. The story also became, from the subject's point of view, more coherent. For example, the supernatural elements were replaced by more mundane episodes. The changes in the remembered versions fell into two categories: (i) the omission of details and themes from the original story, and (ii) the introduction of details and themes that had not been present. These two processes, however, were closely related. Bartlett observed:

> No omission has merely negative import. The story transmitted is treated as a whole, and the disappearance of any item means the gradual construction of a new whole which, within the groups concerned, has an appearance of being more closely organised.[38]

The process of gradual reconstruction to which Bartlett referred was illustrated by the impact on many of the subjects' recollections of themes derived from the Great War of 1914–1918. When some of Bartlett's tests were first carried out, the war was an event still vivid in the collective memory. Thus, the behavior of the young man in the story who declined to accompany the warriors was assimilated to that of conscientious objectors who refused to be conscripted into the army.[39] Other assimilations, evident in many of the subjects' accounts, operated on more general cultural and historical levels. For example, in the final episode of the story, it is recounted that a black substance issued from the mouth of the dying man. The strange event was often remembered as the soul departing the body with the final breath.

Bartlett's main conclusion was that, with repeated recall, the story was increasingly recast into terms familiar to the culture of the person seeking to remember. This process was well known to Cambridge anthropologists. Haddon, one of the leading figures in the field, called it "conventionalization." Haddon had often discussed this phenomenon in publications and lectures, although he usually illustrated it by reference to items of material culture, such as the decoration of artefacts or the design of weapons or boats. Bartlett embraced Haddon's idea of conventionalization, but in doing so he introduced an important modification. Haddon tended to see conventionalization as performing a purely conservative function. Bartlett noted that his experimental data on recall showed that the cumulative effect could be creative rather than merely conservative. Referring to the versions of the story he was analyzing, Bartlett said:

> Nobody, seeing a single reproduction, could predict the remarkable effect which the cumulative loss of small outstanding detail may have. Yet the effect is continuous from version to version, following constant drifts of change from beginning to end.[40]

Bartlett called the technique he had used in this experiment "the method of repeated reproduction," but he also introduced what he called "the method of serial reproduction." He again made use of "The War of the Ghosts," but this time one subject's recollected version of the story was used as the basis for the next subject's recollections. The story was passed from subject to subject as it might be in a rumor or in the passing of a folk story from one generation to the next. Bartlett found that exactly the same processes of rationalization, simplification, and conventionalization were at work. Although he treated the result with caution, it was clear that there was an analogy between the circulation of an idea in the head of an individual and its circulation and modification and conventionalization in a social group.[41]

Bartlett's stress on these two modes of conventionalization and creative elaboration is sometimes misunderstood. The misunderstanding is that the element of creativity in individual recall is mistakenly assimilated to the category of error. From this point of view, Bartlett's results were no more than a demonstration of one of the sources of cognitive unreliability.[42] His results did indeed provide examples of such negative processes, but Bartlett's experiments revealed more than this. Bartlett had demonstrated a mechanism that might underlie creative processes of cultural interaction. This enlarged significance must be appreciated if we are to understand how his approach to the problem of memory fed into his approach to the problem of fatigue.

The Ebbinghaus-Kraepelin Analogy

In Bartlett's eyes, Kraepelin was making the same mistakes as Ebbinghaus. There was, in Kraepelin's work on fatigue, the same belief in scientific simplification, the same assumption that a simple stimulus would produce a simple response, the same emphasis on low-level repetition, the same fixation on a performance decrement, the same imaginative failure to realize that the breakdown of a function could yield something more, rather than something less. The repeated, meaningless drudgery of Kraepelin's addition and crossing-out tests, or the excruciating repetition of the finger flexing in Kraepelin's ergograph, were counterparts to the drudgery of Ebbinghaus's nonsense-syllable tests.[43]

Although Bartlett did not spell out these background thought processes to the Flying Personnel Research Committee, the Ebbinghaus-Kraepelin link did surface retrospectively, after the war, in his 1951 paper "The Bearing of Experimental Psychology upon Human Skilled Performance."[44] Here Bartlett himself made the connection that I have tried to articulate. Unfortunately, the admission was made as the culmination of an overloaded and decidedly inelegant paragraph. The connection is easily missed, so I shall quote the revealing paragraph in full. In the early 1930s, said Bartlett, referring to the publication of *Remembering*, he had

> published a demonstration that in memory studies great preoccupation with "nonsense" material may become more of a hindrance than a help. When, in the course of the second world war we were forced, for practical reasons, to ascertain how people used their skill, both mental and bodily, to track targets, to drive vehicles, to pilot airplanes, or to control them from the ground in concerted flight, we confronted much the same difficulty. The units used in the classic experiments, (simple thresholds of movement; reaction-times having, often enough, succession, but nothing that could be called an ordered sequence; items in the skill separated from everything else, having the accredited form of "stimulus-response," but devoid of any character of accomplishment of an internally varied task) seemed, if they were forced into the everyday world of work and play, to have the same taint of "nonsense."[45]

Stylistically, these were probably the worst sentences that Bartlett ever wrote. But he was making all the connections for which I have been arguing. In the study of fatigue, Bartlett said, he was confronted with *the same difficulty as in the study of remembering*. In both cases, he said, the accepted approach carried the same taint of *nonsense*. Repetitive movement was as devoid of sense as the nonsense syllables. Exclusive focus on such material could not look real life in the face. The nonsense syllables clamped the imagination and memory no less effectively than the ergograph clamped the hand of the subject. But if the insight into the shortcomings of Kraepelin is explained by the prior insight into the shortcomings of Ebbinghaus, what explains *that* insight? Bartlett himself terminated the regress and offered the explanation in the opening pages of *Remembering*. I am referring to his expressions of indebtedness to the neurologist Sir Henry Head (1861–1940).

Positive and Negative Symptoms

Henry Head, like William Rivers, had been a pupil of Jackson's. They had both encountered Jackson when undergoing their medical training in London. Bartlett, by contrast, had no medical training and no direct contact with Jackson. Nevertheless, he interacted closely with Head over a period of two years when Head was composing his 1926 treatise *Aphasia and Kindred Disorders of Speech*.[46] Head's book contained detailed accounts of the brain damage and resulting disorders suffered by many wounded soldiers in the Great War. Bartlett recalled: "We met frequently, and it became his habit to read over to me the chapters of his projected book as they were written."[47] The discussions were primarily methodological, and, in *Remembering*, which appeared some four years after Head's book, Bartlett was lavish in his acknowledgment. He was also tantalizingly unspecific. But one important theme in Bartlett's book can be identified and easily traced back to Head. Bartlett learned from Head the importance of the distinction between what Jackson called "positive" symptoms and "negative" symptoms. Jackson's message was that both forms of symptom had to be identified, and both had to be explained by relating them to an overall picture of the nervous system.

Stated simply, a negative symptom is the detrimental effect on some physiological function that is a direct result of damage done by injury or disease. A positive symptom, by contrast, is the release or intensification of some other physiological function that results, indirectly, from this original deficit. The distinction was rationalized by Jackson's assumption that the nervous system is structured into different levels, where the higher levels inhibit the lower levels. Damage to a vulnerable high-level function not only has a direct and negative effect, it also releases activity from the lower level—and this release gives rise to the so-called positive symptoms. If the idea of levels rationalizes the distinction between types of symptom, it is the identification and classification of the symptoms that gives content to the talk of levels. The combination forms a framework of interpretation. There was no ultimate justification for the framework. Ultimately it was a guess, though it could be supported indirectly, for example by the interpretation of anatomical evidence or by ideas drawn from a picture of the evolutionary development of the nervous system. In practice, the acceptance of the framework by Jackson's pupils largely depended on their trust in Jackson's experience and clinical insight. The attributes of authority and confidence were frequently accorded to Jackson—and were also evident in the ebullient and imaginative Head.

The traditional (non-Jacksonian) approach to aphasia was to map the precise location of any brain lesion and then correlate it with some precise speech deficit believed to result from the damage. Head said that this approach achieved little and generated much that was incoherent. Its practitioners focused on negative symptoms while ignoring or misunderstanding positive symptoms. Head called for a more holistic and functional approach. He noted what his patients could do, as well as what they could not do. He was impressed, for example, by the fact that patients who had lost the power of voluntary speech frequently retained the power to utter expletives. The inability to produce a name or a proposition was a negative symptom; the power and inclination to swear was a positive symptom. It was presumed to express the release and continued operation of some lower and cruder level of verbal functioning.

By observing Head at work, Bartlett learned how to look at his memory data in a similar way. He was emulating the diagnostic flair of Head and Jackson when he asserted that no omission in recalling the "War of the Ghosts" had merely negative import.[48] The omission of relevant detail was a negative symptom, but the intrusions due to conventionalization were positive symptoms. The role of "levels" was given a psychological interpretation in terms of the cultural stereotypes that emerged as the subjects attempted to retrieve the story. The cultural stereotypes, say of battle, or of the refusal to engage in battle, are to be thought of as schema embedded deep in the minds of the experimental subjects by their culture. Like the well-rehearsed habits of coordinated movement, or the reflex of swearing, they have become automatic. They are therefore ready to be released when, due to the passage of time and the press of events, we struggle to remember or to understand. Bartlett's reading of the documents generated by his experimental subjects in *Remembering* thus operated on Jacksonian principles. His criticism of Ebbinghaus was a

Jacksonian criticism, and, as he willingly told his readers, he learned it from Henry Head.

Back to the Problem of Fatigue

We can now see how Bartlett could confidently tell the Flying Personnel Research Committee that a new approach to the problem of fatigue was possible. Bartlett had seen a problem of this form before, and he knew what needed to be done. In the new line of experimentation that Bartlett proposed to the committee, the exotic Spitfire cockpit was the methodological equivalent of the exotic folk story in his memory experiments. In the case of fatigue, a whole array of positive and negative symptoms might be awaiting interpretation—and this is indeed what he found. Craik's new fatigue machine was designed to expose both kinds of symptoms, while the old fatigue machine, the ergograph, was just a negative-symptom machine. An experimenter using the Cockpit could monitor a wide range of pilot reactions. All manner of omissions and commissions were rendered visible, and, potentially, both were revealing. We can now see that Bartlett's question to Craik about work was a Jacksonian question. If Jackson was right, fatigue might release an increased capacity for work from a lower level—even if it was inefficient or misguided work. But was Jackson right? The Cambridge Cockpit gave Bartlett his answer, and it was a positive answer.

In this chapter I have identified the role played by the Jacksonian themes that Bartlett discussed with Head in 1926—themes that fed into *Remembering* and then into the Cockpit work. However, when Bartlett recalled the scope of those discussions, he also mentioned that not all of them were immediately prompted by the chapters that Head was composing for his forthcoming book on aphasia. There was also, he said, "much resulting discussion" of Head's "earlier work on afferent sensitivity."[49] What was this earlier work? Bartlett gave no indication of why he and Head had discussed afferent sensitivity, nor did he reveal what they were saying about it. In the next chapter I shall show that there was good reason both for the discussion of this theme and for Bartlett exercising a certain reticence about it. I shall also explain how these two facts connect with Bartlett and Craik's later analysis of the Cockpit experiments.

3

A Skeleton in the Cupboard?

In 1903, Henry Head and William Rivers began an important physiological experiment on afferent sensitivity in which they claimed to demonstrate the reality of Jackson's functional levels. This work, and its subsequent development by Head and Rivers during and after the Great War of 1914–1918, led to a sharp controversy. For some, the Head-Rivers development of Jacksonian ideas was seen as a genuine contribution to the growth of psychological thinking; for others it was to be treated as unscientific speculation deserving rejection and ridicule. Given that Bartlett had acquired his Jacksonian orientation while studying at the feet of Rivers and Head, this sharp divergence raises significant questions: How should the disagreement be understood, and how does it bear upon the Jacksonian analysis that Bartlett had given to the Cockpit results in his 1941 Ferrier Lecture?

A Heroic Experiment

The Head-Rivers experiment has been called "heroic," and the description is apt.[1] It took four years to perform and involved severing a nerve in Head's left arm. The origin of the experiment lay with a puzzle that troubled Head in his encounter with the steady stream of patients presenting themselves at the London Hospital with injuries to peripheral nerves. (Head does not say so, but these will have been working people injured in factories and workshops.) For Head, the problem was that "it had long been recognised that the consequences of injury to a peripheral nerve could not be adequately explained on any accepted theory of its structure and function." How was the problem to be investigated? His hospital patients, he said, could not be relied upon to give revealing answers to questions about their sensations, so detailed testing was impossible. Such reports "could be made fruitful by the personal experiences of a trained observer only."[2] Head volunteered to be that trained observer and reliable narrator. He reported, tersely, that on April 25, 1903,

> the radial (ramus cutaneous n. radialis) and external cutaneous nerves were
> divided in the neighbourhood of my elbow, and after small portions had been
> excised, the ends were united with silk sutures. . . . This operation produced
> loss of all forms of cutaneous sensibility over an extensive area on the radial
> half of the forearm and back of the hand.[3]

In lay terms, a cut of some six-and-a-half inches (16.5 cm) was made that bisected the fold of the elbow. The skin was folded back, the muscle was hooked outwards, and the nerve was divided. The wound was closed, and the limb was put in a splint so that the hand was free for testing.[4]

Over the following months, Rivers employed a variety of techniques to trace the return of sensibility. Cotton wool was used as a test of light touch; pins were used to test for sensitivity to pain; a pair of compasses were used to test for two-point discrimination; and narrow, flat-bottomed metal tubes, containing hot or cold water of known temperature, were placed against the skin to reveal thermal sensitivity. Rivers also employed von Frey hairs, a graded series of stiff bristles used to deliver a punctiform stimulus of known intensity. The process was sometimes explosively painful, but Head undertook to describe where and how strongly he could detect these stimuli.

From his background in psychophysics, when he had worked on sensory psychology in Foster's Cambridge physiological laboratory, Rivers knew of the danger that Head might unwittingly describe what he knew or expected, rather than saying exactly what he was or was not actually feeling. This phenomenon was traditionally known as the *Kundgabe* fallacy. For this reason, and to aid concentration, Head was prevented by a blindfold from seeing what Rivers was doing to his arm. Rivers also took care not to give him any inadvertent auditory clues as he picked up and put down the experimental instruments from the table around which they sat. The tests and record-keeping carried on until December 1907 (see fig. 3.1).

The experimenters reported that the healing process fell into two stages. During the first stage, the more fully anesthetized central zone became smaller, so that—gradually—the entire area inside the boundary was insensitive to touch but hypersensitive to pain and to strong heat or cold. Head could not accurately locate a painful stimulus. A pinprick to the forearm might be felt as a diffuse pain in the thumb. For more than a year, the experimenters reported that both forearm and hand remained completely insensitive to light touch. More than two years after the operation, the hand had not completely

FIGURE 3.1. Henry Head (1861–1940) and William Rivers (1864–1922). In 1903, Head and Rivers began a lengthy experiment on cutaneous sensibility. A nerve was severed in Head's arm, while Rivers's role was to test and map the return of sensitivity. The photograph shows the two men preparing for these tests in Rivers's rooms in St John's College. They claimed that their results corroborated Jackson's hierarchical model of the nervous system. Source: Wellcome Collection, London. CC by 4.0, https://creativecommons .org/licenses/by/4.0/.

regained its sensibility, when tested with cotton wool and with compasses.[5] The second stage of recovery involved the gradual return of a finer, qualitatively different kind of sensibility. Head increasingly found he could localize the stimulus. Now the extreme reactions were quieted, and the experimenters reported a greater proportionality between the strength of the stimulus and the strength of the felt response. Eventually, the functioning of the afferent side of the sensory nervous system returned to normal, and the experiment was terminated.

Head and Rivers argued that the two phases of the recovery indicated the existence of a hitherto-unknown double system of afferent nervous sensibility. These newly discovered systems were in addition to the well-known system that subserved deep-pressure sensibility (which had endured unchanged throughout the experiment). As Head put it:

All these facts would seem to show that we are here face to face with an un-discovered form of sensibility, capable of producing qualitative changes in consciousness, but incapable of causing a quantitative change apart from the extent of area stimulated . . . To this form of sensibility we propose to give the name "*protopathic.*"[6]

The first phase of recovery was understood to be mediated by the growth of this protopathic system, while the subsequent return of the fine discriminations and graded responses indicated the growth of what they called the "epicritic" nervous system. The protopathic system regenerates more rapidly than the epicritic, hence the two phases of recovery.

The disappearance of the protopathic responses was not interpreted as the simple disappearance or dissolution of the protopathic system but, rather, its *inhibition* by the epicritic system. The protopathic system was still there, but its expression was suppressed; it lay in wait for circumstances in which it would find expression in feeling and behavior. The release from inhibition would arise when the restraining epicritic processes were disabled by injury, disease, or fatigue. For Rivers and Head, their newly discovered protopathic and epicritic systems exemplified Jackson's theory of levels. The protopathic system was low level; the epicritic operated at a higher level.

The Irritation of a London Surgeon

The first negative responses to Head and Rivers's discovery claims were published in 1909 and 1913 by Wilfred Trotter and H. Morriston Davies.[7] Trotter and Davies were both surgeons from University College Hospital in London. They argued that the broad claims made by Head and Rivers were questionable because they rested on too narrow a basis. There was only one "observer"—that is, Head—and only one nerve had been sectioned. To rectify this state of affairs, Trotter and Davies took turns to perform the roles of observer and recorder. They experimentally imposed no less than seven nerve divisions upon themselves. The largest nerve they cut was the internal saphenous at the knee. This section "rendered anaesthetic nearly the whole of the inner aspect of the leg." To reassure those disposed to prosecute researches in this area of physiology, they insisted that the procedure produced "nothing but the most inconsiderable discomfort." No more was required of the investigators, they insisted, than "a certain degree of temperamental stability."[8]

Some of the London results were similar to those generated in Cambridge. Thus, in both Cambridge and London the immediate result of cutting a nerve was a zone of total anesthesia surrounded by a zone of hypoesthesia. But there were four points of difference in their reported findings. First, and most important of all, Trotter and Davies could not detect the two-phased process of regeneration reported in Cambridge. They found that light touch, cold, and pain sensibility tended to be restored at the same time, while Head reported that they emerged sequentially. Secondly, Rivers could find no two-point threshold in the affected area, whereas Trotter and Davies reported no difficulty in establishing such a threshold. Thirdly, with regard to thermal stimuli, Trotter and Davies noted an effect apparently passed over by Head and Rivers. Within the affected area there was an increase in the so-called indifference zone. Normally it is difficult to detect temperature differences within plus or minus 5°C of the skin temperature. After the nerve was cut, the indifference zone of London nerves almost doubled in size. Fourthly, Head had reported the constant presence of hyperalgesia as the (postulated) protopathic nerves reestablished themselves. This finding did not correspond to the experience of Trotter and Davies. They reported a sequence of random comings and goings of pain sensitivity.

Could subtle differences in experimental technique explain these points of divergence? Head and Rivers located the boundaries of the areas affected by the lesion by means of pinpricks and von Frey hairs, that is, by punctate stimuli. Trotter and Davies used the subject's own fingers to stroke the skin and report on where the anesthetized area began and ended. They claimed this method was more sensitive and that it revealed that a greater area of skin was affected than previously noted. Then there was the question of the two-point threshold. Which nerves were responding? Could it be the deep-pressure detectors rather than the regenerating cutaneous system? Perhaps the London experimenters were putting too much pressure on the compass points and thus generating a false reading. In reply, Trotter and Davies insisted that they could always tell, subjectively, which system was producing the response.

As well as differences in technique, there were also differences in professional orientation that informed the two pairs of experimenters. As surgeons, familiar with patients who experienced postoperative discomfort, Trotter and Davies expressed surprise that Head and Rivers treated all the sensations that were reported during their experiment as having a bearing on nervous regeneration. Trotter and Davies observed that in the Cambridge work

the whole problem is dealt with on the assumption that all the phenomena consequent on a section of a nerve whether they occur before or during regeneration are entirely those of loss of function. That the destruction of the nerve after it has been severed from its centre produces no effect upon the tissue in which it lies except the loss of sensibility and that the reinvasion of the affected area by the regenerating nerve produces no effect upon the invaded tissues except a restoration of sensibility seems to us a large and serious assumption which should not be made without critical enquiry.[9]

The simple physiology of local conditions, such as edema, venous congestion, and anoxemia, can generate many symptoms.[10] The London surgeons took for granted that some of the strange sensations felt after the sectioning of a nerve would derive from the irritation of the wound. Using Jacksonian categories: What Head and Rivers treated as a "positive" symptom, Trotter treated as a "negative" symptom of the nervous lesion. What Head and Rivers attributed to the allegedly ancient but newly discovered protopathic system, Trotter and Davies attributed to the itchy feeling of getting over a relatively minor operation.

Who Was Right?

The Cambridge and London participants in this controversy cannot both be right because their claims contradict one another. In principle, both could be wrong because a failed replication is always an equivocal signal.[11] Both sides in the dispute used experimental techniques that, today, would not be deemed adequate because they admitted unacceptable degrees of subjectivity into the procedure. Thus, both sides took for granted the concept of a sensory "threshold." Later, during and after WWII, grounds emerged for suspecting that there were no such things as sensory thresholds.[12] In addition, there were the interpretive problems arising from the object under study, that is, from the complexity of the nervous system itself. This complexity meant that a cloud of uncertainty hovered over the claims of both the rival camps. Careful practitioners in the field of nervous anatomy knew that it was difficult to establish exactly what underlying nervous processes were implicated in the division of a nerve. Because of the ramified structure of the nervous system, any given area of skin would be served by a number of different fibers. When a single nerve was divided, it was still unknown what other nerves would contribute to the remaining sensibility of the affected area. Sharpey-Schafer, in Edinburgh, made this criticism of *both* the experiments of Head and Rivers *and* those of Trotter and Davies.[13]

Head and Rivers's experiment, and their claims about the protopathic and epicritic systems, found its way into the Cambridge psychology curriculum through the second edition of Myers's *Textbook of Experimental Psychology*. It thus became textbook material in Cambridge at almost exactly the same time that Trotter's counter-experiments were published. Myers's treatment was cautious. At present, said Myers, "we have no evidence that to these two systems of cutaneous sensibility correspond two separate systems of peripheral nerve fibres." However, a single physiological system "may conceivably dissociate into separate psychological systems." This possibility, he noted presciently, introduced further complications into the inferences necessarily made in the interpretation of the experimental results.[14]

Other developments that bore upon the Head-Rivers experiment took place in Cambridge but outside the Psychology Department. In the neighboring Department of Physiology, E. D. Adrian was pursuing his pioneering electrical research into nervous conduction.[15] In his 1928 book *The Basis of Sensation*, Adrian showed a degree of sympathy with the Head-Rivers approach. Adrian recommended Sir John Herbert Parsons's *Introduction to the Theory of Perception*.[16] In this book, Parsons had cautiously expounded and applied a version of the protopathic-epicritic distinction to the operation of the retina. By 1935, however, in his *Mechanism of Nervous Action*, Adrian was telling his readers that "the distinction between the epicritic and protopathic systems has not been generally accepted."[17]

Ultimately it was the failure of skeptical anatomists such as Sharpey-Schafer to locate distinct protopathic and epicritic fibers, and the increasingly sophisticated classification of the behavior of different types of nerve fiber, that led to the demise of Head and Rivers's Jacksonian approach to cutaneous sensitivity. In 1949, looking back on the affair, Adrian jokingly declared that Head's theory, unlike most theories in the field, "had the grave defect of being open to experimental test"—and it had failed the test.[18] The "test" was the assumed necessity to find actual fibers that sustained the alleged modes of sensibility. Myers's awareness that there were alternative functional interpretations of the data was passed over. But the immediate point to be extracted from Adrian's humorous sally is that, though he came to disagree with Head and Rivers, Adrian had accorded the Head-Rivers position the status of a genuinely scientific contribution. Their theory now appeared to be false, but that was just the hurly-burly of scientific life—Head and Rivers were simply on the losing side of a routine, scientific skirmish—that is, a skirmish *between* scientists and *within* science. Casting Adrian's stance into Kuhnian terminology: Adrian was saying that the whole affair was "normal science" in action.

Generalizing the Argument

These scientific arguments ran in parallel to another, related but different level of controversy. It will be useful, for purposes of exposition, to distinguish the two. This second level consisted of a more general and speculative application of the Jacksonian approach by Head and Rivers. Their attempt at generalization provoked a correspondingly more general rejection—a rejection that called into question the status of the Jacksonian approach as objective and rational science. The debate was now presented by their critics, not as one between scientists but as one between a scientific and a nonscientific standpoint.

During the Great War of 1914 to 1918, both Head and Rivers put their medical expertise at the service of the war effort.[19] Head acted as a civilian medical expert treating soldiers with brain and spinal injuries, while Rivers was accorded officer rank in the Royal Army Medical Corps and assigned to the duty of treating "shell shock." Shell shock was the name given to a strange condition, identified as a species of "hysteria," that involved forms of paralysis with no detectable physiological basis.[20] For both Head and Rivers, their war work reinforced their commitment to the Jacksonian understanding of the nervous system. For some time, Head had suspected that the relation between the optic thalamus, located in the brain stem, and the upper part of the brain, the cerebral cortex, was analogous to that between the protopathic and the epicritic systems. Rivers believed that he could use Head's clinical conclusions about the optic thalamus to create a bold, Jacksonian synthesis of the fields of biology and psychology—a synthesis that would cast light on shell shock. After the war, Rivers set out these ideas in his book *Instinct and the Unconscious: A Contribution to a Biological Theory of the Psycho-Neuroses*. The book was published in 1920, with second editions and reprints in 1922 and 1924.[21]

The optic thalamus is a large ganglion which was said to act as a "relay station" between the spine and the cortex. Rivers introduced an evolutionary perspective into the discussion and noted that the "optic thalamus represents the dominant part of the brain of lower vertebrates, while the cerebral cortex . . . developed much later." When the cerebral cortex is out of action, because of disease or injury, stimulation of the skin produces a peculiar quality of sensation that Rivers described as having an "absence of objective character" and an "over-weight of the affective aspect of sensation." The symptom complex—referred to as the "thalamic syndrome"—included spontaneous pain and emotionality.[22]

While working in the Empire Hospital in Vincent Square, London, Head encountered many soldiers whose spinal cord had been completely divided by bullet or shrapnel. This enabled him to study "the functions of the lower end of the spinal cord when isolated from the rest of the nervous system." Here the thalamic syndrome assumed a more extreme form. Stimulation of the limb or trunk below the site of injury produced widespread effects that were independent of the locality of the stimulus or the stimulated part. The responses that were provoked were conflicting and violent and involved powerful bilateral flexion of the legs. In addition, "the movements of the limbs and trunk muscles are accompanied by sweating and contraction of the bladder." The "thalamic syndrome" was now characterized as a "mass reflex" and was seen as characteristically protopathic.[23]

Rivers advanced the evolutionary conjecture that the mass reflex "would form an excellent answer to noxious stimuli in the lower animals."[24] It would, he thought, be a good, though primitive, defense mechanism. The unfortunate patients with spinal injuries were thus said to be evincing motor reactions of a kind that might have been offered in self-defense by a primitive organism. This analogy with the supposed responses of the lower organisms was a significant step in the construction of Rivers's generalized Jacksonian theory. With protopathic sensibility on the afferent (input) side, and now the mass reflex on the efferent (output) side, argued Rivers, here was an outline picture of the overall workings of a primitive organism. Our evolutionary ancestors were protopathic. The mysterious character of "shell shock" could then be seen as the result of a trauma that has caused regression to the most primitive and extreme of all fear reactions, namely, a paralyzing immobility.

Allegory and Ideology

Trotter's stance in the original controversy over afferent sensibility had been supported by the polemically formidable clinical neurologist Francis Walshe (later Sir Francis Walshe FRS). Walshe, who had been a pupil of Trotter's, was an admirer of Jackson, but his admiration was highly selective, and he had no time for Rivers's particular extrapolations of Jackson's basic position.[25] In the early 1920s, Walshe authored a number of contributions to the *Medical Research Abstracts and Reviews* in which he dealt scathingly with Head and Rivers. Walshe's campaign continued long after Rivers's death in 1922. Indeed, Walshe published a particularly vehement denunciation in the 1940s after Bartlett had given his Ferrier Lecture.[26] Walshe deplored the appeal to speculative evolution on the one hand and speculative psychology on the other.[27] The result, in his opinion, was "the cultivation of a rank growth of verbiage."[28] In a review of the second edition of Rivers's *Instinct and the Unconscious*, Walshe complained that psychologists simply did not know enough biology to see what had gone wrong with this line of speculation—and he had some powerful points to make.[29]

First, consider the so-called thalamic syndrome. For Rivers and Head, the thalamus represented the protopathic aspect of sensibility. Walshe said that this interpretation did not square with the anatomical facts. The nervous fibers running down from the cortex to the thalamus are very close to those running upward from the thalamus to the sensory cortex. Any severance of the top-down connection would unavoidably involve a corresponding severance of the bottom-up connections. "Cortical control," said Walshe, "could hardly be removed without equally complete loss of the cortical aspects of sensibility."[30] But Head's own clinical records, drawn from his war work, showed no evidence that the apparent "release" of thalamic functions was accompanied by a corresponding disturbance of cortical sensibility.

Secondly, Walshe focused on Rivers's claim, in *Instinct and the Unconscious*, about reflex immobility, that is, the instinctive response of staying very still in the presence of danger. Rivers said this reflex was analogous to hysterical paralysis, but Walshe insisted that the analogy was false. The fearful and immobile animal was not in a state of paralysis. On the contrary: the animal was adopting a defensive posture, and that posture was complex and active. "Immobility," Walshe said, "is a reaction by posture, just as flight is a reaction by movement." Posture, like flight, naturally involved elements of inhibition, but the inhibition in this case played a role different from the one imagined for it in the Jacksonian inhibition-versus-release model.[31]

Thirdly, Walshe rejected the entire picture of evolutionary development employed by Head and Rivers. Like Jackson, Head and Rivers appeared to operate in terms that were closer to Herbert Spencer's version of evolution than to Darwin's version.[32] They characterized evolution, as Spencer did, in terms of a teleological progress from the crude, simple, and rigid to the refined, complex, and flexible. The process involved a shift from homogeneity toward heterogeneity. Controversially, Spencer also believed that progress and development worked through the inheritance of acquired characteristics. The result (according to Walshe) was that Spenser treated levels of evolutionary development as if they were piled on top of one another. Jackson saw that this approach pointed to the conclusion that each progressive step involved the inhibition of the layer below. The implication was that the lower level continued to exist, lying in wait for an opportunity to express itself. Walshe declared that evolution did not work in this way. Jackson's many clinical insights therefore had to be prized apart from these dated Spencerian formulations. Jackson needed to be rescued from his own unfortunate admiration of Spencer. There were, insisted Walshe, no known examples of processes of inhibition that last for long ages and yet permit what is inhibited to spring forth, ready for action but unchanged. An organism that evinced the protopathic mass reflex in response to danger would not survive.

Walshe treated the essential core of Jackson's clinical insights as scientific, but he denied this accolade to Rivers, Head, and Spencer. Rivers's book *Instinct and the Unconscious* was intended as a contribution to knowledge, but, said Walshe,

> we believe that his book should be regarded rather as a bold essay in specula-
> tion, as a biological allegory, rather than as a contribution to knowledge.[33]

Walshe's accusation deserves careful analysis. If Rivers was offering a biological allegory, then it must be an allegory *of* something: It must carry a message about something other than biology. What, then, was Rivers's real message? And how might Jackson's work have provided Rivers with a vehicle for his allegedly allegorical stories? Consider the following passage from Jackson's Croonian Lecture of 1884 on the "Evolution and Dissolution of the Nervous System." Jackson was discussing the nature of an epileptic seizure and used a metaphor to convey his meaning. The higher nervous arrangements, Jackson wrote,

> evolved out of the lower, keep down those lower, just as a government evolved
> out of a nation controls as well as directs that nation. If this be the process of
> evolution, then the reverse process dissolution is not only a "taking off" of the
> higher, but is at the very same time a "letting go" of the lower. If the governing
> body of this country were destroyed suddenly, we should have two causes of
> lamentation: 1, The loss of services of eminent men; and 2, the anarchy of the
> now uncontrolled people.[34]

Jackson's picture of a hierarchical society was meant to convey a picture of a hierarchical nervous system, and the causes of social lamentation divide neatly into his "negative symptoms" and "positive symptoms." But the comparisons can be read and put to use the other way around. The picture of the hierarchical nervous system might indicate the naturalness of a hierarchical social order. Then the social and political message, conveyed allegorically, was that the lower orders must be held in check to avoid anarchy. Here then were the rhetorical mechanisms by which Rivers and Head might be deemed guilty of merely peddling ideology.

Walshe did not use the word "ideology," but this implication of his argument was made explicit by Jonathan Miller in his book *The Body in Question*.[35] Miller suggested, boldly and generally, that the credibility of the Jacksonian picture of the nervous system derived not from scientific evidence but from the appeal of its political analogue—the allegory. Miller said, plausibly, that Rivers and Head were precisely the sort of people who might be attracted by the hierarchical social picture used by Jackson. Such a picture would serve the interests of the class to which they belonged. They had reason to fear the mob—the "uncontrolled people" of Jackson's metaphor. Could those social interests have influenced the way that Head exercised his judgment when he responded to Rivers's psychophysical tests? Miller suggested that Rivers and Head

> were unconsciously interpreting their own findings in a metaphorical rather than a scientific manner: they had started out strongly biased in favour of Jackson's evolutionary theory of nervous illness, and their convictions were so firmly held that Head at least unconsciously reshaped his own feelings until they confirmed the theory which had moved him to embark on the experiment in the first place.[36]

To embrace Jackson was to embrace Spencer's account of evolution, but, said Miller, Spencer's account of evolution was one that "even a moderate amount of thought would have shown to be bankrupt." Head and Rivers's work—or so the argument went—was a distortion of the normal, critical operation of the scientifically trained mind, and the distorting factor was the politics of a class-ridden society.

Evaluating the Critical Arguments

What is to be said about the Walshe-Miller analysis? Let it be granted that Rivers and Head were in trouble over their protopathic-ancestor assumption. From a Darwinian, rather than a Spencerian, perspective, this idea makes little sense, and even their Cambridge friends rapidly made that fact clear to them. Myers deftly upended this aspect of Rivers's position when, in reviewing the argument of *Instinct and the Unconscious*, he said:

> Because a man walks with a certain characteristic gait when certain nervous paths are affected by disease, it would be fallacious to assume that this gait illustrates a primitive mode of progression.[37]

In order to frame a general response to Walshe and Miller, suppose we grant the critics everything they ask. Let it be accepted that a political message of some kind was metaphorically encoded in their reading of Jackson's theory, and that this message made the theory attractive to Rivers and Head. (There can be no doubt that Rivers and Head were imbued with the routine class snobbery of their day.)[38] Suppose we also grant that, in the experiments, Head had, wittingly or unwittingly, evaded some of Rivers's elaborate methodological safeguards. Ideological considerations might then have unconsciously biased his responses and thus played a role in the conclusions that were reached in Cambridge. Even granted these possibilities, there is still a glaring defect in the critic's contrast between the (allegedly) tainted basis of Head and Rivers's work and the (allegedly) unsullied character of Trotter's conclusions. What Walshe almost certainly knew but chose not to say, and what Miller should have known but apparently didn't, was that Trotter was himself a prominent political ideologue. Once this fact is acknowledged, it becomes evident that the Walshe-Miller analysis is based on the systematic employment of a double standard.

Herd Instincts in Peace and War

Trotter wrote a best-selling book with the title *Instincts of the Herd in Peace and War*.[39] It was first published in February 1916. There were a second and third impression in 1917 and a second edition in November 1919, followed by fifth, sixth, and seventh impressions in 1920, 1921, and 1922. The book was war propaganda.[40] Nevertheless, Trotter's underlying idea was, in itself, a perfectly good one: Humans are by nature social animals. There is a social instinct, or what Trotter called a "herd instinct." The herd instinct is fundamental to all social life. It is the foundation of the best human institutions as well as the worst human behavior.[41] This insight is then replaced by crudity. There were, claimed Trotter, different sorts of herds. First, there was the aggressive gregariousness of animals that hunt in herds, such as dogs and wolves. Secondly, there was the defensive gregariousness of sheep or bison. Thirdly, there was what Trotter called the "integrated gregariousness" of the ant and the bee. German culture—so the argument went—embodied the aggressive gregariousness of the wolf, while British culture was the expression of the peaceful gregariousness of the busy bee. Exactly how these remarkable biosocial alignments arose and were sustained was never adequately explained.[42] But one thing is evident. If anyone deserved to be accused of writing "allegory," it was Trotter. While castigating Rivers for his allegories, Walshe maintained an expedient silence regarding Trotter's contribution to this literary genre, and Miller followed suit.

For present purposes, the important point is that Trotter rejected a Jacksonian analysis of the status of the herd instinct—and in this way aligned his social picture with his neurological picture. Trotter denied that, in peacetime, the herd instinct is suppressed. Nor, he said, was the herd instinct released by war in a way that suggested that it had, previously, been held in check or inhibited. Trotter accepted that Jacksonian processes of this kind were often *assumed* in popular discussions about herd instincts. Though widespread, this assumption was wrong. The outbreak of war in 1914, said Trotter, prompted expressions of the usual view that primitive instincts normally vestigial or dormant are aroused into activity by the stress of war, and that there is a process of rejuvenation of "lower" instincts at the expense of the "higher."[43]

What Trotter called "rejuvenation" is, of course, Jackson's "release." Such Jacksonian views, Trotter declared, were theoretically unsound, and he dismissed them as uninteresting.[44]

A Symmetrical Analysis

Coming back to the disputes over afferent sensibility, we see that both sides made claims based on experimentation while, at the same time, their positions were surrounded by a penumbra of ideology. In both cases the ideology corresponded to, and potentially supported, the science, just as the science could be used to support the ideology. But if (as Walshe and Miller maintained) the alignment of the experimental and the ideological is, in and of itself, sufficient to compromise the objectivity of the experiments performed by Rivers and Head, why did it not compromise the objectivity of the experiments of Trotter and Davies?

To maintain consistency in the application of their own stated standards, Walshe and Miller should have concluded that *neither* party to the controversy was behaving in a way that should count as scientifically acceptable. Both parties were open to the charge of peddling allegories and ideologies that could have influenced their thinking and the conduct of their investigations into changing thresholds. But there is an alternative analysis of the situation. Rather than concluding that, in both cases, the science was suspect, consistency could have been restored in the other direction. Rather than inventing unhistorical and unreal ideals of science and objectivity, why not treat science as a broad category of social behavior which, by its very nature, may include controversies where both narrow and broad perspectives are involved? It would then be possible to generalize Adrian's symmetrical conclusion from the narrower to the broader debate and once again conclude that the contributions of both Trotter and Davies and Rivers and Head count as scientific.

Jackson sans Spencer

How did Bartlett understand Jackson when he presented his "cryptic" lecture on the Cambridge Cockpit? He undoubtedly made use of the Jacksonian mechanism of release, but was he thereby subscribing to Jackson's positive estimation of Spencer? There is a clear answer to this question, and it is negative. Bartlett did not share Jackson's understanding of evolution as a remorseless but progressive drive from simplicity to integrated complexity and from rigidity to flexibility. Just as Walshe saw the possibility, and took the liberty, of prizing Jackson and Spencer apart, so Bartlett did the same.

When Bartlett reviewed Head's *Aphasia*, he insisted that the process by which lower levels in the nervous system became disinhibited must not be understood as one of literal retreat back down the path of evolution. He was full of praise for Head's treatise but firmly expressed his disagreement with Head over this reading of the Jacksonian approach. When the level of nervous activity is lowered, said Bartlett, for example by disease, drugs, fatigue, or sleep,

> the more highly integrated responses are thrown out of action and historically simple forms reappear. Again it must be stated, however, that they never appear exactly, but in a form which is specific to the general development and state of the organism at the time.[45]

The same qualification was present in the Ferrier Lecture, when Bartlett carefully chose his words and said that the fatigued "operator" was moving toward a stage of behavior "which has many of the characteristics of an early phase in the history of increasing neuro-muscular guidance." To share many of the characteristics of something is not to be exactly the same as, or identical to, that something. The implication of Bartlett's remarks was that the different levels in the nervous system evolve together and alongside one another. The stages of development do not simply pile up, one on top of another, as Jackson's words seemed to imply.[46]

Perhaps Bartlett should have said more on the subject, but he said enough to do the job that needed to be done. The mechanisms of release assumed by Bartlett were not shackled to Spencerian assumptions about the crudity of the early responses. Bartlett decoupled the use of Jackson's ideas from Jackson's association with Spencer—a historically real connection that Walshe always tried to play down but was happy to impute to others.[47] We can now make an educated guess about the content of those discussions that Bartlett had with Head in 1926, and where, he told us, the talk went beyond the theme of aphasia and dealt with Head's earlier work on afferent sensitivity. The two friends were almost certainly discussing Spencer and Darwin. Bartlett will have been exploring the criticism that he was later to stress in his review of Head's work.[48]

To repeat: for Bartlett, illness, injury, shock, and fatigue do not cause a literal retreat down the evolutionary path until one reaches the stage of the protopathic ancestor. Any mistaken view of evolutionary origins then becomes irrelevant. Nevertheless, the polemical history of this issue may explain why, in his Ferrier Lecture, Bartlett was economical in his references to Jackson. Bartlett left his audience in no doubt that his analysis of fatigue and his criticism of Kraepelin were Jacksonian in character, but Bartlett wanted Jackson sans Spencer. The need to proceed with care, given this history, may also explain why the claim to have made an advance in the study of fatigue focused primarily on the new empirical work he had carried out with Craik and Drew.[49]

Craik's Military Metaphor

In 1943, in a draft for the book that he did not live to complete, Craik compared the structure of the brain to a military command structure.[50] He used the military abbreviation of "C.-in-C." for "Commander in Chief," and wrote:

> For instance, the C.-in-C., Fighter Command, presumably says: "We want a sweep carried out over such and such an area"; he does not have to add: "This means that Spitfire No. so-and-so on such and such a station must have so many gallons of petrol in its tanks and care must be taken that its plugs are clean and its guns loaded." These latter details are delegated to subordinates.[51]

Likewise, he went on, for the psychologist, the rapidity and certainty of basic patterns of action, such as looking, walking, grasping, or balancing, were delegated to lower levels in the nervous system in a manner that can be rapidly turned on or off by the higher centers. There are obvious similarities between Craik's metaphor and what Jackson had written a century earlier when he compared the different levels in the nervous system to the different levels in society. The reason for the similarity is simple. As Oliver Zangwill recalled:

> Kenneth had read Hughlings Jackson and Henry Head . . . Indeed, I think he saw in the idea of levels of function a key concept in trying to build a general theory of the nervous system.[52]

In the years immediately before WWII, Zangwill had worked closely with Craik on visual adaptation and other perceptual processes (see fig. 3.2). In 1939, they had published a joint paper on the perceptual threshold of the contours of small figures enclosed within the closed lines of a larger figure. Craik had constructed the complex apparatus that was needed. This esoteric topic was of interest because it enabled the two young researchers to test some of the claims of the influential German school of Gestalt psychologists. The experiment was focused on certain claims made by the respected professor Kurt Koffka.[53]

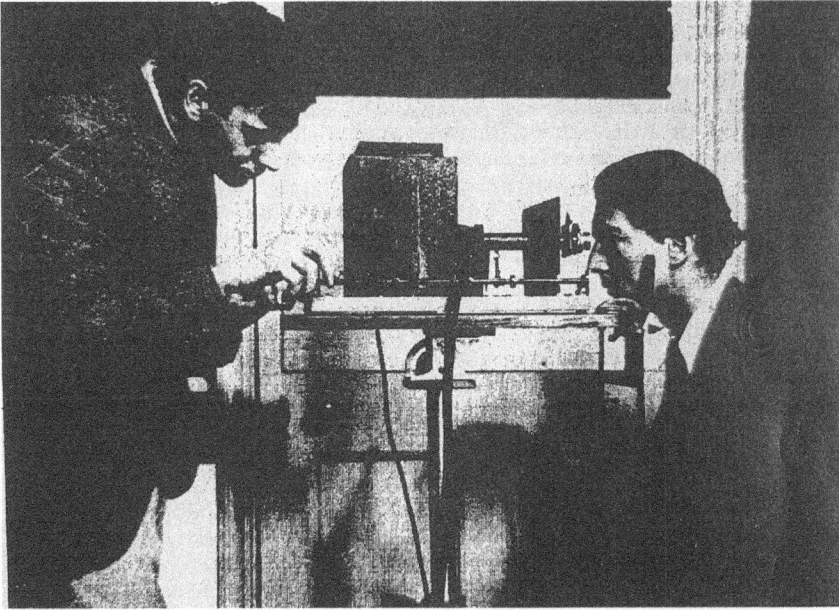

FIGURE 3.2. Kenneth Craik (1914–1945) and Oliver Zangwill (1913–1987). Craik, standing on the left, was a keen constructor of ingenious apparatus. The photograph shows Craik at work in 1938, with Zangwill (*right*) acting as a subject in a study of the effects of visual adaptation on differential brightness sensitivity. In 1952, Zangwill became Bartlett's successor to the chair of psychology in Cambridge. © Department of Psychology, Cambridge.

Gestalt theorists emphasized the brain's tendency to identify and create simple and satisfying forms or patterns (*Gestalten*). Zangwill and Craik's findings, however, told against some of the predictions derived from Gestalt theory. The weakness in Gestalt theory, concluded the Cambridge experimenters, lay in the neglect of Jacksonian principles. Thus:

> Prof. Koffka's theory of perception, despite its brilliant synthetic qualities, seems to us severely limited by its author's persistent refusal to countenance any principle of "levels" in the perceptual process.[54]

Admittedly, the social metaphor that Craik used to characterize the levels in the nervous system differed slightly from Jackson's social metaphor. The difference concerned the relation between the levels and, particularly, the characterization of the lowest level in the hierarchy. Craik depicted the lower levels as reliable and competent. They were likened to well-trained military personnel at the squadron level, not to purveyors of anarchy and disruption. Those occupying the lower levels in the hierarchy of command might be subordinate, but they could be trusted to get on with their job while the higher command got on with its job. We hear nothing of the threatening mob. Fatigue, in Craik's metaphor, was modeled by lapses in directive commands, particularly the timing of the commands, while the lower-level commanded activity still retained its inherent organization. When the upper levels of the social system depicted by Craik begin to break down under stress, control will weaken, but that weakening will result in subordinates doing the right thing at the wrong time. There were no protopathic pilots to earn Walshe's scorn.

If there was a skeleton in Craik's cupboard, it was not ideological but temperamental. Bartlett had hinted at a darker side to his outwardly cheerful colleague. There were gloomy passages from Craik's personal jottings where he reflected on life's illusions and on war as part of life. Around 1942, Craik recorded the following fragmentary thoughts:

> War is a normal part of life; it has been in the past; those who say it ought to be unnecessary cannot prove their case from history. Not a commendable part; but commendable is a word derived from a partial, and wishful, view of life—an unclear wishful thinking that ignores the consequences of granting the wish.[55]

Despite these depressed ruminations, Craik always maintained his commitment to a scientific world view, though he sometimes expressed a pessimistic awareness of its limitations.

Evolving Social Images

There were three main differences between Craik's military metaphor and Jackson's own social metaphor. First, Craik was making a claim about the here-and-now organization of the brain and did not link it to assumptions about the brain's origin or development. Obviously, any such here-and-now claim must be compatible with what is reliably known about developmental and evolutionary processes, but that is all. The psychologist has the job of detecting and explaining the existing regularities in behavior; the evolution of that behavior is somebody else's scientific problem.

Secondly, we have seen that Head and Rivers's failure to identify actual protopathic and epicritic fibers had proven fatal to their early excursions into Jacksonian territory. However, by 1940 experimental psychology was more established as an autonomous discipline than it had hitherto been. Rivers's desire for a synthesis of psychology and biology had ceased to be an obvious disciplinary goal. It might still have been a legitimate distant goal, but that was all. The fate of the earlier Jacksonian campaign and the demise of the conjectured protopathic-epicritic system had provided a salutary warning against the temptation to engage in premature physiological speculation. A form of functionalism could now be asserted without psychologists being open to the charge of evasion. The psychologist discovers what job the nervous system must be doing; it is then the physiologist's task to find out how it does it. Neurologists, such as Walshe, might confidently assert that functions presuppose structures, but it does not follow that knowledge of structure is a precondition of knowledge of function. Psychologists in Cambridge were becoming convinced that it was the other way around. Until function is understood, it is often impossible to make sense of structures. Like engineers, who first draw up a block diagram, psychologists can insist that this "design" part of the project has priority. It was a matter that should be settled before commitments are made about its implementation and material embodiment.[56]

Thirdly, while Craik's military metaphor was well and truly in the Jacksonian tradition, the tradition itself had evolved and was rapidly losing whatever ideological overtones it might once have had. In Cambridge, there was no longer any serious temptation to invent expediently crafted pictures of human nature to help keep order in society or justify positions of social dominance. For Craik it would be evident that social order was secured by policemen, taxmen, schoolteachers, and factory managers, not to mention military commanders, rather than by a moralized conception of nature. This more ideologically neutral, or ideologically indifferent, conception of specialized knowledge was the result of the increased division of labor and the professionalization of science.[57]

To mark the distinction between the earlier and later uses of Jackson's work in Cambridge, it might be advisable to refer to Bartlett's and Craik's commitments as "neo-Jacksonian" rather than use the unqualified label of "Jacksonian." Or we could simply speak of the "Craik-Bartlett" model. The Craik-Bartlett model was indeed informed by the ideas and clinical practices formulated by John Hughlings Jackson, but there was no Spencerian skeleton in this particular Cambridge cupboard.[58]

4

The Landing-Accident Anomaly

In chapter 1, I described Craik's construction of the Cockpit and Drew's experimental report to the Flying Personnel Research Committee. In chapter 2, I outlined the argument of Bartlett's Ferrier Lecture, "Fatigue Following Highly Skilled Work." I identified the way in which Bartlett utilized his past work to provide a template for a new approach to the problem of fatigue. In the course of that "cryptic" lecture, Bartlett utilized the ideas of the neurologist John Hughlings Jackson to rationalize Drew's results. I took the opportunity in chapter 3 to trace the controversial evolution of the Jacksonian idea of "levels" in the nervous system as it was used by Cambridge psychologists. This story culminated in what I referred to as the Craik-Bartlett model. Subsequent chapters will deal with the further elaborations of this model. I now want to come back to experimentation and look at further work using the Cockpit. It is clear that something was troubling the experimenters. Were Drew's results reliable? Could they be replicated? Was the apparatus creating artefacts? The most significant problem confronting the Cambridge workers, however, derived from a statistical study of landing accidents that seemed to contradict the conclusions drawn from the Cockpit work. I shall refer to this problem as the Landing-Accident Anomaly. First, though, I shall look at what the Cambridge experimenters found troubling about the Cockpit results themselves.

A Glimpse Behind the Scenes

Psychological experiments designed to test and refine the Craik-Bartlett model would be difficult to carry out in the best of times, and 1940 was close to the worst of times. The subjects that were required to perform in the Cambridge Cockpit were needed for the completion of their training or for operational flying. While the psychologists were putting their theories on the line, their subjects were putting their lives on the line. Bartlett was all too aware of this fact. In June 1941, in report 325, he had to tell the committee that, since Drew's original account had been circulated, the plan had been to draw more experienced subjects from the fighter and bomber squadrons that were based near Cambridge.[1] "It is obviously impossible to obtain such pilots for experiment during periods of intense operations" said Bartlett, and that was why the supply of subjects had completely dried up. "No objection can be raised for this," Bartlett added, because the experiment itself was fatiguing and no one would want to submit any pilot to the experiment "who might be going on operations in the ensuing 24 hours." There was not yet enough data to draw conclusions about the difference between, say, trainee pilots and experienced pilots, though Bartlett made the guess that the picture of fatigue already given to the committee would not be substantially altered.

Bartlett also revealed something else in report 325. It concerned the attitude of the subjects to the experiment. How did the pilots experience the Cambridge Cockpit and their role as experimental subjects? Bartlett was always conscious that a psychological experiment was, itself, a passage of social interaction and a potential phenomenon of interest in its own right. It is therefore surprising that he did not make more of what he was about to report. Bartlett told the Flying Personnel Research Committee that "a very interesting, consistent and marked difference in attitude has been exhibited as between fighter and bomber pilots." (Clearly the experimental subjects were now more experienced pilots, not just cadets.) The bomber pilots who visited the laboratory apparently showed great interest in the experiment. They wanted to see and understand the mechanism that Craik had contrived. They also tried their hardest when performing the experiment—which is always important because this effort helps to prevent fatigue effects becoming confounded with lack of motivation and effort. The fighter pilots, by contrast, behaved very differently to the bomber pilots. They showed less interest in the machine and asked very few questions about how it worked. They treated the situation as artificial and, said Bartlett revealingly, "most of them are very intolerant of the experiment."[2]

> In a number of cases this has gone so far that the pilot has given up in disgust before he has time to become tired, or just as he is beginning to develop the normal symptoms, and has flatly refused to continue.[3]

Bartlett did not say how many pilots refused to cooperate in this way, nor did he draw any conclusions about the meaning of this phenomenon for the interpretation of the experiment. He did, however, note that all the experienced pilots in question had come from the same two squadrons. This sampling limitation made it impossible to know whether the difference reflected the different temperaments of bomber and fighter pilots as such, or whether it was the culture of their respective squadrons that was displaying itself. And there was more bad news to come.

Was Something Wrong?

Doubts about the Cockpit results must have been circulating behind the scenes for some time, but they did not surface, officially, until report number 488. This report had the title *Fatigue in the Air Pilot* and represented a general summary of the work done on the Cockpit up to August 1942.[4] Drew's original experiment, said Bartlett, showed beyond question that "under the conditions imposed," pilot fatigue occurred, and in most cases the pilot did not know what was happening. What did Bartlett mean by "under the conditions imposed"? The use of the qualification quietly acknowledged that the scope of the experimental result might have presented a problem. If the conditions imposed had been unrepresentative, then the result might have been an uninformative special case. In the report, Bartlett addressed three different possible sources of error in the Cockpit experiments, located, respectively, in (i) the experimental subjects, (ii) the experimenter, and (iii) the apparatus.

To begin with, he said, "it was thought possible that the results obtained might be peculiar to, or exaggerated by the rather inexperienced type of pilot that was being used."[5] The experiment was therefore repeated with a number of operational pilots who, Bartlett said, had prolonged fighting records. Bartlett insisted that there was "no substantial change in the nature of the results."[6] He said:

> The operational pilots who submitted to the cockpit test showed just the same characteristics, but in their case also the onset of the signs of fatigue was deferred for 2, 3, 4, or even more hours. There was one particularly interesting case of a fit, keen bomber pilot, who became exceedingly interested in the experiment. He discussed the plan of the test, and volunteered for an eight hour flight . . . throughout the whole eight hours, he maintained an admirable, steady performance, with no increase in faults either in frequency or amount.[7]

The implied similarity between the replication and the original experiment is said to be based on those pilots who "submitted" to the cockpit test. This phrasing is an oblique reference to the fact that some pilots refused to cooperate. It is not known what the results would have been if more pilots had been cajoled to participate. The breakdown in cooperation raises the question of the relation between the experimenter and the subjects, and this was Bartlett's second point. Was there an "experimenter effect"? He announced that the experiment was repeated using a "second independent" experimenter but reported that it confirmed the original results. There was, however, still the third potential source of difficulty: the apparatus itself. "It was considered possible" said Bartlett, "that the results might have been affected by the design and fittings of the cockpit."[8] Was Craik's fatigue machine generating artefacts? The introduction of a machine called the Silloth Trainer allowed this third possibility to be tested.

The Silloth Trainer was named after RAF Silloth in Cumbria, the base where it was located. It consisted of the fuselage of a Wellington bomber aircraft. The machine did not fly, but it was fitted out with the full cockpit and communication equipment. It was not meant for performing experiments but was used for training pilots and aircrews in the lengthy cockpit checks and procedures that were involved in flying such an aircraft. It even provided realistic noise levels. Although the Silloth Trainer could not register every control movement in the way that Craik's apparatus could, it permitted what Bartlett described as "practically similar observations" to be made on certain aspects of pilot performance. The comparison with the Cambridge Cockpit was somewhat compromised because a fighter was being compared with a bomber, but it was worth a try. The maneuvers demanded in the original experiment were replaced by those appropriate to a bombing mission.

The subjects were seven experienced bomber pilots who were accompanied and guided by their navigators. They were required to simulate a raid on Wilhelmshaven. The pilots had to fly for a period of five hours and were required to change course a specified number of times, as happened in the real raid. Finally, they (notionally) returned to a base in Norfolk. "Readings of airspeed, altitude, rate of climb and dive, and compass course were recorded regularly," said Bartlett.[9] The result of the comparison was declared to be a vindication of the Cockpit. As Bartlett summed up:

> Once again there was no important difference in the character of the results secured, except that as in the earlier cockpit experiment, with operationally experienced personnel, the period of onset of fatigue was delayed.[10]

This declaration may have a reassuring sound to it, but Bartlett was taking for granted a certain perspective and a very particular way to read the results.

Another Way to Read the Results

Without challenging Bartlett's scientific judgment, historians and sociologists can scan these conclusions for information about what Bartlett *meant* when he said that there was no important difference between the findings of the original experiment and the attempted replications. For example, what did Bartlett mean by the sameness of the "character" of the respective results? When, in relation to the earlier replication, he said that there was no substantial change in the nature of the results, what was he identifying as their "nature"? In making these judgments, Bartlett was exhibiting what was central to his concerns and what was peripheral, what was essential and what was accidental. The answer to these analytical questions is that apparently, in Bartlett's eyes, the timescale of the onset of fatigue was *not* the central feature of fatigue. What was central and definitive of the character of fatigue was the sequence of qualitative phases through which it was observed to progress. For Bartlett, it was not the time interval between the onset of these phases that mattered but their causal sequence.

Notice how, in the paragraph quoted above from report 488, Bartlett gave the time of onset of fatigue as being two, three, four, or even eight hours, but he still asserted that this variation in no way compromises the essential similarity between all the results. It must not be assumed that Bartlett was being careless or evasive about the spread of these numbers. The correct conclusion must be that he took this range of variation to be truly consistent with the essential sameness of the findings. We can see that, once again, Bartlett was following the pattern of reasoning he used in his criticisms of Ebbinghaus. In framing that criticism, Bartlett paid little attention to the precise time between the different attempts to recollect versions of "The War of the Ghosts." His interest was in the qualitative character of the changes in the recollected versions of the story, not in the quantitative time-gaps of the sequence. The speed at which the process of recall moved through this sequence was a mere contingency. It was accident, not essence. In the same way, Bartlett considered the time of onset of fatigue to be a matter of secondary rather than primary importance.

If these considerations make Bartlett's stance intelligible, they did not necessarily make it tenable. Tenability depends on other contingencies, such as the aim and purposes and context of the research. The reading I have suggested points to an emerging tension between Bartlett's scientific interest in the phenomenon of fatigue and the practical significance of his findings. The nature and character of fatigue was the focus of Bartlett's attention, but it was the number of hours before fatigue became dangerous that mattered to those who planned bombing raids, selected targets, and had to send exhausted aircrews into danger. The facts that would stand out from the page when senior RAF officers read report 488 would be item 5 of the summary, in which these practical parameters were identified by Bartlett as follows:

> While naturally there are considerable individual differences in resistance to fatigue the evidence is that fit and experienced pilots are fairly likely to show signs of tiredness after about 2½ to 3 hours flying, but very likely indeed, after about 4 hours flying, and are almost certain to do so, unless special efforts are made when they near home after a long and exciting flight.[11]

The crucial question for the military authorities was whether these parameters could be relied on. Both of these themes, that is, both the Cambridge scientific approach and the threatening gap between theory and practice, need to be explored further. I shall pursue this question by looking at the way Bartlett undertook the analysis of accidents in relation to fatigue, and the way in which Bomber Command decided to put his approach to the test.

Bartlett's Approach to Accident Analysis

The Royal Air Force had its own statisticians who kept records of flying activity and aviation accidents as well as losses of aircraft and personnel. It also had an Accident Investigation Branch. From the outset, it was clear that the causes that led to a pilot crashing an aircraft, whether through enemy action or pilot error, would be a matter of concern for the Flying Personnel Research Committee. It was widely accepted within the RAF that fatigue could be a cause of accidents, and the desire to understand accidents was obviously central to the justification for performing the Cambridge experiments. After Drew's report on mental fatigue had been submitted, the Flying Personnel Research Committee charged Bartlett with the tasks of examining the RAF's data on accidents, and planning ways in which the understanding of accidents could be improved.

The result was Bartlett's FPRC report 226, *Some Notes on R.A.F. Flying Accidents*.[12] Bartlett reviewed the monthly accident statistics from March to September 1940, two interim reports from the Flying Accident Committee, and a report on night-flying accidents of September 1940. Despite the fact that some 80 percent of accidents were said to derive from pilot error, he found that the reports had little to say "that bears on the human functions in which I am interested."[13] The writers of the reports would invoke errors of judgment but failed to explain their nature or causation. There would sometimes be a gesture toward fatigue as a cause. This suggestion, said Bartlett, was probably right, but no evidence or further analysis was offered in these reports. There was also a tendency in the reports to blame the inadequate training of the pilots, and Bartlett was disinclined to accept this move. He felt that it ought to be possible to be more specific regarding causation, and he therefore proposed an entirely new approach:

> My opinion is that no further important advance will be made until a different kind of analysis is attempted, in terms of those functions of a well-trained skill which are most liable to suffer disturbance, of when they are most likely to suffer, and of what preventive measures can be taken.[14]

A good place to start would be to look at landing accidents, he suggested, because these accounted for 40 to 50 percent of the total number of accidents. The kind of analysis he had in mind was one that was open to the wide range of contributory factors that might be in play. It must not be assumed, he cautioned, that the failure of any single function would account for landing accidents. "There is almost certainly no 'one best way' of landing," he argued, and "no single master function which can be found in terms of any simple sensorial process."[15]

The rejection of the idea that there is "one best way" of performing a task was a familiar Cambridge theme. It can be traced back to the formula used by both Bartlett and Myers when they argued against the exponents of "Taylorism," a fashionable but pernicious American movement in the field of industrial psychology. The alleged increase in productivity to be derived from time-and-motion study was said by Taylor and his followers to rest on discovering the best way to perform a task. Contrary to the theory of these self-appointed efficiency experts, skilled workmen knew better and would perform a task differently on different occasions. Forcing the activity into one repetitive pattern would be self-defeating. Bartlett had originally broached these themes in *Psychology and Primitive Culture* and had directed the reader to the criticisms of Taylorism published independently by his colleagues C. S. Myers and Eric Farmer. For Bartlett there was no single best way to land an aircraft because there was no best way to perform any skill, just as there was no "one best way" to remember something. Taylor and company were perpetrating an error analogous to that originally committed by Ebbinghaus: namely, that of imposing a method in which there was only one way to get a right answer. Bartlett was again making the same counter-move.[16]

Bartlett's positive suggestion was that every effort should be made to document the conditions that might have caused the pilot to lapse into error when making a landing. One such factor, he suggested, was that a pilot might actually be in the grip of some theory about the best method of landing. The pilot might have been taught a rigid routine and have lost flexibility. What was urgently needed was more concrete information such as

(i) the total number of flying hours before the accident, (ii) the distribution of those hours, (iii) the date of the last leave and its duration, (iv) the duration of the actual flight before the accident, (v) the general resting conditions on the air base, (vi) details of the sleeping conditions, and (vii) the stresses on the pilot in the preceding twenty-four to forty-eight hours. Bartlett's project was characteristically causal, but he acknowledged that, at least initially, the requisite investigations were to be of a qualitative nature. Such investigations required expertise in the observation of what he referred to as "human functions." Who, he asked, "is well trained to think in terms of human functions?" The answer was: a Cambridge psychologist. The best person I know to do this job, said Bartlett, would be Mr. Drew.[17]

Bartlett returned to these themes in April 1942 when he submitted a report with the title *Some Notes on the Investigation of Flying Accidents.*[18] He reiterated his concern that accident reports should be related to a background understanding of recurrent and characteristic types of error, but this time he suggested that the Cambridge-based psychologists who were best suited to establish these facts were Mr. E. C. Chambers and Mr. E. Farmer.[19] Chambers and Farmer, who were industrial psychologists financed by the Medical Research Council, had contributed significantly to long-running discussions over the reality of the condition of "accident proneness"—an idea that Bartlett had invoked and endorsed in his previous note on accidents. Bartlett also took the opportunity to sharpen up his commitment to the role played by fatigue in accident causation. He drew attention to the existence of an opposing belief, namely, the belief that "the expert pilot is not liable to fatigue."[20] Bartlett did not name any exponents of the no-fatigue hypothesis but declared that it was widespread. It was clearly a threat to the Cambridge program. He added:

> I have never actually met any operational pilot, just off operations who subscribed to this belief, or whose behaviour agreed with it; nevertheless it continues to be held and hinder research.[21]

What was needed was a study that would dispel the no-fatigue hypothesis and reaffirm what the Cambridge Cockpit work had surely established, namely, the vital importance of fatigue and its proper understanding. When Bartlett had asked the Flying Personnel Research Committee for the remains of a Spitfire aircraft to build the Cambridge Cockpit, he had immediately been given what he wanted. Now he was asking for a new way to understand flying accidents and a new way to banish those who doubted that fatigue was a danger to the expert pilot. This time he did not get what he asked for. The committee undertook a large-scale study of landing accidents, but it was not the kind of study that Bartlett had requested, and it was not conducted by Drew, Chambers, or Farmer, or any other Cambridge psychologist.

The Landing-Accident Study

At the twenty-seventh meeting of the Flying Personnel Research Committee, on September 3, 1942, the committee passed a resolution to the effect that Professor Bartlett, Dr. Austin Bradford Hill, and Wing Commander G. O. Williams should study landing-accident statistics.[22] Bradford Hill had been co-opted onto the FPRC in 1941.[23] He was an epidemiologist and statistician who had been a pilot in the Royal Naval Air Service during the Great War. Between the wars, he had authored a widely used textbook on statistical method for medical researchers and had written a number of reports for the Industrial Fatigue Research Board. He was later to achieve eminence for his role in establishing the causal link between cigarette smoking and lung cancer.[24] In the event, Bartlett dropped out of the study, although the reasons were not minuted or further discussed. It was Bradford Hill, not Bartlett, who would take charge of the next large-scale investigation of fatigue. It was designed to test the temporal parameters for the onset of fatigue that Bartlett had acknowledged in report 488.

The rationale for Bradford Hill and Williams's study of landing-accident statistics was simple. Landing an aircraft, especially a heavy bomber, is a difficult task. After a lengthy sortie over enemy territory, a pilot may be fatigued, and the longer the flight the greater the likelihood of fatigue. On the basis of these commonsense considerations, a simple prediction can be derived. All other things being equal, landing accidents will occur more frequently after long flights than after short flights. This prediction does not depend on the supposition that fatigue will manifest itself in a strictly regular way over time. During a flight, bouts of fatigue might come and go, although, ultimately, fatigue must show a tendency to get worse over time or it would hardly be recognizable as fatigue at all. Even if fatigue consists of sporadic lapses, the implication is that, as the pilot tires, these lapses will occur with increased frequency, or increased length, and thus increase the probability of pilot error. The prediction about landing accidents will still hold. It may be acknowledged that pilots will vary in their susceptibility to fatigue. One pilot's "breaking point" will differ from that of another pilot. But in a sufficiently large, random sample of pilots, this individual variability will not subvert the prediction. It will still be true that the performance of a large group of pilots making long flights should generate a greater number of landing accidents than a group of the same size making shorter flights.

After a preliminary look at the records, the time period selected for study was from April 1940 to March 1942. The investigators also limited themselves to accidents after night sorties. It was necessary to gain access to the documentation that recorded the number of landing accidents during the period of the study and the length of the prior flight. In order to provide a measure of the risk—that is, the probability of a landing accident—it was necessary to know how many sorties of a given length had been carried out during that time period. Then the number of accidents after a sortie of a given duration could be divided by the number of sorties of that duration. Accident numbers were recorded by the statistical department of the Air Ministry. The original data was divided into categories such as heavy landings, collisions, undershooting or overshooting the runway, whether the undercarriage was or was not lowered and properly locked, etc.

Unfortunately, these statistics only rarely recorded the duration of the preceding sortie. There was, however, another crucial document. RAF form 541 was used to record the times of takeoff and landing of all operational flights along with the type of aircraft, the names of the pilots, and their squadron. Bradford Hill and Williams had to coordinate these two sources of information. They then used a random sample of one in ten of the (nearly) fifty thousand operational night flights for which relevant documentation existed. This sample allowed them to arrive at an estimate of the number of sorties of a given duration during the chosen time period. The documentation also enabled them to rule out all the landing accidents that had been attributed to mechanical failure, bad weather conditions, or damage to the aircraft caused by enemy action. These contingencies were counted as special circumstances that merely served to obscure the effects of fatigue. Once such accidents were eliminated, all that remained were those accidents for which pilot fatigue alone might be responsible. This procedure finally gave a sample of around three hundred landing accidents that had taken place in the course of what were described as ordinary, straightforward landings.

The Landing-Accident Results

Bradford Hill and Williams drew up their results in FPRC report 423 of April 1943.[25] They reported that, over a broad spectrum of flight times, there was no significant difference in the landing-accident rate. Longer flights did not terminate in more landing accidents than shorter flights. The prediction derived from the theory of fatigue was false. There were some qualifications to this overall conclusion, but the two analysts said that they did not upset the main result. There was a high accident rate for very short flights of less than two hours, and also for very long flights of ten hours or more. The investigators suspected that the short-flight result was because many of these were early returns and were due to mechanical faults or bad weather conditions. They allowed that the accidents after very long flights might have been caused by fatigue but noted that there was counter-evidence even to this assumption. The problem for the fatigue hypothesis was that the data for very long flights showed a sudden and abrupt increase in accidents for flights of ten hours or more. If fatigue had been the cause of the accidents, then, for a large sample, surely the overall effect should have been building up gradually over increasing lengths of flight. There was no such progressive increase. By way of illustration, one of their main numerical findings, which exemplified their general conclusion, is set out in my table 4.1, below.

Bradford Hill and Williams's overall result was both striking and counterintuitive. The result stood in contrast to the opinions expressed in interviews with one hundred instructors from Bomber Command Training Units who had themselves completed operational tours of duty in Wellington, Whitley, and Hampden bombers. The interviews showed that the instructors all believed in the reality and significance of the fatigue generated by long flights.[26] But opinions, even well-informed opinions, are one thing, and the statistics were telling a different story. Bradford Hill and Williams broke down their data in a variety of ways, for example by aircraft type, season of the year, and type of accident. They found different accident rates for different aircraft (Blenheim bombers had a particularly high accident rate), and they found that there was a higher rate for winter compared with summer months, but they found no consistent, overall link between either accident rate or accident type and the length of the preceding sortie.

TABLE 4.1. Landing accidents in relation to duration of sorties

	Under 2 hours	2–	4–	5–	6–	7–	8–	9–	10 hours or over
Estimated number of sorties flown	1,080	3,730	4,860	9,020	10,240	5,960	2,570	1,070	410
Number of accidents recorded	13	27	34	53	48	32	14	5	8
Accidents per 1,000 sorties	12.0	7.2	7.0	5.9	4.7	5.4	5.4	4.7	19.5

Note: In column headers, "2–" means a flight of two hours or more but under four hours; "4–" means a flight of four hours or more but under five hours; etc. Data is combined for Wellington, Hampden, and Whitley aircraft (from table 8 of FPRC report no. 423, 1943).

Bartlett's Silence

What was Bartlett's response? The place to look is the minutes taken at the meeting of the Flying Personnel Research Committee at which Hill and Williams formally presented their findings. The meeting in question, meeting 31, was held on July 30, 1943. The meeting began with a report announcing that certain instrument-lighting problems had been resolved. Then there was a minute noting some statistical problems detected by Bradford Hill in an attempt to assess pilot temperament and neuroticism. The assessment had been made by Derek Russell Davis, a medical psychologist from Cambridge. Russell Davis was the man who had taken over from Drew to see whether there had been an experimenter effect at work in the original Cockpit report. At meeting 31, it was decided that Russell Davis should consult with Bradford Hill before proceeding with the research on pilot temperament. Finally, in minute 296, note was taken of the submission of the report by Hill and Williams, FPRC report 423, *An Investigation of Landing Accidents in Relation to Fatigue*. The minute book is worth quoting in full:

> Dr Bradford Hill stated that this investigation arose out of the findings from the cockpit experiment that under laboratory conditions, fatigued individuals after 3 or 4 hours carried out their tasks rather inaccurately. Other work suggested that if the subject could correct this state, he would be able to carry out his task efficiently. An endeavour was made to ascertain whether pilots do compensate for their fatigue, so that when returning from long bombing sorties they can avoid a crash on landing. Such accident rates have been calculated for sorties of different durations and show a practically unchanged level for the great majority of sorties flown, i.e. from 2 to 10 hours' duration. A slightly higher rate is apparent at each end of the curve, i.e. in very short or very long sorties.
>
> Professor Bartlett said that analysis of records of accidents in a number of industrial operations during the past year confirmed that a peak of accidents occurs about 2 hours after the start of the operation. He maintained that if individuals are aware of the liability to accidents from fatigue, they will be able to correct the tendency. He wished to avoid the spread of false ideas about fatigue in the R.A.F.

The best that can be said is that Bartlett's immediate, minuted response was opaque and of questionable relevance. More bluntly, it might be dismissed as obviously inadequate. Was Bartlett simply lost for words?

Bartlett and Bradford Hill had crossed swords on the committee on a previous occasion, and then, by contrast, Bartlett had been far from reticent.

The previous confrontation had been over a line of Cambridge research concerning "pre-selection" tests. This unglamorous but important work was designed to identify, at an early stage, which candidate pilots would cope with flying training and which ones should be rejected at the outset. Bartlett had submitted a number of reports describing the prosecution of the research.[27] One line of investigation that interested the psychologists was the different way in which candidates tried to solve problems. Bradford Hill expressed his doubts. He had visited the Cambridge Psychology Laboratory to talk with Bartlett but had concluded that their experiments were not valuable. Bartlett was displeased and did not mince his words. In February 1942, Bartlett told the committee that, while he had no quarrel with Hill's statistics, the point at issue was a matter for psychologists, not statisticians. Bartlett's case was that Bradford Hill did not understand the exploratory character of the psychological research that was involved: "He is forcing on the whole investigation a fixity and uniformity in all its phases to which it has never pretended and which is certainly never reached."[28] Perhaps, in Bartlett's eyes, the landing-accident investigation exhibited a similar rigidity.

The Bomber Command study had done exactly what Bartlett said should *not* be done. What Bartlett saw as the *nature of skill itself* had been ignored. With the exception of a passing remark about compensating for fatigue, the character of the changes caused by the onset of fatigue went unaddressed. Rather than conducting a scientific investigation into the nature of fatigue, the landing-accident study was confined to the timescale of the onset of fatigue—the matter of least scientific significance as far as Bartlett was concerned. The questions Bartlett had formulated about the background of accidents, such as the pilot's amount of sleep and the conditions and stresses prior to the accident, had been rendered irrelevant by virtue of being placed outside the scope of Bradford Hill's investigation. All Bartlett's scientific concerns had been sidestepped by the methodology of the Bomber Command study. No attempt had been made to address the nature of the psychological functions that mediated, or facilitated, or inhibited the skilled but varied activity of landing an aircraft.

All that mattered, as far as Bradford Hill and Williams's field study was concerned, was the final outcome, the success or failure of the landing, not the whys and wherefores of the accomplishment, or the different, skillful ways the aircraft had actually been brought back into safe and controlled contact with the runway. All the detailed differences had been swept into the single, undiscriminating category that the analysts referred to as ordinary, straightforward landings. Ebbinghaus had counted up the number of nonsense syllables successfully remembered. Bradford Hill counted up the number of landings successfully achieved. The methodology of the investigation had truncated the curiosity of the investigators. As a result, in both cases product, not process, had been pushed into the foreground.

But however narrow the question, the fatigue theory had given the wrong answer. Was this Ebbinghaus's revenge? Was the ghost of Kraepelin laughing? Had Muscio been right all along that the concept of fatigue caused more trouble than it was worth? The choice seemed stark. Bartlett appeared to be in the position of having to accept what Bradford Hill and Williams had found, and abandon the Cockpit results as misleading, or discover some way to challenge the result of the landing-accident study. In the crucial committee meeting on July 30, 1943, Bartlett did neither of these things.

What Could Bartlett Have Said?

There were three relevant and potentially important observations that could have been made about the landing-accident result. They were all of a methodological character. First, Bartlett could have explored a weak point that Bradford Hill himself admitted in his report. Despite the apparently extensive trawling of data, the number of accidents studied was, from a statistical point of view, relatively small. Less than three hundred accidents had been analyzed and, when the data was broken down into the systematically different accident rates for the different types of bomber aircraft, the numbers were very small indeed. Doesn't this fact make the overall negative conclusion suspect? Bradford Hill and Williams worked on a sample of 10 percent of the available data. Bartlett could have voiced the concern that their study was statistically underpowered. He could have suggested that Bradford Hill needed to repeat the analysis using all the available data rather than a 10 percent sample. A full-frontal criticism of Bradford Hill's statistical methodology, however, was hardly to be expected from the temperamentally nonstatistical Bartlett. In any case, given Bradford Hill's widely recognized expertise, it would have been a decidedly risky strategy. A safer strategy for Bartlett would have been to revisit his previous criticism and focus not on the statistics but on the psychology.

Second, and in accord with his previous strategy, Bartlett could have gone back to first principles and compared the methodology of the Cockpit and landing-accident studies when understood as psychological investigations. The aim of the Cockpit experiment was to place the pilot-subjects in a situation where it was reasonable to suppose that fatigue would ensue, so that the nature of the effect could be examined. Bradford Hill's test did not share this open-ended character. He began with a reasonable supposition about fatigue, but it could be argued that what was really under test in the landing-accident study was more than just the hypothesis that fatigue would be proportional to time on task. In the background there was a nexus of assumptions, both explicit and implicit, that sustained this focus and simplification. When no landing fatigue seemed to manifest itself, the problem then became one of locating the source of the error in the prediction. The negative result, as such, did not indicate where the error lay. It could have lain in the hypothesis nominally under test, or it could have lain in one of the numerous background conditions of the test. Bradford Hill's negative result constituted a genuine puzzle, but it did not furnish a decisive refutation of the positive result of the Cockpit experiment. Bartlett's predicament was therefore not as stark as it might at first appear. Nor, indeed, was the problem one for Bartlett alone. Scientifically, both researchers were challenged by the need to understand the relation between their two respective findings, and both were equally under an obligation to seek the explanation.

Thirdly, and more specifically, Bartlett could have put critical pressure on the character of the data on which the entire landing-accident study rested— that is, on the crucial Air Ministry statistical data, or the status of the information in the RAF form 541. Bartlett might have asked: *Who* provided this data and *who* filled in these forms? Indeed, *when* were the relevant forms filled in, and *when* was the accident data submitted and assigned to various descriptive categories? The significance of these questions is implicitly signaled in Bradford Hill and Williams's own description of how they designed their study. Their original intention had been to examine a more extensive time period beginning in March 1940 and ending in August 1942. However, at the time their inquiry began, in November 1942, it transpired that the relevant forms 541 for the period April to August had not yet been submitted by most of the bomber squadrons. The result was that the final study was confined to the period April 1940 to March 1942. There was sometimes a considerable time-gap between the dramatic events on the runway and the description of those events that finally came to rest in the statistics that constituted the institutional memory of Bomber Command.

Cambridge psychologists had already raised issues of this kind with the Flying Personnel Research Committee in connection with the reconnaissance reports compiled by the aircrews and intelligence officers of Coastal Command. The question of *when* a report was compiled was central to an investigation carried out by Bartlett's Cambridge colleague and friend, psychologist Gwilym Cuthbert Grindley. Over a period of some two weeks, Grindley flew with the reconnaissance crews of Coastal Command based at the Leuchars airfield on the east coast of Scotland. His assignment was to study how the crews reported the results of their observations of maritime activity. How did they distinguish fishing boats from military shipping and one military vessel from another? In opposition to the current practice, Grindley recommended that crews should be debriefed immediately on landing rather than writing up their reports after a significant period of rest after the long flights. They needed to get their reports down on paper more rapidly, that is, before their background assumptions and taken-for-granted sense of the relevant probabilities had done their creative work. In making this recommendation with the aim of increasing accuracy, Grindley made explicit appeal to Bartlett's findings in *Remembering*.[29]

The general form of the third question that could have been asked about the landing-accident study therefore concerned the formal, and informal, understanding of the purpose of the accident data when it was originally compiled. What were the perceived consequences attached to the assignment of this or that cause to a landing accident in the minds of those responsible for making the statistical records? Some complex psychological and social process generated the data used in the landing-accident study, so the question arises of what effect that process might have had on the final conclusion. Sophisticated investigators such as Bradford Hill and Williams would certainly have been well aware of these considerations, but being aware of a problem is not the same as overcoming it. In the event, they were not called upon to explain where they stood on such questions.

Bartlett's strategy could have been to respond to the landing-accident study by once again following Boas and seeing the bureaucratic paper-trail as the "autobiography" of the institution—that is, of Bomber Command. Bartlett could have invoked the principle: Do not explain what happens by appeal to the claims made in the statistical records or on form 541, but explain what appears in the records, and on the form, by reference to the happenings that ultimately generate this data. The problem, of course, was to find out what was really happening—and that was precisely the purpose behind Bartlett's proposals for gathering more extensive and detailed accident data, proposals that the Flying Personnel Research Committee, collectively, chose not to follow.

As far as can be judged by the minutes of the crucial meeting that received the report of the landing-accident study, Bartlett raised none of these issues. Of course, the processes that generated these minutes themselves raise a number of questions, and they do so for exactly the reasons that have just been laid out regarding the accident data itself. What lay behind those minutes? What purposes, and whose interests, do they serve? Do they perhaps conceal as much as they reveal? Do we really know what went on at that meeting? To use Boas's metaphor once again, the minutes are the "autobiography" of the Flying Personnel Research Committee, but are they to be relied upon? They can be read in a number of ways. Was Bartlett conceding defeat? Had he, under the pressure of events, "forgotten" his own ethnological perspective that might have prompted some of the methodological considerations sketched above? Or was he just keeping his powder dry?

Flying Neurosis, Radar, and Pavlov's Dogs

After Bradford Hill's study, the experimental program originally aimed at problems of fatigue acquired two new aspects. First, the Cambridge Cockpit work changed direction. Bartlett handed over the Cockpit to Derek Russell Davis (1914–1993), and the theoretical focus of the research shifted from flying fatigue to flying neurosis. Whether this shift was a direct result of the discordant character of the findings in the landing-accident study compared to those of the Cockpit study remains unclear, but it seems highly probable. There was, however, no discussion of any change in research policy minuted in the proceedings of the Research Committee. Had Bartlett lost confidence in Craik's Cockpit as an instrument with which to study the vicissitudes of high-level skill? The second change in the field of fatigue research was that the topic of skill fatigue was supplemented by a parallel concern with another sort of fatigue. There was a shift to the study of perceptual fatigue, in particular the perceptual fatigue caused by the protracted attention demanded by the use of radar apparatus. In this chapter, I shall discuss both of these new aspects to the Cambridge approach to fatigue. I shall start with the work of Derek Russell Davis, who challenged the idea that it really was fatigue, rather than some other phenomenon, that was being manifested in the Cockpit.

Flying Neurosis and the Cambridge Cockpit

Russell Davis had trained in medicine at Clare College, Cambridge, and had become a specialist in the field of psychiatry. He was said to be a man who was happy to challenge professional authority.[1] He admired Bartlett's humanistic approach, but this admiration did not lead him to accept Bartlett's analysis of the Cockpit work. After the war, Russell Davis was appointed to the lectureship in psychopathology in Cambridge, and later to a chair in Bristol. He was an eclectic theorist and even invoked the ideas of the American behaviorists. He spoke of the need for a "liaison officer" to mediate between the psychiatric clinic and the psychological laboratory.[2] He saw himself as playing precisely that role.

Russell Davis argued that it was incorrect to see the Cambridge Cockpit experiments as studies of fatigue. Initially (when acting as a stand-in for Drew to test for the possibility of an "experimenter effect"), he appeared to have accepted the fatigue interpretation, but he soon moved to a position more like that of Muscio. He came to believe that the concept of fatigue had no explanatory power. That the pilot's behavior became disorganized was beyond doubt. But in a 1946 report for the Medical Research Council, Russell Davis said: "It was no explanation of the disorganisation to attribute it to fatigue, and further experiments had to be carried out to discover the reasons for it."[3] Later, in the 1948 publication *Pilot Error*, Russell Davis noted that the attempt to explain the errors by appeal to fatigue was, as he put it, purely hypothetical: "No evidence was obtained to show with what tissue changes the errors were associated."[4] The fatigue hypothesis, he argued, was untenable. Significantly, Russell Davis cited Bradford Hill and Williams's landing-accident study in support of his conclusion that a change in the direction of the research project was needed.[5]

In his criticism of the fatigue interpretation, Russell Davis noted that the disorganization of flying skill in the Cockpit often revealed itself early in the tests. Sometimes it would appear after a mere ten or fifteen minutes. If it had really been a manifestation of fatigue, said Russell Davis, this disorganization would have occurred late, not early. Furthermore, if it were really fatigue that was being observed, then its effect would develop in a uniform and progressive way throughout the test. Clearly, Russell Davis was happy to work on the basis of at least some a priori definitions rather than treat the nature of fatigue as an open question, as Bartlett and Drew had. This definitional stance led Russell Davis to cast a skeptical eye on some of Drew's other primary data. The graphs showed that directional errors in the flying increased uniformly throughout the test. That certainly looked like fatigue at work, but what about the graph for sideslip that revealed the pilots' handling of the ailerons? That was much more irregular. To Russell Davis, it did not look like a graph of fatigue at all. He therefore reanalyzed Drew's data by breaking down the errors into small, medium, and large. He found that small errors died away in the second half of the test, while large errors increased. This finding suggested to Russell Davis that two different modes of response were at work, and he set out to identify and explain them.

Above all, why did performance in the Cockpit fall off so rapidly and catastrophically? Russell Davis pointed to the fact that Craik had made the Cockpit unstable and had simplified some of the responses to the controls. He conjectured that the mechanical unreality of Craik's machine meant that subjects would be unclear about what counted as an adequate performance. How seriously would the subjects have taken the aircraft-like nature of the apparatus? How were they to have known whether they were acquitting themselves well or badly? The standards that the pilots would have acquired during their flying training might have told them that they were doing badly, but they did not really know if these were the standards they should have been meeting. Russell Davis concluded: "The rapidity and intensity of the deterioration probably depended upon the lack of means of judging what relative standards were being attained."[6] Thus:

> Many pilots are unable to control the machine with the degree of accuracy which their training as pilots might lead them to expect of themselves, and they had no means of judging what relative degree of success they were achieving . . . The experimental conditions were of a kind which might readily lead susceptible individuals to an increase of anxiety.[7]

The appeal to anxiety provides the key to Russell Davis's standpoint. Bartlett would have called this appeal the "great, unverified guess" that runs through Russell Davis's analysis of the Cockpit results—and it was a different guess from that which informed the original Cockpit experiments. Russell Davis cited Bartlett's Ferrier Lecture but studiously avoided any mention of Jacksonian ideas. Superficially, the account he gave of Bartlett's position might have passed as accurate. But all that was most characteristic of Bartlett and Craik's thinking—for example, regarding the role of Jacksonian levels and the process of inhibition and release—was rendered invisible. Thus, Russell Davis said that the central characteristic of a pilot's behavior was that it was *organized*. It was not a simple sum of component actions because the components of the skill would have become integrated. The nature of this organization, he said,

> has been discussed in detail by Bartlett (1943) who has suggested that errors arise from a change in the relationship, within the whole activity, of components which are themselves correctly performed, and especially from a disturbance of the timing of the components in relation to one another. Thus, the *disorganization* of the skill results in errors.[8]

At first glance, this passage seems like a respectful nod to Bartlett's Ferrier Lecture, but in fact Russell Davis bypassed Bartlett's actual approach by

failing to note that, for Bartlett, the "organization" in question was a hierarchical system of levels. Nor was it revealed that the main form of "disorganization" discussed by Bartlett was described as a higher level failing to
control a lower level. In its stead, and under the cover provided by the general
category of "organization," Russell Davis appealed to a very different reper-
toire of concepts. For him, the question was whether the overall organization
was "stable." His belief was that the factors that determined this stability were
emotional and motivational. Unstable people had unstable skills. Russell Davis also spoke of "anticipatory tension" and gave anxiety the status of a "drive."
Whatever reduced this drive would provide "reinforcement," and hence any
anxiety-reducing course of action would be prone to repetition. These statements reveal an informal and non-rigorous, though perfectly legitimate, ap-
propriation of the ideas of American learning theory.

On a commonsense level, and putting behaviorist terminology aside, Russell Davis was talking about fear, the susceptibility to fear, and the desire to
escape from fear. Craik had not built a fatigue machine; he had built a fear
machine. Russell Davis did not, however, construe fear in a narrow way. Pilots in battle certainly know fear, but, Russell Davis asserted, they did not
only fear injury and death: They also feared letting their squadron down.
They feared failure and shame. A similar fear of failure, he said, came into
play when the pilots were subject to unfamiliar demands such as those posed
in a psychological laboratory. This is why tempers sometimes flared. Anxiety
could explain why some subjects refused to cooperate and complete the experiment, and why performance levels in the Cambridge Cockpit were often
so poor.

Given Russell Davis's medical training, it is natural that he wanted to relate his work to that of the medical branch of the RAF. The medical officers
of the RAF had the job of identifying and treating aircrew whose behavior,
under stress, was judged to have become pathological and a danger to themselves and their comrades. Russell Davis began to use the Cambridge Cockpit
as a device for studying the forms of pathology that had been given the name
of "flying neurosis." Flying neurosis took two related forms, although a state
of acute anxiety was seen to be at the root of both. Sometimes it expressed
itself in agitated activity, in which case it was known as "anxiety neurosis."
The condition was accompanied by loss of sleep, irritability, loss of concentration, and, predictably, a loss of confidence and flying skill. Neurosis also
expressed itself in apathy and withdrawal, in which case it was referred to as
"hysterical neurosis." Hysterical neurosis led to the "conversion" of the anxiety into a range of apparent physical symptoms such as digestive problems,
headaches, and visual disturbances. Such symptoms can be the side effects of

physical illness, but "neurosis" was the diagnosis when the symptoms had no identifiable cause such as bacterial or viral infection. In lay terms, the pilots "broke down."

The basis of flying neurosis was assumed to lie within the character of the individual, whether innate or learned. The problem for squadron medical officers was how to identify and prevent flying neurosis or treat it once it was established. Russell Davis did not claim to have the answers, but he was confident that his research could contribute to the solution of these problems. His central idea was that the Cockpit exposed embryonic neurosis. The detailed record of control movements could be read as revealing outer behavioral signs of the problematic inner states of mind:

> Hence it may be argued that the type of error concerned is a particular form of a general class of behaviour, which appears in its most marked form in neurosis.[9]

Degrees of disposition to neurosis would express themselves as degrees of disposition to error, and the two different kinds of neurosis would be reflected in different kinds of error. Anxiety neurosis was assumed to be expressed by overactivity in the Cockpit, and hysterical neurosis by inertia. Here, then, was a new theory and a new research program. But it was also an old Cambridge preoccupation. Russell Davis was reactivating the concerns of Rivers and Myers when they had studied shell shock during the Great War. The same broad diagnostic categories were still in use. The old contrast between anxiety neurosis and hysterical neurosis was still employed, only now the symptoms were identified through the telltale details of control movements recorded by Craik's machine.

Testing the Neurotic-Disposition Theory

Russell Davis gained access to a large number of experimental subjects by participating in a collaborative research program known as the "Harrogate Investigation." This was an ambitious (though never completed) study involving one thousand bomber pilots at an intermediate stage in their training. It derived its name from the RAF Personnel Reception Centre no. 7 at Harrogate, which was the administrative center of the project. The overriding aim was to discover the personality characteristics of the reliable and successful bomber pilot. The military authorities wanted to predict who would break down under the stress of operational flying. Those with a disposition to flying neurosis could then be excluded from such roles. To achieve this goal, it was necessary to make a preliminary assessment of any latent predisposition toward neuroticism. It was then necessary to show that the assessments were consistent and validated by subsequent performance. Doing so involved following the sample of pilots through their lengthy training as they moved from initial flying training to advanced flying training, and then to operational training. The ultimate validation of the assessments would be provided by their operational performance. Given such validation, the procedure could then be given practical application as a selection procedure.

The assessments were to be made by RAF psychiatrists who would interview each pilot for about forty-five minutes, noting any history of mental instability, depression, tendencies to nightmares and sleeplessness, and other personality traits deemed relevant. The pilots were then assigned to one of three categories: (i) zero disposition to neurosis, (ii) slight disposition, or (iii) moderate disposition. To have a disposition to neurosis is not the same thing as having a neurosis or displaying neurotic symptoms or behavior, but the aim was to identify potential problems that were not easily visible. In the event, the war ended before the study was complete, though enough work was done to establish one frustrating negative result. The degree of imputed neurotic disposition did not correlate with any subsequent tendency to have flying accidents.

Russell Davis acknowledged this negative finding in his postwar *Pilot Error* but could not have known of this outcome when he began his research during the war. When he aligned his research with the Harrogate Investigation, that negative result lay in the future. Nevertheless, the initial psychiatric discriminations at the start of the Harrogate Investigation gave Russell Davis an opportunity to cross-check the assessments with the performance in the Cambridge Cockpit. Russell Davis studied a sample of 355 of the pilots who had been interviewed and evaluated by the psychiatrists. His question was: Would those assigned "nil" disposition to neurosis behave differently in the Cockpit tests from those alleged to have a "slight" or "moderate" disposition to neurosis? Would the nil-disposition pilots be less inclined to be overactive or underactive in their handling of the controls? Russell Davis identified the levels of activity in terms of behavior that could be read off the recorded traces of the control movements and, in this way, operationally defined the three categories of "normal," "overactive," and "underactive" levels of pilot activity. Overactivity was plausibly associated with anxiety neurosis, while hysterical neurosis and lapses into an inert state represented a withdrawal from the field of motivational conflict. This picture was broadly confirmed by a standardized test of forty-five minutes of "flying" in the Cambridge Cockpit. A simplified but representative version of his findings, taken from *Pilot Error*, is given below as table 5.1.

TABLE 5.1. Association of grade of predisposition and test class

Grade of predisposition	Test class			Total	Percentage abnormal
	Normal	Overactivity	Inertia		
Nil	130	13	12	155	16
Slight	113	32	9	154	27
Moderate	25	14	7	46	46

Note: Pilots with no identifiable predisposition to neurosis are less likely to be either overactive or underactive (from table 10 of Russell Davis 1948).

TABLE 5.2. Test results of patients and fit pilots

	Normal		Overactivity		Inertia		Total
	Number	Percentage	Number	Percentage	Number	Percentage	
Fit pilots	268	75	59	17	28	8	355
Patients	13	33	11	28	15	38	39

Note: The proportion of fit pilots in the normal class is significantly higher than that of the patients (from table 9 of Russell Davis 1948).

In a second study, Russell Davis used a sample of thirty-nine pilots who were already being treated in an RAF psychiatric clinic in the spa town of Matlock, Derbyshire. These pilots, who sometimes had impressive operational records, had broken down under stress. They were deemed to have become neurotic, rather than just being disposed to neurosis. Russell Davis accepted the clinical diagnosis used in the hospital and compared the Cockpit performance of these patients with those of fit pilots used in the previous experiment who had turned in a "normal" performance. The number of neurotic patient-pilots was small, but the result went in the predicted direction (see table 5.2).

Three-quarters of the non-neurotic pilots turned in a normal performance in the Cockpit, while only one-third of the neurotic pilots handled the controls in a way that could be rated normal, that is, neither over- nor underactive. The data in table 5.2 is again taken from *Pilot Error* and is presented in a simplified form.

Accidents and Suspensions

Russell Davis sifted through the RAF records to find out what had happened to the pilots who had been tested in the Cambridge Cockpit as they moved through the Advanced Flying Units and on to the Operational Training Units.

TABLE 5.3. Test category and outcome of training course

Test category	Courses entered	Courses completed	Suspensions	Fatal accidents
Normal	369	350	13	6
Overactivity	75	70	3	2
Inert	34	24	6	4
Total abnormal	109	94	9	6
Total	478	444	22	12

Note: In the case of both suspensions and fatal accidents, the proportion in the abnormal classes was significantly higher than in the normal class (from table 11 of Russell Davis 1948).

He wanted to know whether these pilots had successfully completed the advanced courses, or had been suspended, or had suffered a fatal accident. Where records existed, it became clear that the proportion of suspensions and fatal accidents was much lower for those whose Cockpit performance was deemed "normal" as compared with those whose tests revealed the pilot to be overactive or inert (see table 5.3).

Summarizing the data, Russell Davis stated:

> Altogether there were 12 fatal accidents in the 478 courses entered at advanced flying and operational training units. A relatively large proportion of these pilots had been placed in the abnormal classes. In the normal class there was 1 fatal accident for every 58 courses completed; in the overactive class there was 1 for every 35 courses completed; in the inert class alone there was 1 for every 6 courses completed.[10]

In making his scientific case, Russell Davis did not confine himself to the medium of tables and numbers; he also provided a less formal but more dramatic narrative accompaniment to the statistics - an accompaniment in which he told the story of individual cases. To convey the full impact on a reader of his *Pilot Error*, it is worth quoting at some length a section dealing with an episode associated with the Harrogate Investigation. At this point in *Pilot Error*, we might say that Russell Davis was simply recounting a war story. Some future Boas might record it under the title "The War of Two Pilots." Like "The War of the Ghosts," the story dealt with the fate of two young men of very different temperaments.

"The War of Two Pilots"

Here is the full text of the story as Russell Davis told it:

> The skill of one pilot proved in the Cockpit Test to be very easily upset. His performance during the practice period was quite normal. Early in the test

itself, however, he started a series of manoeuvres at 7,500 feet, as part of the prescribed exercise, and, losing control of the machine, he allowed it to lose height without making any real attempt to retrieve the situation. The test was stopped when o feet was reached. The pilot was plainly distressed and emotionally upset, but he seemed as if paralysed and incapable of any organized activity by which at any time in two or three minutes he could have regained control.

He was killed in an accident three months later when his aircraft crashed shortly after taking-off on a night exercise at an Advanced Flying Unit. What happened could not be known for certain, but the investigating officer interpreted the evidence of several eye-witnesses thus: "After taking-off and climbing some distance the pilot failed to rely on, or concentrate on his instruments and allowed the aircraft to turn to the right as he was levelling out. On seeing the airfield lights on his right instead of on the left beam, he must have panicked and for some reason began to put his flaps down. This panic would account for the see-saw up and down seen by the eye-witnesses. In an attempt to turn to the right again opposite the end of the runway, having flaps going down, he allowed the nose of the aircraft to drop on the turn. The aircraft then dived into the ground under considerable power."

The test performance of another pilot was good; he made very few errors. However, he sustained ten accidents during an operational tour in an extraordinary series of mishaps, for none of which was he to blame. The nose of his aircraft was cut off by collision with another aircraft over enemy-occupied territory and two of the crew were killed, but he brought the severely damaged aircraft back to his base and landed it successfully.—"A very good show," as his Squadron Commander remarked. Once his aircraft was damaged by enemy action; and on seven occasions, once when flying low, he encountered engine failure for reasons outside his control each time. On every occasion he behaved appropriately and effectively, showing unusual ability to deal with every emergency without disorganization of skill. He was never injured.[11]

Whether an anecdotal account of this nature should be accorded the status of scientific evidence is perhaps a moot question—and the tight-lipped "good show" of the squadron commander gives the narrative a distinctly dated feel. Nevertheless, the overall figures for suspensions and accidents supplied by Russell Davis indicate that the Cambridge Cockpit tests had some claim to predictive power regarding the danger confronting the trainee pilots of the RAF. And it is well to be reminded that pilot training, as well as operational flying, was highly dangerous. It is reassuring to think that psychologists could at least hope to identify a class of persons who were especially at risk. But whether or not it counts as evidence, I suspect that it is this episode in Russell Davis's *Pilot Error* that was the basis of the story that still circulated among students in the Psychology Laboratory some twenty years after the war, and which perhaps still circulates today. It was not true that Cambridge psychologists could, or even tried, to predict who would be shot down. They had no such sinister predictive power, and nor did they aspire to it. That aspect of the story was simply false, though no one who has read through all the versions of "The War of the Ghosts," as Bartlett set them out in *Remembering*, would be surprised that the final iteration of "The War of Two Pilots" should take the simplified and dramatic form: *They knew who would be shot down.*

A Gap in the Argument

Whatever Russell Davis's role as the generator of arresting war stories, there was a gap in his argument in *Pilot Error*. Russell Davis had established that there was a link between neurotic predispositions and abnormal Cambridge Cockpit results. He had also shown that there was a link between abnormal Cambridge Cockpit results and failure and fatality on subsequent advanced and operational training. If A causes B, and B causes C, then, if there is no ambiguity in the middle term, it is reasonable to conclude that A causes C. Causation, unlike, say, mere similarity, is a transitive relation. The problem was that if A = neurotic disposition and C = accident proneness, the relation did not hold. That much at least had been established in the aborted Harrogate Investigation. Russell Davis fully acknowledged that there was an inferential problem:

> It would be expected, therefore, that neurosis would make an individual especially liable to accidents. Although not conclusive, there is considerable evidence that this is so. There is also evidence to the contrary, the most weighty of which is an absence of association between the psychiatrists' grades of disposition in the subjects of the Harrogate investigation and the occurrence of accidents. In fact, whatever the grade of predisposition, the accident rate proved to be similar, a finding which weakens the theory.[12]

The finding weakened the rationale for using the Cockpit as an instrument for studying the emotional cause of accidents. A possible explanation for the logical gap, said Russell Davis, is that the predictions of different psychiatrists are based on different understandings of the term "neurosis." He had criticized Bartlett's theory because its central concept of fatigue was vague and not grounded in physiology; now he had to admit to the same weakness in his own theory.

There was also another, and perhaps deeper, problem. Whatever their theoretical explanation, Drew's empirical results still stood, and so did Bradford Hill's. If the landing-accident statistics were a basis for saying that Drew's results could not be explained by fatigue, they posed a similar problem for Russell Davis. If fatigue explains the Cambridge results, then the Bomber Command statistics implied that there was more fatigue generated by "flying" the Cockpit than by flying a real bomber aircraft. Can this be true? But on the anxiety theory, these same statistics committed Russell Davis to saying that "flying" the Cambridge Cockpit was more emotionally demanding and neurosis-inducing than flying a bomber. Is that any more plausible?

Bartlett Draws a Boundary

Bartlett insisted that fatigue was a normal rather than a pathological phenomenon. He said so, emphatically, in the opening sentence of his Flying Personnel Research Committee report 488 of 1942. Bartlett also believed that there was nothing abnormal about everyday anxiety. Ordinary anxiety, he insisted, was one thing, and neurosis was another. Bartlett accepted the reality of character and temperament, and he was prepared to divide pilots into overactive and underactive types, but he insisted that this division reflected a differentiation within a normal population.[13] Bartlett also accepted Russell Davis's finding that performance in the Cockpit was associated with subsequent failures in flying training. But note the qualification when, in discussing the Harrogate Investigation, Bartlett said that the Cockpit "was found to be a more accurate predictor of the accident prone individual (not the potential psychoneurotic) than any known method, including the psychiatric interview."[14]

In his Ferrier Lecture, Bartlett had said that fatigue revealed itself in an increased irritability. The pilot tended to blame other people and other things for his own failings. Russell Davis had confirmed this tendency to project the cause of difficulty onto external sources. However, Russell Davis interpreted this tendency as evidence for the claim that fatigue-like symptoms were really expressions of incipient neurosis. Bartlett tried to head off this interpretation in report 488. He noted:

> Since the projection of personal characteristics upon other people or inanimate objects is found in a good many neurotic patients, there is a danger that it may be taken as a sign of a neurotic trend when it occurs under fatigue. For this there is no justification whatever. Both the increased irritability and the projection are normal symptoms. They disappear rapidly once the resting period is reached.[15]

Bartlett's discomfort with the psychiatric interpretation of the Cockpit results was expressed again in the course of the twenty-ninth meeting of the committee on February 23, 1943. The minutes record that

> Professor Bartlett thought that there was a case for telling aircrews something about symptoms of normal fatigue, but he was most anxious to prevent the idea from spreading that fatigue was a symptom of psychoneurosis.[16]

Spreading the idea that fatigue was a symptom of neurosis was, arguably and in effect, just what Russell Davis was doing.

Two other interventions in meeting 29 show how serious Bartlett was in his desire to keep his distance from psychopathology. The first of these further interventions was prompted by the receipt by the committee of a series of detailed medical reports by Air Commodore (later Air Vice-Marshal) Charles Symonds. These reports were referred to as "the FPRC 412 series." They were made public after the war, in 1947, in a three-hundred-page book issued by the Air Ministry under the title *Psychological Disorders in Flying Personnel*.[17] Commenting on Symonds's reports, before their publication, Bartlett expressed his concern that the sheer bulk of material created the wrong impression. The minutes read:

> Professor Bartlett said that the size of the FPRC 412 series of reports by Air Commodore Symonds could give the impression that cases of neuroses were common in the R.A.F., whereas, in fact, the incidence was low. Exposure to the sort of conditions obtaining in the Air Force does not produce psychoneurosis unless there is a constitutional predisposition.[18]

The psychologically normal person, as Bartlett understood normality, was assumed to possess the resilience to withstand the stress of wartime flying duty.

The second of these interventions involved Bartlett having a tilt at a report (FPRC 508) by Squadron Leader D. D. Reid. The report bore the title *The Influence of Psychological Disorder on Efficiency in Operational Flying*.[19] Reid (who would later hold a chair in epidemiology at the London School of Hygiene and Tropical Medicine) used medical records and clinical interviews to argue for a link between neurotic disposition and failures in operational flying. His theory was similar to that adopted by Russell Davis, who cited Reid's work in support of his own position. Bartlett, however, evinced skepticism about the value of Reid's approach to accident data. In the minutes, it is reported:

> Professor Bartlett thought that there was a tendency in FPRC 508 to tie up accident proneness with some sort of psychological disorder, whereas the former was a notion arrived at by statistical analysis of data collected from

a normal population. It was important that if individuals suffered flying accidents it should not be assumed that they are cases of flying stress. It was essential that the medical officer should be able to distinguish between cases of
mental disorder and one whose manner is a slight variation from the normal.[20]

If Bartlett's first intervention, asserting that fatigue was normal, sounded reassuring, the second carried a sting in the tail. Bartlett was saying that accidents should not be excused by designating them as symptoms of stress.
There should be no routine inference from accidents to an underlying pathology. For Bartlett, even the fatigued pilot was still responsible. The implication
was that squadron medical officers must not override commonsense moral
distinctions and notions of responsibility.

Although Bartlett had handed over the Cambridge Cockpit research to
Russell Davis, Bartlett kept his distance from all references to neurosis and
pathology. Like Rivers, Myers, and Head during the Great War, Russell Davis wanted to build bridges between experimental psychology and medicine.
Though Bartlett admired Rivers and Myers, he possessed none of their medical expertise, and there was every sign that, in World War II, he opposed this
strategy of rapprochement. In his work for the Flying Personnel Research
Committee, the closer medical concerns came to the territory of experimental
psychology, the more cautious Bartlett became. He wanted to anchor psychology in the normal rather than pathological.[21] When Davis rejected the appeal
to fatigue because nobody could demonstrate the relevant "tissue changes,"
Bartlett will have remembered the old dispute over the protopathic-epicritic
distinction. The critics of Head and Rivers had demanded proof of the reality of epicritic and protopathic nerve fibers. They wanted to see the relevant
tissue. The response that the distinction was functional had been brushed
aside. But even in 1940, it was still difficult for psychologists to separate what
was valuable to them from what was questionable in the Jacksonian tradition.
With Bradford Hill's landing-accident results on the one side and Russell Davis's psychiatric results on the other, Bartlett's neo-Jacksonian stance was not
easy to maintain, and the problem was about to get more difficult.

Radar Watchkeeping

New fatigue-like effects revealed themselves in the course of WWII when radar operators had to scrutinize radar screens for long periods of time. Bartlett
described the strange character of these fatigue effects by saying:

> Although the operator continues as intently as ever to look . . . or to carry out
> whatever other sensorial responses may be required, there are items which he

fails entirely to register, even though they are directly within the field to which he is all the time vigorously attending.[22]

In British terminology, "radar" was originally known as Radio Direction Finding, or RDF, and the British spoke of monitoring a radar screen as an exercise in "watchkeeping." This task became the focal point of a new species of fatigue research.

During WWII, the Canadian-born Norman Mackworth, originally a medical doctor, had worked closely with Bartlett in Cambridge on a number of projects. The projects were commissioned by the FPRC and financed by the Medical Research Council. The two men had studied the layout of RAF control rooms, and they had also investigated the way bomb-aimers were influenced by the random, spatial distribution of marker flares when estimating their aiming points.[23] Mackworth recalled that it was toward the end of 1943 that "the Royal Air Force asked if laboratory experiments could be done to determine the optimum length of watch for radar operators on anti-submarine patrol, as reports had been received of overstrain among these men."[24]

The radar operators flew with the aircraft of Coastal Command, and the flying conditions were often atrocious, but the task was vital. Large numbers of British ships were being sunk, and the consequences were potentially catastrophic for the British conduct of the war.[25] Even with the recently improved radar equipment, the efficiency of watchkeeping deteriorated rapidly. Mackworth and Craik went down to the Coastal Command station at Chivenor in Devon, which was the base for many of the patrols across the Bay of Biscay. They wanted to look at the problem firsthand, but the reasons for the deterioration proved obscure and called for systematic study. Mackworth confronted the question experimentally and reported his findings to the Flying Personnel Research Committee in May 1944. Mackworth's FPRC report no. 586 bore the title: *Notes on the Clock Test—A New Approach to the Study of Prolonged Perception to Find the Optimum Length of Watch for Radar Operators.*[26]

Mackworth asked his experimental subjects to keep watch on a display that resembled a clock face rather than a radar screen. (No synthetic training devices for the new air-to-surface-vessel ASV Mark III radar sets were available, so it was necessary to improvise.) The important thing, said Mackworth, was to devise a laboratory task that captured the essence of the problem. Five conditions had to be realized: (i) real-life submarine search mainly amounted to "waiting for nothing to happen," so the required responses would be very infrequent; (ii) like the radar operator in an aircraft, the subject would be isolated except for some form of telephonic intercommunication; (iii) there was no way for the subject to check on their own efficiency; (iv) like the elusive blip on a radar screen—the real object of the subject's search—the experimental signal should not be easy to see; and (v) there should be limited time for making a response, so the stimulus had to be brief as well as infrequent, just as a radar contact with a rapidly submerging submarine might be lost after a few moments. All these conditions were met by the Clock Test. Mackworth wrote:

> The subject sat in a wooden cabin in a room entirely by himself and looked straight ahead at a black pointer 6 inches in length which normally moved like a second hand of a large clock in front of a vertical white surface. Every second it jerked on to a new position, one hundred of these movements making up the full circle. There were no scale markings or reference points of any kind on the background.[27]

The subjects were told that every now and then, the pointer would make an unusually large jump forward and move twice the normal distance. This double jump would be a signal that required the subject to press a response key. All that the subject would hear during the test would be the hum of the apparatus. There would be no auditory cue to the double jump. A practice period of five minutes was provided in which the experimenter showed the subject examples of the usual background movements and the unusual double jumps that constituted the signal. The experimenter accompanied the subject's practice responses with expressions such as "That's one!" or "Now!" The subsequent experimental session lasted two hours.

When analyzing the results, the two hours of watchkeeping were broken down into four half-hour sessions, and the subject's performance was scored in terms of the number of signals missed in each of these sessions. These scores provided four data points for the graph that registered change in the efficiency of the watchkeeping. Within each of the four half-hour segments of the watch, the subject was given just twelve double-jump signals by the clock. The eleven intervals that separated the signals were (in minutes): ¾, ¾, 1½, 2, 2, 1, 5, 1, 1, 2, 3. This sequence took twenty minutes, and the remaining ten minutes of each half-hour session were devoid of any signals. All of the half-hour sessions followed on from one another without a break, repeating the same sequence of twelve signals with the same intervals between them. The subjects were 170 RAF cadets from a nearby Initial Training Wing. The repetition of the sequence was not known (or noticed) by the subjects.[28] The results clearly exhibited a stark decrement that, in an operational flight, would mean missed opportunities to detect a submarine and, as a consequence, perhaps the loss of a merchant ship, its cargo, and its crew. The graph of decreasing efficiency is shown in figure 5.1.

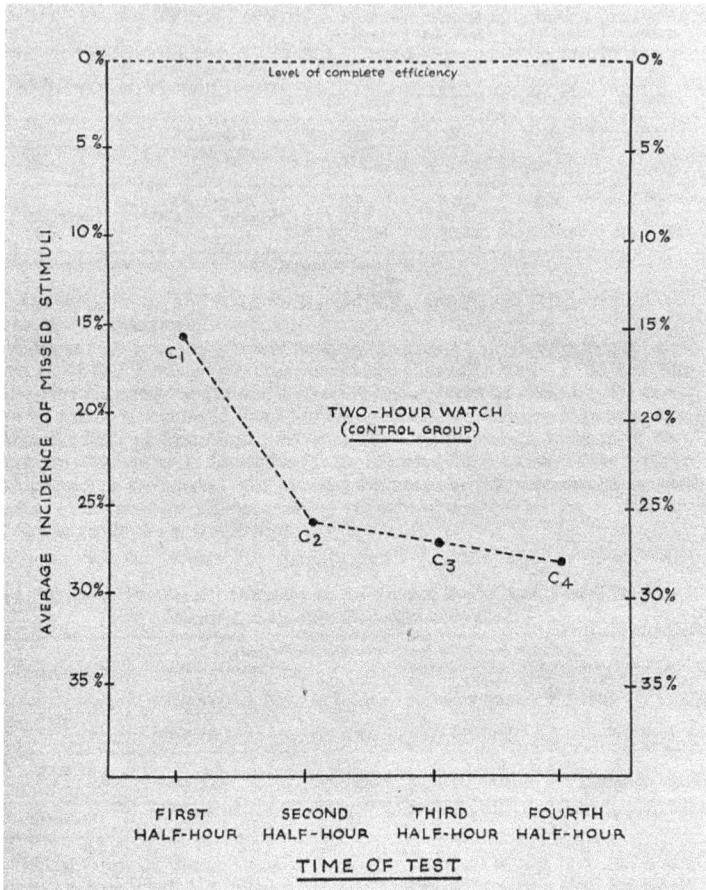

FIGURE 5.1. Decline of vigilance in radar watchkeeping. Using his Clock Test, Mackworth plotted the percentage of missed signals for each half hour of a two-hour watch. His graph showed the marked increase in the percentage of missed signals after the first half hour. The implication was that a radar watchkeeper should only work for half an hour before being given a rest. From Mackworth 1950, 15.

Mackworth's summary was stark:

> For every ten signals missed during the first half hour, between 17 and 18 signals were missed during each of the second, third and fourth half hours of the test.[29]

The indications were that the optimum length of a watch was about thirty minutes. Mackworth then conducted a number of further experiments to see what might be done to mitigate the situation. Could the rapid fatigue decrement be prevented? Using the same experimental arrangement and the same sequence of signals, he showed that two radar operators, alternating

half-hour spells of watch over the two hours, would sustain the initial, relatively high, level of vigilance. A rest of half an hour restored the efficiency of
the watchkeeping. By contrast, he found that general motivational talks or
further instructions by an officer before the test had no beneficial effect.

Other modifications had more unexpected effects. Using twenty-five further cadets, Mackworth introduced a telephone into the experimental room.
The subjects performed the experiment in the usual way except that they
were told to listen out for a telephone call. They were to answer the call and
then they would be given further instructions. The Clock Test proceeded in
the standard fashion until near the end of the second half-hour session, when
the telephone rang in the middle of the ten minutes that were devoid of signals. The subjects were not in fact given new instructions but only a standard
message requesting that they do even better in the remainder of the test. The
first noticeable result of the modification was that listening for the phone call
lowered the efficiency of watching for signals. If this finding was predictable,
the second part of the result was surprising. The rate of signal detection after
the brief telephone call returned to the level of the first half hour. The beneficial effect again lasted about half an hour, but the improvement more than
counterbalanced the decrement due to listening out for the call.

Mackworth found that 10mg of Benzedrine prevented the vigilance decrement over the full two hours of the Clock Test. Placebo tablets proved to have
no effect. He also studied the effect of telling the subject, after every signal,
whether or not it had been detected. He would say "You got that" or "You
missed one there." This feedback mechanism also prevented the decline in
efficiency. Such feedback has no counterpart in real submarine hunting, but it
was a potentially revealing result. Mackworth also varied the apparatus. Craik
had now built a mock-up of the ASV Mark III radar set, and Mackworth
repeated his Clock Tests using this more realistic display. He got the same
results, thus confirming his original insight into the essence of the problem.[30]
Mackworth's wartime report to the Flying Personnel Research Committee
contained reliable experimental facts and sound practical recommendations
but no theoretical analysis. One large and obvious question presented itself.
How did the fatigue generated in watchkeeping and vigilance tasks relate
to the fatigue apparently manifested in tasks demanding skill? Mackworth's
original report to the Flying Personnel Research Committee was entirely empirical. It was only after the war that he published a theory to account for the
result. That theory was seen as problematic, and the relation between perceptual or vigilance fatigue and other forms of fatigue was one of the themes that
informed the postwar research program of Cambridge psychologists. What,
then, was Mackworth's understanding of his own results?

Radar Screens and Pavlov's Dogs

At the first meeting of the new Experimental Psychology Group, held in the library of the Cambridge laboratory in October 1946, Mackworth proposed a theory to explain the vigilance decrement he had observed. His theory was based on the experiments described in Pavlov's classic treatise *Conditioned Reflexes.*[31] Pavlov's work was widely recognized as a contribution to a truly hardheaded, physiological standpoint from which all traces of subjectivity had been expunged. Utilizing Pavlov's ideas, Mackworth interpreted the initial instruction given to his subjects in the Clock Test as a form of conditioning. The period of the experimenter's instructions and encouragement was thus seen as a period in which certain responses were rewarded. The subsequent experimental run, by contrast, was a sequence of responses that were not rewarded in the same way. As Pavlov had demonstrated in his experiments on dogs, a natural or "unconditioned" response, such as salivation in the presence of food, can become "conditioned" by association to appear at the sound of a bell. But if the response is repeatedly evoked in this way but not followed by the reward of food, the response will diminish and cease. Pavlov called this phenomenon "experimental extinction."

For Mackworth, the watchkeeping decrement was to be seen as an instance of Pavlov's experimental extinction because most of the acts of watching the radar screen were unrewarded—there was no signal that betrayed the presence of the sought-for target. Mackworth then followed Pavlov in identifying experimental extinction as a form of inhibition. By this, Pavlov meant that the tendency to make the response had not been destroyed by the lack of reward but had merely been suppressed. This explanation supported Mackworth's finding that efficient detection reappears after a rest—the buildup of inhibition dissipates with time. The return to efficiency after a rest can therefore provide a measure of how rapidly, and by what laws, the inhibition dissipates. Pavlov also reported that a sudden, surprising stimulus, such as a loud noise, could dispel inhibition and make the conditioned response return. Here, argued Mackworth, was the reason why, in his Clock Test, the sudden telephone call generated an improvement in detection.

Rereading Pavlov

The appeal to Pavlov might seem surprising in Bartlett's Cambridge, but it is easily rendered intelligible. Cambridge physiologists respected Pavlov, and in the interwar years the university had granted him an honorary degree.[32] What was important to the physiologists was bound to attract the notice—even if not the unqualified agreement—of the psychologists. Pavlov's conditioned reflex had even featured, albeit briefly and rather obliquely, in the pages of *Remembering*. Pavlov's conditioning experiments, conjectured Bartlett, might be a way of explaining why certain details stood out from a scene when it came to be recalled.[33] A more general and more searching link to the Cambridge interest in Pavlov in the interwar years, however, was revealed by Mackworth's postwar citation, and use, of Carolus Oldfield's thoughtful 1937 paper "Some Recent Experiments Bearing on 'Internal Inhibition.'"[34]

Oldfield did two things in his 1937 paper. First, he pointed out certain potentially revealing similarities in the experimental methods and findings of a number of recent Cambridge experiments—studies that at first sight seemed very diverse. The studies included ones conducted by Rawdon-Smith (on experimental deafness), by Drew (on satiation of drives in experimental animals), and his own (on eyelid conditioning in humans). Oldfield noted that these studies all dealt with cases where the experimental subjects, whether animal or human, were exposed to a low level of background stimulation and where, set against that background, a sudden stimulus was found to have had a striking effect. In commonsense terms, the stimulus seemed to have "surprised" the subject, but it was the general effect of the surprise that interested Oldfield. The surprising stimulation could restore old threshold settings, jaded appetites, and conditioned responses. Oldfield pointed out that all of these reported results could be seen as beginning with a process of experimental extinction—that is, the growth of internal inhibition and then, crucially, disinhibition by a sudden stimulus standing out against a uniform background. Mackworth cited Oldfield's paper because he wanted to add his wartime Clock Test results to Oldfield's prewar grouping of experimental findings.

Oldfield also pursued a second, rather more critical, theme in his paper—a theme that did not fit so easily with the use that Mackworth now wanted to make of it. Oldfield suggested that Pavlov's experiments were not quite what they seemed to be. Pavlov's stress on the conditioning process was misleading. The experiments might seem to be about the conditioned reflex but, in Oldfield's opinion, they were really about something else. They were about the process and nature of inhibition itself. This was the real topic of the research. Pavlov was studying inhibition in a response *that happened to have been built up by a process of conditioning.*[35] In doing so, he said, Pavlov had studied inhibition in certain "exceptionally simple and favourable circumstances."[36]

Whether independently or prompted by Oldfield's 1937 paper, Craik had made a similar observation in 1943 in *The Nature of Explanation*. In fact, Craik generalized the point made by Oldfield. The interest of Pavlov's experimental technique, Craik said, was that it was a simple way of investigating the general phenomenon of the modification of a response. The conditioned reflex, said Craik,

> is thus an excellent experimental situation for studying modification of response and the mechanisms involved in it; it is not itself, as a phenomenon, the kind of unit from which a theory of thinking and action should be built.[37]

Pavlov, in contrast to Craik, thought that the conditioned reflex was precisely the unit out of which a general theory of thinking and action should be built. For Pavlov, all instances of behavior, and all habits, are simply linked sequences of conditioned reflexes. It is "obvious," Pavlov said, "that the different kinds of habits based on training, education and discipline of any sort are nothing but a long chain of conditioned reflexes."[38] But what was obvious to Pavlov and what may have seemed inviting to Mackworth was not obvious to all of Mackworth's Cambridge colleagues.

After the war, Donald Broadbent dramatically deepened the methodological doubts raised by Oldfield and Craik about the real character and point of Pavlov's work. Broadbent was by then a newly appointed, and still junior, member of the staff of the Applied Psychology Unit. On leaving school, during the war, Broadbent had volunteered for the Royal Air Force. After a preliminary course in engineering in Cambridge, he was trained as a pilot in the RAF. After the war, though still flying in the reserve, he left his engineering behind him and retrained as a psychologist in Bartlett's laboratory. In a brief autobiographical paper, Broadbent explained how, as his final examinations approached in 1947, he had been hoping to get a job after graduation as a psychologist in industry. Such work, he felt, would satisfy his desire to do something of practical use in the world. Alas no jobs showed up in industry,

> but the word began to get around the lab, shortly before the exams, that the navy wanted someone to work on effects of noise. As it happened, I had been very impressed by some notions of Carolus Oldfield . . . relating to internal inhibition (Oldfield, 1937); and it was clearly a practical problem.[39]

Broadbent did not explain the link he saw between Oldfield's paper and the problem of noise but, I shall suggest, what attracted his attention was Oldfield's stress on the role played by uniform and boring levels of background stimulation. If working against the background noise of a ship's engine led to the accumulation of internal inhibition, where did the intensity of the noise fit into this picture? Perhaps the research the navy wanted would be intellectually interesting as well as practical because it would provide an occasion to fit these pieces of the jigsaw together. A disinhibiting stimulus, Oldfield noted, depended for its effect on providing an element of surprise against a uniform background. The stimulus therefore had to possess what, in everyday language, would amount to an attention-getting novelty. When Mackworth introduced Pavlov's work into this nexus of problems, he invited a comparison between the predicament of the radar watchkeeper and the predicament of Pavlov's dogs.

The Paradoxical Pavlov

Broadbent did not directly challenge Mackworth's Pavlovian reading of the Clock Test results. Instead he did something that was potentially far more interesting—he offered a radical rereading of Pavlov's entire experimental corpus. In his 1953 paper "Classical Conditioning and Human Watch Keeping," Broadbent inverted the analogy that Mackworth was seeking to employ.[40] Rather than seeing the lapses of attention of the radar operator in terms of conditioning, Broadbent proposed that Pavlov's conditioning experiments were to be seen in terms of the focusing and lapsing of attention. Pavlov's dogs did not illuminate the behavior of the radar watchkeepers, but the behavior of the watchkeepers could illuminate the little-remarked oddity of Pavlov's enterprise.

"We reach here the central peculiarity of classical conditioning," said Broadbent:

> that the dominance of the situation by the experimentally controllable stimulus makes it possible for us to follow the movements of the animal's attention, using the learned response as a detecting instrument.[41]

This formulation deepened what Oldfield and Craik had said. They had both offered a reading of Pavlov in which conditioning was seen as a detecting instrument used in the investigation rather than being the object of the investigation. It was true that Pavlov's experiments encouraged the dog to learn that there was a link between the ringing of a bell and the presentation of food but, beyond that, learning played no further role—just as learning played very little role in radar watchkeeping. Pavlov claimed to have illuminated the essence of the learning process, when, in reality, learning played a small and incidental role in his work. "Paradoxical though such a position may seem," added Broadbent, Pavlov's experiments deliberately *minimized* the role of learning.[42] The experimental technique developed by Pavlov was a technology for controlling and focusing the short-term and fluctuating attention of his canine subjects. Seen from this perspective, the technology was highly effective. Experimenters using human subjects can ask the subject to attend to X and ignore Y. Pavlov had found a way to do the same with non-verbal animal subjects. This need to control and focus his subjects' attention explains the soundproofed laboratory; the harnesses constraining the dogs; the exploitation of their dependable natural responses; and the stillness, silence, and simplicity of the experimental procedure on which Pavlov insisted. And then, against this dull and uniform background, came the intrusive, attention-getting ring of the bell that functioned as the conditioned stimulus.

If Pavlov had found ways to study and manipulate the dog's attention, what is "paying attention"? The concept of "attention," with the implied selectivity and focus of a person's conscious awareness, is a mentalistic construct taken from ordinary language. It might look as if Pavlov could be accused of dressing up common sense in scientific-sounding words, but Broadbent was not making this claim. Like Pavlov, Broadbent acknowledged the need to sharpen up mentalistic everyday terms (and if necessary abandon them) in order to provide a shared vocabulary of concepts suitable for scientific communication. Attention is a real phenomenon but one that needs to be understood in objective and measurable terms. An important part of what we call "paying attention," said Broadbent, involves some kind of "filtering" process where some stimulation from the environment is allowed to pass into the nervous system of the organism, while other stimulation is blocked by the filter. Stimuli that are blocked fail to generate a response.

Now we can see what it was about Oldfield's paper that caught Broadbent's attention. The common factor in the experiments Oldfield described was the uniformity of the perceptual background, and this uniformity was one of the conditions that would prompt a filter to spontaneously switch to another channel of stimulation. The scientific task was to find the other principles of filtering. To set the process in motion and begin to turn the metaphor of "filtering" into a scientific theory, Broadbent suggested that the filter, whether in the canine or the human nervous system, operated according to three simple principles. The probability of a stimulus passing the filter depended on (i) the intensity of the stimulus, (ii) the novelty of the stimulus, and (iii) the biological importance of the stimulus. If these three principles could be shown to be empirically plausible, the theory could then be subject to further refinement.

To drive his point home and demonstrate the plausibility of the three principles, Broadbent systematically reanalyzed the wide range of results reported in Pavlov's *Conditioned Reflexes*. He showed that, in each case, they could be rationalized in terms of the operation of a perceptual filtering process. Mackworth's findings in the Clock Test then made sense without any need to mention conditioning or the accumulation of inhibition. The supposed time taken for inhibition to accumulate or dissipate becomes information about the time schedule of spontaneous filtering, and filter-switching, processes. The initial instructions to the subject in Mackworth's experiment correspond, in the nonhuman case, to the stimuli having a given biological importance. Biological importance has now been reinterpreted to include social importance. The striking decrement in the efficiency of watchkeeping is explicable because, with the passage of time, the filter begins to pass novel stimuli at the expense of the increasingly familiar task stimuli. The effect of the intrusive telephone call derives from the intensity of the stimulus that resets ("disinhibits") the filter.

In a veritable tour de force, Broadbent identified result after result in Pavlov's treatise and showed how they could be rationalized as a case of filtering. Broadbent's analysis covered the phenomena that Pavlov describes as (i) "extinction," (ii) "spontaneous recovery," (iii) "disinhibition," (iv) "external inhibition," (v) "positive induction," (vi) "irradiation," (vii) "partial reinforcement," and (viii) the so-called "paradoxical phase."[43] On Broadbent's reading, Pavlov had written an empirical treatise describing the psychological phenomenon of filtering.

Pavlov's Questionable Premise

Broadbent's reanalysis of Pavlov's achievement was not a case of saying the same thing in different words. Pavlov and Broadbent invoked different hypothetical mechanisms. Broadbent's stress was on perceptual processes, while Pavlov's was on response processes. In Pavlov's analysis, however, the crucial role of perceptual processes was hidden by all the complex experimental procedures. It was this experimental aspect of Pavlov's work that commanded Broadbent's respect, but not the theory. In fact, in Broadbent's eyes, Pavlov's theory was completely at odds with Pavlov's experimental methodology. Pavlov's theoretical analysis was based on the assumption that all the stimuli that impinge on the sense organs ultimately make a cortical contribution to a response. The cortex was influenced by "countless stimuli," Pavlov insisted. "Every one of these stimuli produces a certain effect on the animal, and all of them taken together may clash and interfere with, or else reinforce, one another."[44] Suppose that two stimuli, one strong and one weak, impinge on the sense organs. The stronger stimulus may mask the weaker, but Pavlov assumed that the weaker stimulus still proceeds into the nervous system and plays a role in generating the final response. Pavlov declared: "Although the effect of the weaker stimulus when tested singly is invisible, it nevertheless plays an important part in the stimulatory compound."[45] Broadbent, by contrast, argued that there was compelling evidence that most stimuli are blocked by the perceptual filter and do not penetrate beyond the perceptual system and journey further into the nervous system.

In the next chapter, I shall look further into the origin of Broadbent's Filter Theory and indicate its relation to the wartime Cockpit experiments.

Dismantling the Cockpit

Despite the failure of the landing-accident study to detect pilot fatigue, Bartlett never conceded that the Cambridge Cockpit results were wrong. But nor did he ever give a direct reply to Bradford Hill. A degree of uncertainty therefore hangs over the question of how the two bodies of work stand in relation to one another. I shall argue that, notwithstanding Bartlett's silence in the crucial FPRC meeting, the intense, postwar research program in Cambridge psychology can be read as a response to the Landing-Accident Anomaly—that is, to the uncomfortable conjunction of the Cockpit results and Bradford Hill's results. Cambridge psychologists did not themselves describe their postwar program in this way, but there are revealing hints in Bartlett's postwar publications that point to this conclusion. Of course, there were also other important research themes that commanded attention, such as the impact of noise on skill. But the work from circa 1945 to circa 1958 can be seen as a retrospective justification for Bartlett's refusal to back down in the face of Bradford Hill and Williams's negative report. Had Bartlett been less tenacious or had less impact on his students, that later, postwar work, with its increase in scope, might never have been pursued in the way that it was. In this chapter, I shall describe some of the salient features of the postwar research and identify their connections with the Cambridge Cockpit experiments.

Broadbent's Student Memories

For Bartlett, the end of the war in 1945 brought the sadness of Kenneth Craik's death. This loss was not only a personal one: it meant that the newly created Applied Psychology Unit needed another director. Bartlett took over the position on an interim basis with Norman Mackworth as his deputy. Meanwhile, the Medical Research Council had sent Russell Davis on a visit to Germany to report on what he could find out about the state of that country's wartime research in applied psychology—a theme that I shall take up later.[1] The end of the war also brought the resumption of normal teaching duties to Cambridge University and the influx of a new cohort of psychology students. I have already mentioned that Donald Broadbent was one of those students. A number of years later, he was to assume the role originally occupied by Kenneth Craik as director of the Applied Psychology Unit. A revealing glimpse into the immediate postwar Cambridge scene is provided by Broadbent's recollections of his student days in Bartlett's laboratory in 1947.[2]

Broadbent admits that the teaching he received was patchy, but this patchiness was offset by "a constant thread of excitement that came from being at a frontier where things were really happening."[3] Dan Berlyne was on the staff and gave students a rigorous grounding in American learning theory. The backbone of the teaching, though, was provided by Alan Welford, whose lectures furnished well-ordered accounts of a range of major topics, including the current state of research on motor skills. There were cheery lectures by G. C. Grindley on animal behavior (though, because of the war, there were still no animals in the laboratory). Students diligently read *Remembering*, and beyond that, they also had Bartlett's own lectures. As Broadbent recalled:

> He would give one lecture a week, not of course following any particular syllabus, but pouring out a steady stream of ideas, odd references to things he had recently read, general concepts that seemed to him widely applicable, and so on. On these occasions everyone was too enthralled and too occupied with taking notes to interrupt or discuss the lectures but he was incredibly approachable.[4]

Bartlett impressed on his students the importance of applied psychology.[5] Craik's spirit, added Broadbent, "had communicated itself throughout the building," and the students "met for the first time a startling new approach to human beings, one that was not available in any textbook or journal article and one that had potentially revolutionary implications."[6] Despite the intellectual richness of the Cambridge Department of Psychology, the postwar years both inside and outside universities were undeniably ones of economic austerity. The students had to await their turn to take notes from the department's single copy of Woodworth's prewar textbook *Experimental Psychology*.[7] Broadbent was the proud owner of a rare copy of Hilgard and Marquis's recently published *Conditioning and Learning*, but that was only because it had been sent to him by an American friend he had made when completing his flying training.[8] The austerity of the time was the product of the war and the sudden termination of American economic cooperation.[9] The demands made on the British population in the cause of postwar reconstruction were, however, in conformity with Broadbent's own morally serious view of the world.[10]

Looking back, many years later, Broadbent set out to articulate the "principle which I was taught by Bartlett, and which was in my youth taken for granted."[11] The principle in question was the neo-Jacksonian approach used by Bartlett in his Ferrier Lecture. But the immediate problem for Broadbent's generation was how to express those traditional Cambridge insights in combination with the exciting wartime developments in information theory—a mathematical framework devised by communication engineers in order to understand the limits of material systems in transmitting information. Because of the struggle to bring all these themes together, Broadbent admits: "Looking back again, I can see that the full explosive quality of the Bartlett-Craik approach began to be watered down almost at once."[12] Broadbent's explanation was that the Craik-Bartlett model was too complex. The model embodied a potentially powerful insight, but it lacked "a structure of data and technique" to link it with established knowledge and practices.[13] Lacking that supportive structure, it needed to be dismantled and broken down into simpler, more tractable problems. "To create manageable research out of this approach," said Broadbent, "it was necessary to carve individual problems out of it." The compromise was to focus on the upper level of the Jacksonian hierarchy and analyze it in information-theoretic terms. Hence, "the restrictions on the upper level of the process were examined separately, giving rise to 'single-channel theory.'" Even now, he concluded, "thirty years later, we are only just regaining the full complexity of the Bartlett-Craik view."[14]

Although the nonmathematical Bartlett kept his distance from information theory, he acknowledged the problem created by the complexity of the Cockpit investigation and the price paid in the attempt to make experimentation realistic. After describing the wartime work in "Some Growing Points in Experimental Psychology" (a paper that appeared in April 1945, the same month that Craik died), Bartlett admitted:

> Some simpler situation must be found which nevertheless preserves the essential characters of skilled behaviour. This is much less easy than may appear, and a fully satisfactory experimental set up has not yet been achieved.[15]

Eventually solutions were found to these problems, and appropriate pieces of apparatus were constructed. These pieces of apparatus were used to probe the different aspects of perceptual-motor skill that had been bound together in the Cambridge Cockpit. One of these was the Five-Light Test, a serial-reaction task; another was a tracking test; and others involved monitoring banks of dials. All of these tests will be described shortly. By this means, the Cambridge Cockpit was methodologically though not physically "dismantled." I now want to look at how Bartlett's postwar papers set the scene for these developments.

The Measurement of Human Skill

In the late 1940s and 1950s, Bartlett produced a series of survey papers describing the work coming out of the Psychological Laboratory and the Applied Psychology Unit. The quotation above, about the problems generated by the need for simplification, comes from one of them. Others bore forward-looking titles such as "Some Problems in 'Display' and 'Control,'" "The Measurement of Human Skill," "Man, Machines and Productivity," "The Effects of Flying on Human Performance," and "Challenge to Experimental Psychology."[16] These publications can give us a glimpse of the material in those discursive lectures that excited Broadbent and his fellow students. They also reveal Bartlett's enduring preoccupation with the Cambridge Cockpit.

In "The Measurement of Human Skill," Bartlett argued that psychologists must engage with factors which were "internal to the skill performance itself."[17] He took the example of a pilot flying on his instruments. How should the pilot's skill be measured? Should the investigator focus on the statistics of achievement or on the psychological processes underlying the achievement? Is the focus to be on the outer product or the inner process? Bartlett's stance was unequivocal: the stress should be on the process—hence the importance of studying the factors internal to the skill itself. Thus:

> The whole performance, from take-off to landing, is a long-drawn-out inter-linkage of receptor and effector functions, forming a veritable series and not merely a succession, each step a step to another step, right to the end. Yet it has often been claimed that an informed observer, sitting by the pilot, knowing nothing, or next to nothing, of the terrific play of his senses and muscles and mind, can criticize, rate, and measure his achievement. More than this is sometimes claimed; for it is said that an observer who never sees either the pilot or his aircraft, but is merely supplied with an objective record of what the machine has done . . . can with reasonable success measure or assess the pilot's skill. Even beyond this, though possibly with some exaggeration, it is often considered that the skill of the pilot can be correctly criticized and judged if nothing more is available than an accurate objective record of the landing.[18]

Who are these observers who, knowing nothing of the inner processes in the mind, nevertheless judge the pilot's achievements? Bartlett does not tell us. Since they are pictured as sitting next to the pilot, this criticism may be a tilt at military flying instructors, though it would be uncharacteristic of Bartlett to criticize the military. More pointedly, who are those who think pilot skill can be judged on the basis of nothing more than an objective record of the landing? Who has ever attempted to do such a thing? Again, no names were mentioned, but this remark must have been aimed at Bradford Hill. He attempted to judge the state of fatigue of the pilot, and to decide whether the pilot's skill had been eroded by fatigue, by using one single measure: the RAF documentation recording whether the pilot successfully landed the aircraft.

If the accusations of ignoring the inner processes of the mind was meant as a criticism of Bradford Hill—and there can be little doubt about the matter—how cogent was it? What, exactly, had Bradford Hill done wrong? In Bartlett's view, he had failed to get inside the skill and probe its internal structure. Merely taking an overall measure of a result leaves the investigator uninformed about changes in the organization of the action. The inner process may have changed even when the outer product remains unchanged when measured by some given range of tests. In Bartlett's opinion, until the psychologist is informed about the nature of these inner changes, the skill has not been adequately understood—at least, not by scientific standards.[19] Every pilot in a squadron might have landed his aircraft without an accident, but how many were close to their limits? How would those limits have been revealed under slightly changed circumstances?

The virtue of Craik's Cockpit was that it delivered multiple, cross-referencing measures of the different components of the pilot's response. As well as giving the subject a wide scope in selecting their responses, it allowed the experimenter to probe the interior of the skill in the way Bartlett wanted. Bartlett was saying that it was a methodological mistake to focus on a narrowly defined achievement rather than on the complex, sensory-effector functions involved in the overall task. This point was equally true regardless of whether the task was pursued successfully or unsuccessfully. Focus on success can lead an investigator into a trap. Sensory-effector functions work together, he said,

> in the pursuit of and achievement of some task which, when it is achieved, seems at first sight as if it can be discussed, criticised, evaluated, and even measured in its own right, apart from the means used to bring it about.[20]

Note the wording. At *first sight* it may look as if success can be measured in its own right. But, for Bartlett, success is not to be separated from failure, or achievement from nonachievement. The one cannot be understood without the other because they are both the upshot of the same underlying interior mechanisms. The argument was essentially the same as that developed in *Remembering*. The achievement of accurate recall cannot be understood by the psychologist apart from inaccurate recall. Of course, experiments in which achievement are treated in isolation from failures can be and are performed. They yield results, but the results only shed a partial and one-sided light on the overall problem. They truncate the phenomenon under study.

The criticism prompted by the landing-accident study was therefore the same as that originally directed at Ebbinghaus. Ebbinghaus had used non-sense syllables as the material to be recalled in his studies of memory because doing so made it easy to decide when the recall was right and when it was wrong. There were no borderline cases or problems of grading degrees of greater or lesser accuracy. Scoring was made easy, and this advantage made it look as if the achievement of successful recall could be measured in its own right, apart from the means used to bring it about. Similarly, the simple classi-fication of "crash" versus "no-crash" disguised the complexity of the processes involved in creating the results that had been given this convenient binary structure. For Bartlett, behind this simplicity there will always be complex-ity. Stipulating that a phenomenon lies outside the scope of an investigation (as was done implicitly by focusing on "straightforward" landings) does not ensure that the actual causalities that are at work are thereby curtailed.

Moving now to the postwar work conducted within this Bartlettian framework, there were three bodies of Cambridge experimentation that are particularly relevant to appreciating the legacy of the Cockpit work. These are (i) Donald Broadbent's theory of the perceptual filter, (ii) Christopher Poul-ton's work on tracking, and (iii) Richard Conrad's work on speed and load. I shall consider each in turn and then show how they relate to the problem of landing an aircraft. I begin with Broadbent, who, it will be recalled, began his research career in the Applied Psychology Unit working on problems insti-gated by the Royal Navy.

The Noise Anomaly

There is an obvious need to maintain the operational efficiency of a crew working in the noisy environment of a battleship or submarine. What is less obvious is why there was a need for more research. The answer is that there were long-standing puzzles surrounding the effects of noise. Everyday experi-ence of factories, foundries, shipyards, railway stations, crowded restaurants, or noisy classrooms may suggest that noise has a bad effect on comfort and the capacity to concentrate, but scientific researchers had repeatedly failed to measure any objective decrements in efficiency. This discrepancy between science and common sense may be called the Noise Anomaly.

During WWII, the problem of noise had been the subject of a well-financed research effort in the Harvard Acoustic Laboratory in Cambridge, Massachusetts. The eminent psychophysicist S. S. Stevens had carried out an extensive range of tests and reported negative results. Admittedly, noise makes verbal communication difficult—for example between members of an aircrew. Stevens and colleagues such as George Miller tackled these problems with success, but while noise impeded communication it did not appear to impair skill.[21] In the United Kingdom, some small-scale tests on the effects of noise had been run using the Cambridge Cockpit. Russell Davis reported on them briefly in *Pilot Error*. Some, but not all, pilots seem to be affected by noise. The elevator-control measurements registered some negative effects, but the results were neither clear nor consistent.[22] Within a relatively short time, however, Broadbent had demonstrated, empirically, that the damaging impact of noise on skill was a reality and not just a popular misconception—as Stevens was inclined to think. Broadbent showed why noise might be expected to have this negative effect and explained why others had failed to find it. Two experiments played a central role in reaching these conclusions: they were known as the Twenty-Dial Test and the Five-Light Test.

The Twenty-Dial Test

Broadbent's apparatus consisted of twenty steam-gauges of the kind found in a ship's engine room. The subject's task was to monitor the dials to see if any of the pointers, which were usually stationary, moved into a danger zone. On detecting this signal, subjects had to turn a knob under the dial that moved the pointer back to below the safety line. The watch lasted one-and-a-half hours, with fifteen signals distributed amongst the twenty dials. The signals were delivered without warning, at unpredictable intervals varying between one and twelve minutes. On some runs, the task was performed in silence; in other runs, it was performed under noisy conditions. Apart from the noise variable, Broadbent's task differed from Mackworth's Clock Test in two respects: (i) there were multiple sources to be monitored, and (ii) the signal stayed in place until noticed by the subject. For the purposes of this experiment, "noise" was defined as 100 decibels and "quiet" as 70 decibels.

The average reaction time to detect a danger signal in this test was over ten seconds, but to get a clearer picture, allowance had to be made for different kinds of response. If the subject happened to be looking at one of the dials when its pointer moved, that was counted as a "seen" signal and contrasted with a "found" signal. A "found" signal was recorded when the subject did not actually see the move but found that one of the dials had moved while he was scanning the others. Broadbent then divided the "found" cases into "quick finds" (< 10 seconds) and "slow finds" (> 10 seconds). He discovered that performing the Twenty-Dial Test in noise produced no alteration in the number of "seen" signals, but it markedly reduced the number of "quick finds." "The effect," he said, "was quite startling, the proportion of signals seen in less than 10sec being 0.36 for quiet, and only 0.22 for noise, a drop of nearly 40 per cent."[23] Here, then, was the noise effect that had eluded the researchers in the other Cambridge.

The Five-Light Test

Watchkeeping tests, said Broadbent, only provide evidence about the state of the subject at the moment the infrequent stimulus is presented. How can the psychologist find out what is happening inside the nervous system between the occurrences of such signals? Here, the use of serial-reaction tasks became important. They allowed the psychologist to bridge the gap between what Broadbent called the "vigilant man" and the "active man."[24] Like the Cambridge Cockpit itself, serial-reaction tasks furnish many signals rather than few signals; the signals arrive in a steady flow rather than intermittently, and responses are likewise numerous rather than sparse. As Bartlett had said, finding a simple apparatus that captured these aspects of the Cockpit experiments had not been easy. Fortunately, Alfred Leonard in the Applied Psychology Unit had found an answer.[25] Leonard's serial-reaction apparatus consisted of a display with five signal lights arranged in a regular pentagon and five similarly arranged metal contact points. The lights were automatically illuminated, one at a time, in random order. On seeing an illuminated light, the experimental subject was required to touch the corresponding contact point with a stylus. Even if the subject touched the wrong contact point, the device would still present the next stimulus.

Leonard's Five-Light Test could be carried out in two ways: namely, "paced" and "unpaced." In the unpaced mode, the subject's response triggered the next signal light. Subjects therefore set their own pace. In the paced mode, the apparatus presented the signal light at a rate predetermined by the experimenter, and it was the subject's job to keep up (if the pace was fast) or to be patient (if the pace was slow). Leonard's apparatus automatically counted the number of correct and incorrect responses and recorded the number of periods of two seconds or more during which no responses were made. Equipped with Leonard's apparatus, Broadbent was able to test the impact of noise on the performance of a serial-reaction task and also see how noise might interact with the difference between a paced and an unpaced task. The results were revealing.

First, Broadbent asked what, if anything, was the difference between the performance on the paced and unpaced versions of the Five-Light task under *quiet* conditions. He used separate groups of subjects in the respective conditions in a test lasting one-and-a-half hours. It emerged that the unpaced group showed no significant decrement in their output of correct responses over this time.[26] There was, however, a statistically significant, and marked, change of another sort. There was an increase in the number of pauses of two seconds or more. Things were very different for the paced group. Broadbent set the pace by making the apparatus deliver one stimulus light every second because this was the average rate spontaneously adopted by the unpaced subjects. Being paced in this way led to a falloff of the output of correct responses after the first ten minutes. Every subject showed this deterioration.

Measured by the output of correct responses, performance in the paced task therefore showed fatigue-like effects while the output of the unpaced performance did not, though it became irregular. The obvious explanation, said Broadbent, was that the attention of all the subjects (whether paced or unpaced) wandered away from their repetitive and boring tasks for short periods of time. The subject's attention would then switch back to the task—the process overall taking about two seconds. The subjects in the unpaced group would be able to speed up between these lapses and so could keep up their average output. The subjects in the paced group, by contrast, could not make up for missed signals. Once missed, a signal, and any chance of a response to it, was gone forever. Broadbent pointed out that there was an analogy between the paced Five-Light Test and Mackworth's Clock Test. Although one was a vigilance test and the other a serial-reaction test, they both have a paced character. The transient signals in the Clock Test arrived at unpredictable times, and if the signal coincided with a lapse of attention, the chance of a correct response had gone forever. This similarity could explain why both the Clock Test and the paced Five-Light Test displayed fatigue-like performance decrements.

The next step was to compare performance on the *unpaced* Five-Light Test in quiet and noisy conditions. Broadbent arranged for eighteen naval ratings to perform the unpaced version of the Five-Light Test for two twenty-five-minute runs: once in quiet (70 decibels) and once in noise (100 decibels). The runs were on successive days. Half the subjects encountered the noisy condition on the first day, and half on the second. The overall result corroborated the previous finding that noise could have an adverse impact on the performance of a skill. Thus, "the average number of errors per subject was 37 in quiet and 57 in noise"—that is, a rise of about 50 percent. In the unpaced condition, of course, the subjects are likely to adopt different speeds to one another, so the relation between the number of correct responses and the number of errors can vary in all sorts of different ways. For example, Broadbent found that the "output of correct responses showed no noticeable effect of noise," but when errors in the second run were compared they were "markedly more frequent in noise."[27] A further, notable feature of the experiment was that "the relative efficiency in noise again declined with time since the beginning of the test." Thus, when performing in noise, "it seems that the length of the task is truly an important condition."[28]

Was it really noise that was having these effects, or was it beliefs about noise? Many people believe noise has a bad effect. What if the subjects shared this belief and it de-motivated them and degraded their performance? To guard against this possibility, Broadbent resorted to a degree of deception. He assembled another group of naval ratings, gave them the same Five-Light Test under the same two conditions, but (aided by a display of convincing graphs), he informed them that working in noise *improved* the performance of operators on new sorts of radar sets that were being designed. The purpose of the experiment, subjects were told, was to help the designers. The result was that noise *still* reduced efficiency and to almost the same degree. On the assumption that Broadbent's subjects believed what he had told them, the conclusion was drawn that the effect of the noise was real. But this conclusion leaves another problem. Why did Stevens's group in the USA fail to detect it?

The Harvard Failure

The Harvard experiments used highly selected subjects and a wide range of tasks that were changed frequently to keep up motivation. The tasks included choice reaction times, card-sorting, translating written codes, judging distances, perspective reversal, and the use of a Pursuit Rotor. The Pursuit Rotor looked rather like a gramophone with a turntable. The subject had to keep a stylus on a target point on the turntable as it rotated. We shall encounter this interesting device again in a later chapter. For the moment, the important point is that none of these tasks showed any detrimental effect of noise on performance. Other tests, such as speed of visual accommodation, dark adaptation, hand steadiness and muscle tension, produced equivocal results. In Broadbent's view, however, the range of tasks used in the Harvard work was badly chosen because it covered "precisely those which we would expect to minimise the effects of noise."[29] They were all well-practiced tasks which involved low levels of information transmission. In contrast, Broadbent described his own use of the Twenty-Dial Test as "deliberately designed to avoid the conditions which the negative experiments had in common."[30] There was, however, one result that emerged in the midst of their negative findings that might have led the Harvard group to attribute bad effects to noise—but the opportunity was missed. The task actually involved aircraft-like controls. As Broadbent explained, the subject used the controls to direct a beam of light at a series of targets:

> This task showed an apparent effect of noise on speed and on accuracy, but it was later decided that the effect might have been due to auditory cues from the relays of the scoring mechanism: such cues would naturally be more help-ful in the quiet condition.[31]

Noise would mask the faint click from the apparatus, so in the noise condition of the experiment, the subjects might be denied useful feedback that was available in the quiet condition. In this case, it would not be the noise itself that made the performance worse, it would be the masking of a useful, additional cue. The apparent ability of noise to reduce efficiency would then be a mere artefact. On this basis, and without further investigation, Stevens discarded the result.[32]

Blocks, Filters, and the Cambridge Cockpit

The postwar noise research may seem very distant from that undertaken on the Cambridge Cockpit, but that impression is misleading. There are some intimate and important connections that can be discerned. Both the noise program and the skill-fatigue studies involved British Cambridge psychologists in a confrontation with an (apparently) compelling negative result—in one case the Bomber Command failure to discern the negative effect of fatigue and, in the other, the Harvard Laboratory failure to discern the negative effect of noise. In both cases, the Cambridge response to the negative results was to advocate a particular and identifiable inductive research strategy. The strategy was embodied in Bartlett's injunction to probe the internal structural changes in the phenomenon by tracing the interaction of both success and failure. I want to explore this nexus of connections so that I can then make the link that I indicated between Broadbent's contribution and those of Poulton and Conrad.

In 1931, the American applied psychologist Arthur Bills had demonstrated that when people are given a continuous and protracted task in the laboratory, they eventually begin to insert periods of nonresponse into their performance. He called these episodes "blocks" and declared that "blocking" was "a new principle in mental fatigue."[33] The blocks, he said, probably served a recuperative function by "giving response mechanisms frequent rests." Notice here the assumption that the locus of the effect was on the process of responding. Later, still assuming that the "block" operated on the responses, he conjectured that the blocks might be the result of local shortages of oxygen.[34] Bartlett alluded to Bills's work in his report *Fatigue in the Air Pilot*, presented to the Flying Personnel Research Committee in 1942, but he noted that the real nature of "blocks" remained obscure.[35] Did "blocks" put a stop to all mental activity or only some of it? Bills's work had also been discussed in Woodworth's prewar textbook (the book that had been in short supply in postwar Cambridge). Woodworth suggested that "blocks" were really shifts of attention, but he was still unsure whether the resulting response was the outcome of stimulus competition or response competition. Given his exposure, both to Bartlett's thinking and Woodworth's textbook, it is hardly surprising that Broadbent should relate his own experimental results, and his own metaphor of filtering, to Bills's ideas. Like Bartlett in his committee report, Broadbent wondered but doubted whether "blocks" meant that the subject's mind was blank. Like Woodworth, he asked whether the "block" was on the perceptual side or the response side. Unlike Woodworth, Broadbent was unequivocal in saying that the data implied that the phenomenon under discussion must be perceptual.[36]

Broadbent's analysis was based on the assumption that the central nervous system was too small to handle all the information that our sense organs might extract from the surroundings. Some sort of selection must always be at work—and this fact, he argued, makes the "block" a perceptual rather than response phenomenon. The filter protects the limited capacity of the decision-making processes in the brain. The next question for the experimentalist was how that selection process is best understood. In his 1958 *Perception and Communication*, Broadbent argued that a practical analogy may be found

> in a radio receiver designed to eliminate impulse interference, and so present the signal to the listener free from such interference. Such a receiver may make use of the fact that the interference possesses frequency components not present in the desired signal: when an incoming signal has such components it is therefore rejected, and only signals without these frequencies are passed to the later stages of the receiver.[37]

Kenneth Craik had previously used a similar analogy, arguing that "the essential part of physical 'recognizing' instruments is usually a filter—whether it be a mechanical sieve, an optical filter, or a tuned electrical circuit—which 'passes' only quantities of a kind it is required to identify, and rejects all others."[38]

A filter switch did not "block" all input but only "blocked" information from certain sources in the environment—sources with a common physical feature. Information from other sources was allowed to pass, and then the filter might switch back. The data gleaned from the Twenty-Dial Test and the Five-Light Test pins down some of the parameters governing the switching operation. A simple and clear picture seemed to be emerging. With regard to noise, as time-on-task increased, the perceptual filter was more frequently captured by the intense, competing stimulus of the noise; that is, it allowed these stimuli to pass more deeply into the nervous system. Once the data was understood in these terms, the conclusion could be drawn that noise caused inefficiency because it was a distracting stimulus. Later Broadbent concluded that, in certain respects, this plausible analysis was false. Noise did not operate as a distracting stimulus but through a very different mechanism. But, at the time, the distraction theory was an important step forward.

Where was the link between Broadbent's noise research, his conclusions about filtering, and the Cockpit? It is to be found in the section of *Perception and Communication* which bears the title "Perceptual 'Fatigue' in Prolonged Work." Broadbent noted that repeated, muscular work, provided it was not too demanding, often did not result in a performance decrement or decline in output.[39] In this respect the sustained performance of correct responses in the (unpaced) Five-Light Test is representative of many activities. But, Broadbent went on:

> In the more complex task of flying aircraft, it has again been found that prolonged work does not affect ordinary measures of average performance. On the contrary (Bartlett 1943; Drew 1940) the deterioration is rather in the forgetting of peripheral tasks such as changing fuel-tanks; or in concentrating on one instrument in the cockpit to the exclusion of others. This result resembles those on industrial "fatigue" and on the five-choice reaction task in that it makes it difficult to uphold the simple theory that repeated responses become less efficient.[40]

The reference to Bartlett 1943 and Drew 1940 was, of course, a reference to the Cockpit work. At the beginning of the quoted paragraph, then, we have Broadbent's acknowledgment that a pilot flying an aircraft for prolonged periods may, under the right circumstances, show no marked fatigue decrement. Although no mention was made of Bradford Hill's report, this acknowledgment amounts to Broadbent's tacit acceptance of the landing-accident result. But what did Broadbent mean when he said that fatigue in flying an aircraft would not show itself in "ordinary measures of average performance"? What are the ordinary measures? This phrase was a reference to Bradford Hill's use of documented safe landings as a performance measure. Broadbent was here following Bartlett's complaint about the use of landing-accident statistics. Such data only deals with overall outcomes and fails to register any inner adjustments between the components of a skill. Ordinary measures deal with end products rather than inner processes.

This understanding of Broadbent's meaning is supported by the next sentence of the quoted passage. In that sentence, Broadbent set "ordinary measures" in contrast with more subtle measures of skill fatigue—measures in which note is taken of the neglect of certain instruments at the expense of other instruments, and the neglect of peripheral as compared with central tasks. By implication, these measures were being counted as "non-ordinary" measures that dealt with the changing relations between the parts of the display or parts of the task rather than end results.

Without an appreciation of its historical context, this passage in *Perception and Communication* cannot easily be understood. The passage ends with an assertion of the link between the Five-Light Test (referred to as the five-choice reaction task) and the central theme of Bartlett's Ferrier Lecture: namely, the untenability of any view of fatigue that reduces it to the simple idea that repeated responses become less efficient. This admittedly compressed passage from *Perception and Communication* shows what Broadbent retrospectively found most significant about the Cockpit results. He was mobilizing the old Cambridge Cockpit, along with the new Five-Light Test, as evidence of the *perceptual* locus of fatigue effects.[41]

I now want to move on from Broadbent's own experiments to two other important contributions relevant to understanding fatigue—from Poulton and Conrad, respectively.

Tracking

Eustace Christopher Poulton (1918–2000) originally came to Cambridge in 1937 on a mathematics scholarship to Trinity Hall, but he switched to medicine and later served in the Royal Army Medical Corps. After discharge from the army on medical grounds, Poulton was employed by the Medical Research Council in Cambridge and turned to psychological experimentation. He became the local expert on tracking and had a formidable reputation as a methodological critic.[42] Poulton indicated, in the opening lines of one of his early papers, that "the experiments are an elaboration of an unpublished experiment carried out by K. J. W. Craik and M. A. Vince in 1945 to study anticipation in tracking."[43] The unpublished experimental work referred to was not specified but was probably the series of reports that Craik and Vince had submitted to the Flying Personnel Research Committee.[44] The string of papers by Poulton that followed included an influential, theoretical synthesis published in 1957 with the title: "On Prediction in Skilled Movements."[45] This paper was described by Broadbent as "one of the most distinguished produced by any British psychologist in the last thirty years."[46]

Tracking is involved in many skills—for example, following a target seen through a gun sight, but also picking up an item from a moving conveyor belt. The typical laboratory apparatus used to study tracking presents the experimental subject with a long, wavy line drawn on a strip of paper that passes at constant speed in front of the subject. The task is to follow or "track" the wavy line by using a pencil that is permitted a limited movement across the paper. The subject must draw a line that, ideally, covers the original line exactly. If this goal were achieved, the tracking would be perfect. The subjects are only allowed to see the part of the original line exposed through a more or less narrow viewing window in the apparatus. The length visible in advance can be varied so that the subject will be given a greater or lesser amount of information on which to anticipate how the target line will behave. As well as offering different degrees of preview, the paper strip can be moved at different speeds, and the line to be tracked can be more or less wavy and more or less predictable.

In his 1957 survey, Poulton was able to show, on the basis of numerous experiments, that every movement and every phase in a tracking task was saturated with anticipation—where every act of anticipation is a form of prediction. Some of the initial movements in a tracking response have the character of what engineers call "open-chain" or "ballistic" responses (or what psychologists would recognize as mere reflexes). Other responses involved in tracking had the "closed-loop" or "negative-feedback" character that was increasingly familiar from the operation of servomechanisms. Above all, it was clear that, when engaged in a tracking task, subjects had the capacity to detect and exploit some of the statistical properties of the ever-changing stimulus. Poulton used the label "perceptual anticipation" to refer to this ability. Provided that the course to be tracked has a degree of predictability, perceptual anticipation may yield remarkably good tracking performances.[47]

One of Poulton's most interesting findings was that anticipation and prediction occurred even when the subjects were denied visual feedback from the tracking task. A naturally occurring case of a denial of feedback arises when subjects blink their eyes. In a joint paper of 1952, Poulton and Richard Gregory recorded the subject's unselfconscious blink-rate during a tracking task and measured its effect on accuracy with tasks of different degrees of predictability. A normal, voluntary blink lasts about 0.25 seconds. Although Poulton and Gregory found a performance decrement during these brief periods of "blackout," as they called them, they also established the reality of successful tracking during such blackout when dealing with predictable tasks, such as tracking a sinusoidal curve. They calculated the tracking error that would have occurred (i) if the subject had ceased to initiate *any* movement during the "blackout"; and (ii) if they had ceased to initiate any *new* movement, "maintaining instead his initial speed of movement." They found that the subject performed better than would be expected on either assumption. In other words, the subject was managing to do something more sophisticated than was envisaged in either of these hypotheses. As Poulton and Gregory put it:

> It seems clear that the subject must have been responding to an acceleration component of the course during his blink "blackouts," as well as to a velocity component. He must in fact have been tracking an expected course during his "blackouts."[48]

These results were confirmed by other experiments, and the time period during which blind tracking was found to be possible was extended. In another 1957 paper, "On the Stimulus and Response in Pursuit Tracking," Poulton reported:

> Tracking for 5.0 sec. with the eyes closed was sometimes as accurate as normal tracking with the eyes open. The fast simple input gave rise to a continuous cyclical response, which appeared to be adjusted only relatively infrequently to improve its match with the input.[49]

This result echoed but slightly modified Craik's earlier idea that the human operator acts as an intermittent correction servo.[50] Poulton's work on anticipation also has an obvious application to Broadbent's Filter Theory. It shows that while subjects are briefly switching attention to new stimuli, they might also continue to respond to past stimuli. The response can be delegated to one mechanism, while the filter passes information to another mechanism. Here again one glimpses the levels of control embodied in Craik's neo-Jacksonian picture. The "lower level" continues to respond while the "upper level" confronts other tasks, decisions, or distractions and only makes intermittent modifications to the anticipations built into the lower-level subroutines.

Before leaving the theme of tracking, it is important to notice that Poulton, like other experts in this field, always distinguished between two forms of the tracking task which were known, respectively, as (i) "pursuit tracking" and (ii) "compensatory tracking." Landing an aircraft under visual conditions, as in Bradford Hill's study, is an example of pursuit tracking, even if the visual conditions are somewhat degraded. In pursuit tracking, then, both the goal and the error signal are known. In compensatory tracking, by contrast, the only signal on the display is the error signal. In simple terms, under these conditions, given the appropriate instruments, the pilot might know from reading the instrument panel how far he is from the runway, but he cannot actually see the runway. Some American experts, such as Franklin Taylor, have challenged the importance of the distinction between the two sorts of tracking. They argued that, in the compensatory case, it was often possible to infer the missing items of information. Poulton, however, always maintained that, in practice, pursuit tracking was more efficient and less prone to error than compensatory tracking.[51]

The Paradox of Speed and Load

Richard Conrad's early experimental work was linked to the theme of anticipation. His experiments, which were originally overseen by Bartlett, did not involve tracking; rather, Conrad presented his subjects with an array of four dials.[52] Each dial had a pointer that rotated at a fixed rate, although the rate differed from dial to dial. Each dial had markers placed at intervals around the perimeter. The experimental subject had to scan the dials and press a Morse key beneath the dial whenever the rotating pointer on that dial was about to pass a perimeter marker. The event of a pointer passing a marker counted as a "signal." It was not possible to learn the order in which the signals would arrive, but an element of anticipation was possible. The problem for the subject, said Conrad, "was not to respond as soon as possible after the signal had occurred, but to anticipate on which dial the next signal would occur before it occurred, and then respond at the moment of coincidence."[53] Using this experimental arrangement, Conrad could vary the number of dials while keeping the total number of signals per minute the same. He could also keep the number of dials the same but vary the number of signals per minute. He referred to the number of dials as the "load" imposed by the task, and the total number of signals per minute as the "speed" of the task.

Conrad's first finding might be considered obvious. As "speed," in his sense, increased, so did the number of missed responses. His second result was not obvious and, at first sight, seemed puzzling. He found that as "speed" increased, so did the rate of correct responding.[54] At higher speeds, not only was the rate of correct responding greater than the rate of correct responding at the lower speed, it could be greater than the maximum possible rate of scoring at the lower speed—a maximum that was typically never achieved. What looked like a ceiling to the performance turned out not to be a ceiling. The challenge of the increase in speed seemed to energize or activate the subjects and increase their capacity to make correct responses. In 1948, in one of his postwar survey papers, Bartlett made an oblique reference to this phenomenon when he said:

> A simple objective increase in urgency—in speed, for example, or in the resistance opposed to movement—will be met swiftly by an almost reflex-like increase in effort. Skill that is threatening to deteriorate can, paradoxically, often be restored by an objective increase in the difficulty of the situation.[55]

Conrad was not the only psychologist to report findings of this kind, but it is particularly easy to link his experimental arrangement to the task of flying an aircraft.[56] The pilot has to respond to information from a number of dials such as the airspeed indicator, the altimeter, the compass, etc. For a given cockpit, the number of dials on the dashboard does not change, so the "load," in Conrad's sense, is constant. However, in preparing a maneuver, such as landing, the rate of signals needing a response will increase. The "speed" of the task in Conrad's sense will change, and this change, as his results show, can "paradoxically" help to offset threatening deterioration.

The three bodies of experimental work that I have just described, by Broadbent, Poulton, and Conrad, were all summarized and synthesized in Broadbent's *Perception and Communication*. The synthesis is captured in the

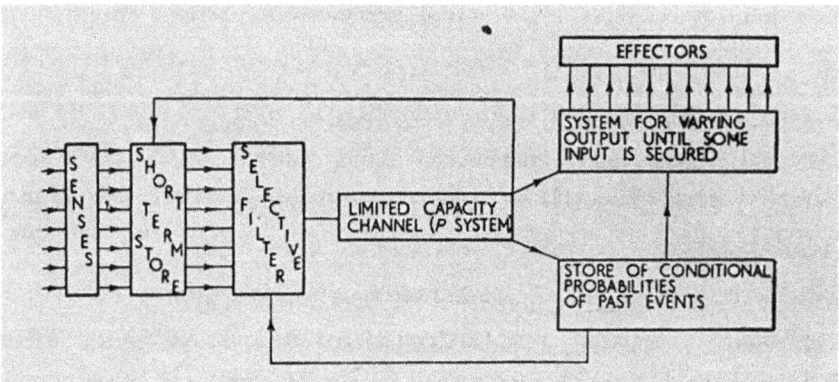

FIGURE 6.1. The flow of information within the nervous system, circa 1958. The figure indicates that a process of filtering precedes cognitive analysis and response. Broadbent emphasized that the diagram was displayed at the end of his book *Perception and Communication* because it summarized the outcome of experimental research, rather than representing a hypothesis constructed prior to experiment. It was an inductive conclusion. From Broadbent 1958, 299.

celebrated flow diagram that was presented in the final chapter of the book, and which is reproduced as figure 6.1. Broadbent was at pains to stress that the diagram was not the expression of an abstract theoretical stance but was an economic summary and interpretation of numerous and diverse experimentally ascertained facts. That is why, following the canons of his inductivist methodology, the diagram came at the end of the book, rather than the beginning.

The Laning-Accident Study Revisited

Can we allow ourselves to say: "*Perception and Communication* was the Cambridge reply to Bradford Hill and the Landing-Accident Anomaly"? No. Such a claim would be a historical distortion, although the distortion would not be great. What can be said, without any distortion, is that *Perception and Communication* contained material that could have been used in a plausible effort to reconcile the results of the Cockpit experiments and the landing-accident study. The reconciliation might be called the Broadbent-Conrad-Poulton Theory of Fatigue, but for brevity I shall refer to it as the Filter Theory of Fatigue. The theory is that fatigue compromises and diminishes channel capacity, while predictability and arousal both sustain channel capacity. Capacity limitations are expressed through perceptual processes as well as through motor processes. The narrowing of attention is a reflex attempt by the filter mechanism to use available capacity more efficiently. Success or failure in landing then depends to a significant extent on the contingent timing of the elements of the task and their interaction with the timing of filter-switching processes.

The flow diagram can be understood as an embodiment of this theory once given the basic principles that Broadbent listed and spelled out for its interpretation.[57] I will now show how these principles provide the ground for the required reconciliation. The most important of the interpretive principles, for present purposes, is stated by Broadbent as follows:

> The effect of prolonged work and also of continuous distracting stimuli will be greatest at the end of the work period, and will consist of intermittent failures in the intake of information. Tasks in which a great deal of perceptual anticipation is possible will therefore show no such effect.[58]

The first sentence identifies the locus of a form of inefficiency that becomes more marked with increased time on task; the second sentence identifies a perceptual process that might serve to counteract these decrements. Tasks that permit anticipation will not show fatigue effects—or will only show them with diminished severity. Further counteracting fatigue is the activating effect of confronting increased demands. The contingencies of the task—and the timing of responses—of course play a vital role in the outcome. Here, the distinction between paced and unpaced tasks becomes a matter of central importance. It will be recalled, from chapter 1, that the task imposed on the pilots in the Cockpit experiment involved a rigorously timed sequence of maneuvers. Bartlett identified the main symptom of the disorganization of behavior that is designated by the word "fatigue" as a failure of timing. In short, the Cockpit task was a paced task, and laboratory studies show that it is paced tasks that can, like the Cockpit task, display dramatic performance decrements. I will now relate these ideas to the different phases of the actual operational flights that provided the raw data for Bradford Hill and Williams's study. Those phases are (i) the return flight, (ii) joining the circuit, and (iii) the final approach.

The ideas current in Cambridge circa 1958, summarized in *Perception and Communication*, imply that, on a return flight after a long sortie, a pilot and crew will manifest some internal effects of fatigue. With increasing frequency, the perceptual filter will block out task data and admit task-irrelevant data. The pilot will be aware of distracting stimuli from outside the cockpit, or sensations of bodily discomfort from within. However, the main tasks for the pilot on the return journey are to maintain straight and level flight at a given altitude, follow a compass course, and make good a specified track across the terrain toward home. The operational flight plan will have specified the turns to be made onto the different legs of the route, but the timing, though important, is not sensitive. At least, it is not so sensitive that brief lapses of attention, a brief filter shift at the wrong moment, will generate errors that cannot be corrected and the situation rectified. The return home is, effectively, a simple and undemanding tracking task with extensive preview. The information-processing demands are relatively low.

All this is true of the fortunate crews who experience a peaceful return flight. The situation is different if the aircraft is suddenly subject to attack. The contingencies of the attack impose their own pace on the order of events. Here a momentary lapse of attention by any member of the crew can have disastrous consequences. The opportunity to issue a warning, or to take evasive action, or to retaliate, may be missed—just as a transitory signal in a vigilance task, or the stimuli in a paced, serial reaction task, may be missed. In such cases the aircraft may never return, though this fact would not register on Bradford Hill's statistical conclusion. It was the number of landings that were counted, not the number of non-returns.

The crews who make it back to the vicinity of their base airfield must now make contact with the control tower and join the circuit in preparation for the landing. The pilot will have work to do modifying the position, speed, and altitude of the aircraft to align his machine with other returning aircraft circling the aerodrome. There will be pre-landing checks to carry out. Throttle settings must be changed and carburetor heaters engaged. Flaps and undercarriage must be lowered and trim adjusted. The pilot must turn into the base leg of the circuit, lose altitude, and then make a difficult turn into the wind for the final approach. All this activity keeps up the level of the information-processing demand. The pilot is now in the position of Conrad's experimental subjects who could be pushed into a higher level of activation by virtue of the increased demands made on them.

When the pilot has been given clearance to land, the problem to be solved requires the visual selection of a point of touchdown near the beginning of the runway. The pilot must aim the aircraft at this point and attempt to fly toward it, down an imaginary straight line. The difficulty is to stabilize the speed, direction, and rate of descent of the aircraft in order to make a safe approach. The pilot will use the elevators to raise or lower the nose to correct deviations from the recommended landing speed. The rate of descent is controlled by adjusting the engine throttles. Lateral deviations, caused by crosswinds, are corrected by the use of rudder and ailerons. The technique is to keep the image of the point of touchdown at a fixed position in the windscreen. If the point shifts up or down, or moves to the right or left, some correction must be made. Ideally, the aiming point should simply grow larger and larger in the visual field as the runway is approached. The final act of judgment, initiating the rollout, a few feet above the ground, involves pulling back the control stick to kill the speed and allow the aircraft sink onto the runway. The gesture needs to be as automatic as putting a coffee cup back onto the saucer.

Poulton's experiment on tracking with the eyes closed provides a simplified model of a pilot on final approach who is "blinking" with fatigue. The model only applies if the course to be tracked is simple, if the track is well known, if there is good preview, and if the person who is tracking is well-practiced. For the operationally trained pilot landing a familiar aircraft on a familiar aerodrome in good weather, with good visibility, these conditions are satisfied. Such landings might be classified as "ordinary" or "straightforward," and these were exactly the landings that Bradford Hill and Williams said they wanted to study. Obviously, blind tracking cannot work where there are complicated, randomized, and fast-moving changes in the course to be tracked. This would be the case if the pilot confronted bad visibility, or turbulent crosswinds, or was flying a machine whose responses were erratic due to mechanical faults or damage. A blink, or lapse of attention, or a filter shift, might now prove disastrous if it coincided with a moment of emergency. Under these conditions, however, the accident, traced through Air Ministry statistics and form 541, might not have been counted as pilot error but instead have been attributed to the weather or a mechanical defect.

In 1953, Broadbent delivered a paper with the title "Neglect of Surroundings in Relation to Fatigue Decrements in Output." The occasion was a symposium on fatigue held at the Royal Air Force College at Cranfield.[59] He summarized for his audience a number of the experiments that I described earlier in this chapter. He concluded that a person who had, say, been driving a car for many miles might handle the vehicle as well as a fresh driver. This is similar to saying that a pilot on a long flight might appear to be as efficient in handling the aircraft as a pilot after a shorter flight. But, said Broadbent, there was a precondition. The equivalence between the two car drivers would only hold if the road were clear and the driver had adequate warning of any crisis. The difference between fatigue and non-fatigue might be revealed only in a crisis, for example, if a child suddenly stepped off the curb. Then the fatigued driver might not notice what the unfatigued driver would notice. In Bartlett's phrasing, fatigued drivers, like fatigued radar operators, might be looking as intently as ever but still fail to register something that was within their field of vision and to which they were, or thought they were, vigorously attending. The same preconditions would hold for two pilots, one fatigued and one fresh.

An Uncanny Ability

The facts that Broadbent marshaled in his 1953 Cranfield discussion, and which he consolidated in later experiments, have implications for the design of any study meant to detect and measure fatigue. They reinforce the points Bartlett made in his postwar papers. Suppose that, in some such proposed study of fatigue, the measure used to judge performance only took account of a broadly defined end result, such as arriving safely at the destination of a journey. Again, suppose that, in a misguided attempt to sharpen the focus of the study, the demands made on the subjects were all predictable—the rate of information transmitted through the subject would then be low, as it was in the Harvard tests on the impact of noise stress. Under these highly controlled conditions, the difference between fatigued and non-fatigued states, though real, might never be revealed. Bad weather and enemy action in Bradford Hill's study corresponded to the child stepping off the curb in Broadbent's hypothetical example.

The conclusion must be that Bradford Hill and Williams had chosen to study a task—the landing of an aircraft—that was saturated with anticipation. In the light of postwar Cambridge research, this choice was unfortunate if the aim was to illuminate the inner psychological mechanisms of fatigue. Bradford Hill and Williams's research design, like Stevens's research design in the Harvard noise study, contained a hidden trap—a trap that only those possessed of the broad sweep of an inductivist sensibility would be likely to notice. Given such a sensibility, the negative result of the landing-accident study might have been predicted, and perhaps it was.

Broadbent and other Cambridge colleagues always said that Bartlett had the

> uncanny ability to predict the results of almost any experiment performed on human beings; he might not be able to tell you how he arrived at his prediction, but he seemed always to know what somebody would do in a certain situation.[60]

Obviously, such laboratory stories cannot be taken literally: Broadbent was not really attributing supernatural powers to Bartlett. Nevertheless, Bartlett might indeed have had a surprising capacity to make correct guesses about experimental outcomes. The use of the word "uncanny" merely provided an appropriate idiom in which honor could be paid to Bartlett's experimental sagacity. It was a way for colleagues, students, and friends to show respect for an elder of the Laboratory Tribe. If we permit ourselves to use this mythical though oddly plausible idiom, we might indulge the imagination and suggest that Bartlett probably "knew" what Bradford Hill was going to find, and "knew" that eventually there would be a way to diffuse the potential conflict between the Cockpit findings and the Bomber Command study. It was just that, at the crucial meeting of the Flying Personnel Research Committee in 1943, Bartlett could not say with any precision how he knew, and why he expected, this outcome. That had to await the publication of Broadbent's *Perception and Communication* in 1958.

Was There a German Cockpit?

In 1927, the high-altitude physiologist Arnold Durig published two chapters on fatigue in Edgar Atzler's bulky *Körper und Arbeit: Handbuch der Arbeitsphysiologie.*[1] (Atzler was the director of the Kaiser-Wilhelm-Institut für Arbeitsphysiologie in Dortmund.) In 1935, Ernst Simonson prepared a further, widely quoted review of the fatigue literature. His "Der heutige Stand der Theorie der Ermüdung" (the current state of the theory of fatigue) appeared in the journal *Der Ergebnisse der Physiologie.*[2] The medical authorities in the German Air Force were well aware of this literature and well aware of the reality of fatigue and its military significance. Did it prompt them to commission a German Cockpit similar to the Cambridge Cockpit? Were they moved to adopt the course of action taken by the Flying Personnel Research Committee? The answer is negative, and the reasons must be investigated. They turn out to be interesting. First, though, I shall explain why the initial negative answer is also surprising.

The Death of the Psychotechnik Movement

There were German precedents for large-scale and realistic apparatus, some of which bore a certain similarity to the Cambridge Cockpit. These pointers suggest that similar steps might have been taken in Britain and Germany when confronting the problem of fatigue. During WWI, German psychologists had constructed moving panoramas to test trainee pilots and observers.[3] In the 1920s and 1930s, they also devised aptitude tests for locomotive drivers where candidates were required to respond to various stimuli by pulling levers and adjusting controls similar to those in the cabin of a locomotive. The aim was to select those who appeared to be the best judges of speed and distance and who displayed the quickest reactions. The analogy between the demands made on pilots and those made on locomotive drivers is evident.[4] These facts are well known to historians of psychology because of the extensive research of Horst Gundlach and others into the "psychotechnics" movement.[5] If psychologists could take over the cabin of a railway engine and use it as a piece of psychological apparatus, then they could do the same for the cockpit of an aircraft. Despite these precedents, I have found no German fatigue machine in the form of a cockpit. In 1928, Hans-Georg Gade of Danzig published his *Zur Psychotechnik des Flugzeugführers* (On the Psychotechnics of Aircraft Pilots), but his apparatus merely took the form of a rotating chair. His main concern was to develop aptitude tests that singled out candidate pilots who could best cope with disorientation and dizziness.[6]

Part of the explanation for the lack of any strict equivalent to the Cambridge Cockpit may be the eclipse, indeed the suppression, in the mid-1930s, of the German psychotechnic movement.[7] After a period of post-WWI enthusiasm for apparatus-based aptitude testing and scientific selection procedures, by the late 1920s the psychotechnic movement had lost momentum. There were professional disagreements about the status and direction of the field. Was psychotechnics a matter for engineers or for professional psychologists? And economic crisis and mass unemployment overshadowed the significance of refined selection procedures. On top of these problems, the post-1933 National Socialist regime purged the discipline of psychology of its Jewish members and generated a polemic against so-called "*jüdische*" *Psychotechnik*.[8]

The result was that in the years approaching WWII, many of those most likely to have instigated the transition from trains to planes ceased to have prominence and influence in the field of applied psychology. The apparatus-based tests of the kind favored by exponents of psychotechnics were increasingly deemed old-fashioned and theoretically superficial. The exponents of the new fashion decreed that psychologists should look for a deeper unity behind the various aptitudes that had previously been the focus of investigation. What mattered was character, and psychologists, particularly those associated with military selection procedures, shifted their allegiance from psychotechnics to characterology.[9] The advocates of characterology were sailing with the right-wing political tide.[10] Nevertheless, some academics, who can be seen as representatives of the older psychotechnic movement, continued their work as they sought some degree of accommodation to the post-1933 regime. For example, there were Walter Moede and Emil Everling at the Technische Hochschule in Berlin-Charlottenburg. Moede had worked on selecting and training lorry drivers for the German army during WWI.[11]

He continued his work on apparatus-based selection procedures but did not work on pilot fatigue. One of his students produced a thesis on the fatiguing effects of oxygen deficiency, but that was all.[12] Everling, by contrast, had built an interesting, cockpit-like object called the *Fliegerdrehkammer*.

The *Fliegerdrehkammer*

In 1936, Everling was appointed to the chair of *Flugzeugbordmeßgeräte*—that is, his teaching and research responsibilities concerned the construction and properties of aircraft instruments.[13] Everling had a special interest in the problems of blind flying and the instruments that could serve to offset the misleading information sometimes generated by the natural human capacity to sense direction, position, and bodily orientation. Were there perhaps ways of selecting pilots who had a particularly good sense of spatial orientation and who would be less challenged by the task of pure instrument flying? Were there ways to strengthen and train our innate capacities? The vestibular system in the human ear was a sort of natural instrument that could be understood in engineering terms. The Cambridge Cockpit experiments were essentially tests in instrument flying, so if anyone was positioned to construct a counterpart to the Cambridge Cockpit, it was Emil Everling.

In 1938, Everling and his student Wolfram Eschenbach constructed an enclosed cockpit-like structure equipped with aircraft-style controls. The device could house a subject and an instructor, sitting side by side, and it could be made to rotate to the right or left about a vertical axis. This was the *Fliegerdrehkammer* (see fig. 7.1).[14] Everling and Eschenbach built it from scratch rather than using an old aircraft fuselage. The device was conceived as an apparatus to test blind-flying ability. It was sometimes used to test different designs of aircraft instruments, but there is no indication that the *Fliegerdrehkammer* was ever used to study fatigue. The "Berlin Cockpit" (if it may be so called) was never employed in a way that made it the equivalent of the Cambridge Cockpit.[15] Had the German Air Force, or the German Air Ministry, requested that the apparatus be put to use to investigate fatigue, there is every reason to suppose that it would have served this purpose—and then the *Fliegerdreh-kammer* might indeed have earned the title of the Berlin Cockpit. But even if Everling's apparatus had been used to study fatigue, because of its circumscribed range of movement it could not have provided the wealth of information about the interrelated responses of the subject that was furnished by Craik's Cockpit. In the event, the authorities made no such approach to Everling. Exactly why they did not remains a problem. Something about the

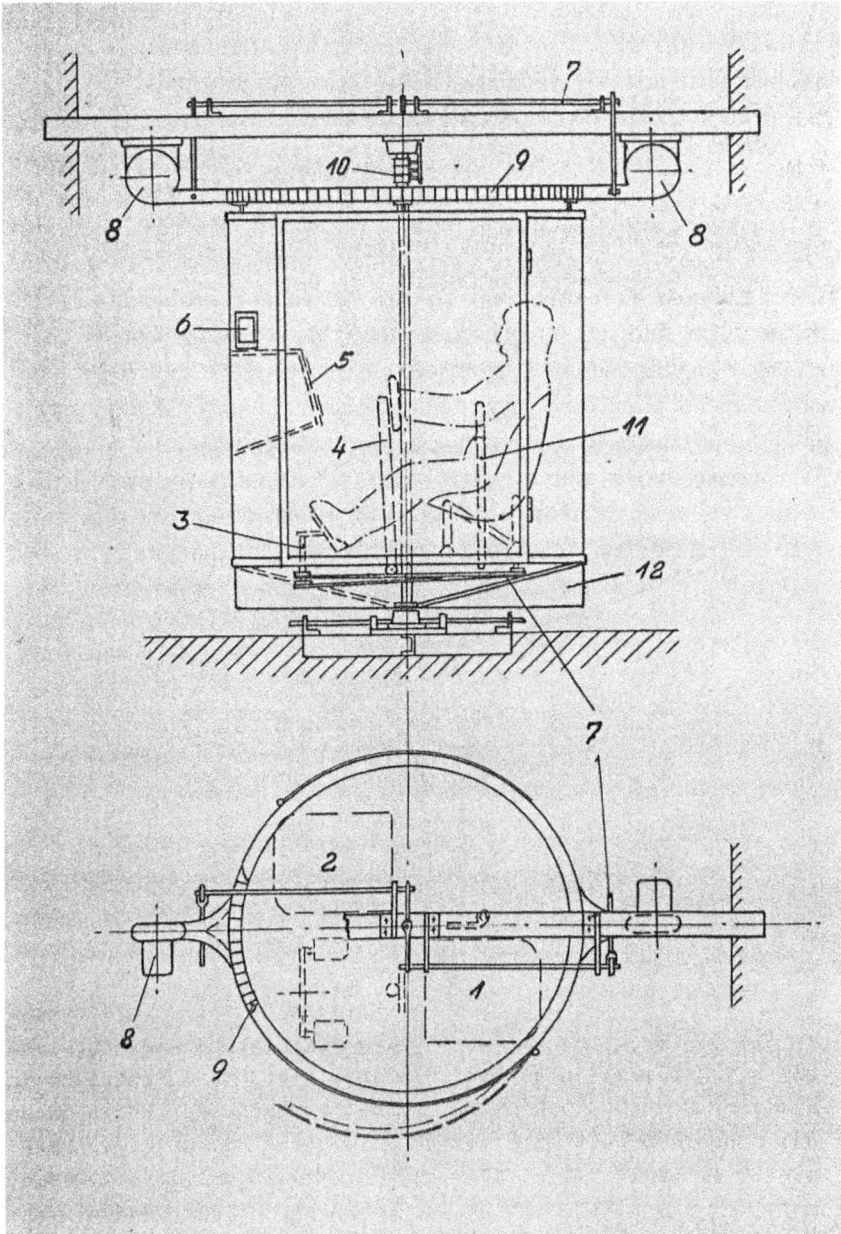

FIGURE 7.1. The Charlottenburg *Fliegerdrehkammer*. Constructed by Everling and Eschenbach at the Technische Hochschule in Charlottenburg in 1938, the *Fliegerdrehkammer* was cockpit-like, but it was only used for a limited range of testing and selection purposes. Unlike the Cambridge Cockpit, it lacked the capacity to record a range of different but closely linked responses to complex situations. From Eschenbach 1938, 32.

prevailing view of fatigue, and the prevailing distribution of responsibilities for dealing with fatigue, seemed to have precluded such a course of action. I shall look further into these matters in the course of the chapter.

Medical Pessimism in 1941

On December 5, 1941, the *Deutsche medizinische Wochenschrift* published a collection of articles on the theme of fatigue and the problem of combating fatigue. The collection was given the title *Ermüdung und Ermüdungsbekämpfung* (Fatigue and Combating Fatigue). It was introduced by a useful and revealing summary paper: *Leistung, Ermüdung, Übermüdung* (Performance, Fatigue, Exhaustion), authored by the physician Professor Albert Johann Anthony.[16] Anthony (1901–1947) held the rank of *Oberartz* (medical consultant) in the German Air Force.[17] His own research dealt with oxygen metabolism.[18] Following the material in Simonson's review, which Anthony duly cited, the overall message was that the scientific understanding of fatigue had fragmented along disciplinary lines. There was no overall and coherent account to be had. As Anthony observed:

> *In dem umfangreichen Schriftum, das sich mit dem Ermüdungsproblem befaßt, findet man zwar ein großes Tatsachenmaterial zusammengestellt, über die allgemeinen Gesetzmäßigkeiten, die den genannten Problem zugrunde liegen, werden aber sehr verschiedenartige, zum Teil widersprechende Anschauungen vertreten. Dies erklärt sich zum Teil durch die tatsache, daß das ganze Problem aus verschiedener Blickrichtung gesehen wurde.*[19]

In the extensive literature dealing with the problem, a great deal of factual material has been assembled regarding the basic laws that underlie fatigue, but there are very variable and sometimes contradictory opinions on the subject. The explanation lies partly in the fact that the whole problem can be seen from very different viewpoints.[20]

The constellation of problems confronting the clinician, the psychologist, the work physiologist, and the military commander were fundamentally different. Anthony complained that there was no accepted scientific measure of the fatigue state. He cited and endorsed the pessimistic conclusions expressed by Durig. As Durig had said in the Atzler volume: "so ergibt sich, daß es überhaupt keine Methode gibt, mit welcher Ermüdung auch nur einigermaßen exakt gemessen werden könnte"—so the upshot is that there is simply no method by which fatigue can be measured even approximately.[21] But if fatigue evaded measurement, Anthony nevertheless took it for granted that fatigue was to be classified as a process of inhibition—rather than, as in Cambridge, a process of disinhibition. Thus:

> *Die bei Leistungen auftretenden hemmenden Einflüsse, die zu örtlicher oder allgemeiner Ermüdung fuhren, werden als* Ermüdungsvorgänge *bezeichnet.*[22]

> The inhibitory influences which arise during the performance which lead either to local or general fatigue are to be referred to as *fatigue processes*.

The problem was that these inhibitory processes were made stronger or weaker by virtue of their relation to many other factors. Anthony explained that there were numerous interacting variables. As well as the level of central arousal or excitation (*Erregung*) and magnitude of the stimulus (*Reizgröße*), there was the susceptibility to stimulation (*Anregbarkeit*) and the internal drive-state of the organism (*Antriebsfähigkeit*). Added to these variables, the mental orientation or *Einstellung* of the responding person and the effect of their determination (*Leistungswille*) also needed to be taken into account. All of these factors entered into the final performance, and they were all possible sites of specific forms of fatigue. Thus:

> *mag es zunächst unmöglich erscheinen, weiteren Einblick in die Zusammenhänge zu gewinnen, und es wird verständlich, daß eine quantitative Ermittlung der menschlichen Leistungsfähigkeit auf direktem Wege nicht möglich ist, solange Erregbarkeit und Antriebsfähigkeit eine quantitativen Messung nicht zugänglich sind.*[23]

> It may appear impossible at first to acquire further insights into the relationships portrayed and it is conceivable that a direct quantitative determination of human performance is not possible, as long as arousal and drive are not amenable to quantitative measurement.

Anthony appeared resigned to the possibility that Durig was right and adequate measurements of fatigue might never be forthcoming.

> *Mit Recht weist deshalb Durig darauf hin, daß es heute keine einzige Methode gibt, welche am lebenden Menschen quantitativ die Ermüdung zu messen gestattet und daß es auch gar nicht zu erwarten ist, daß es solche Methoden einmal geben wird.*[24]

> Durig has correctly pointed out that there is not a single method nowadays which permits fatigue in living human beings to be measured quantitatively and it is not to be expected that such methods will ever be available.

There was no compelling way to draw conclusions about the experimental subjects' capacity to work from their work performance. A decrement in the output of work does not permit any direct inference to a decrement in the capacity to work. There are too many other, unknown factors in the situation. Furthermore, Anthony argued:

> *In diesem Zusammenhang ist Leistung kein absoluter Begriff, sondern erhält nur Sinn durch Bezug auf eine bestimmte Arbeitsform, auf eine bestimmte Leistungsart.*[25]

> In this context, capacity is not an absolute concept, but gains its meaning only by reference to a specific form of work and a specific mode of performance.

Because the concept of capacity is relative to specific types of work, it is impossible to draw inferences about the performance of complex tasks from data about simple tasks. Like Bartlett, Anthony was clear that

> *Es ist vor allem nicht statthaft, aus dem Verhalten bei primitiven Arbeitsform ohne weiteres auf das Verhalten bei anderen komplitzierten Arbeitsformen zu schließen.*[26]

> Above all, it is not permissible to draw any immediate conclusions about complicated forms of work from simple forms of work.

Unlike Bartlett, however, Anthony appeared to accept this situation and made no proposals to overcome the evidential gaps he had identified: for example, by initiating a new research program devoted to the nature of fatigue in complex and skilled tasks.

Despite these problems on the scientific front, Anthony argued that there was still room for the exercise of informed clinical judgment. He said that certain "simple basic laws" were at work. By "laws" here he meant not scientific generalizations but medical regimes and a clinical awareness of the dangers posed by fatigue. He pointed out that the best pilots may be those whose character inclines them to push themselves beyond the limits of their capacity. The remainder of his paper was devoted to these clinical concerns, such as the need to draw distinctions between healthy and "harmonious" forms of fatigue and pathological states of exhaustion that generate physiological damage. Pathological symptoms take the form of high blood pressure, irregular heartbeat, and disturbances to the autonomic nervous system. Anthony was critical of the use of performance-enhancing drugs such as Pervitin and advised against their use except in emergency. Clearly, Anthony's main concern was the clinical and physiological aspects of extreme fatigue—not the detailed ways in which skilled performance might degrade during a demanding and protracted task. Neither the methodological pessimism nor the clinical orientation that was displayed in Anthony's contribution would be apt to lead anyone to the construction of an object like the Cambridge Cockpit.

Pervitin, Encephalography, and Anoxia

If Anthony's stance represents one particular professional perspective on fatigue, namely, that of the clinician, there were other standpoints adopted by other German researchers—indeed, that very diversity was seen as the problem with the overall research effort. However, some of these specialized approaches undoubtedly led to an active scientific involvement with fatigue phenomena and generated local pockets of confidence that scientific advance was possible.[27] Recall Anthony's warning about misuse of the stimulant Pervitin. The drug was meant to overcome fatigue, so research into Pervitin counts as research into fatigue.[28] Pervitin was known to be addictive, and there were worries lest its users become stimulated to the point where they would exhaust all their natural reserves. Pervitin could be dangerous, and there had been accidents, which was why the German Air Force, which issued emergency packs of Pervitin to its aircrews, demanded that any use of the drug be officially recorded. Predictably, given its extensive use, there was intense study of this drug in Germany during the war. For those involved, there was a corresponding sense of contributing to advances in psychopharmacology, which was viewed as a field that was especially cultivated in Germany and (in the view of one expert) had been "introduced and developed by Kraepelin" and "based on Wundt's psychology."[29]

Electrophysiology provided another, technically sophisticated, approach to fatigue. The attempt was made to correlate fatigue states with identifiable patterns of electrical activity in the brain and thus to exploit the recent advances in electroencephalography that had been made by German researchers.[30] Perhaps it would be possible to identify the characteristic brain waves that expressed or heralded the onset of fatigue. This project was undertaken by Prof. A. E. Kornmüller at the Kaiser-Wilhelm-Institut für Hirnforschung in Berlin-Buch. Kornmüller believed that fatigue correlated with the slowing of the alpha frequency. His idea was that all members of an aircrew could be wired up to a device that would analyze their brain waves and automatically inform the pilot if anyone was displaying fatigue symptoms.[31] The war ended before Kornmüller's apparatus could be assessed for its practicality, but postwar attempts by the US Navy to build an alertness indicator, based on Kornmüller's work, were not a success. The American researchers found that "the range of individual differences in the amount and frequency of the eyes-open alpha rhythm in a group of normal people was quite remarkable and seemed to be relatively independent of alertness." They concluded (not entirely seriously) that a simpler solution would be a harness that positioned a sharp spike under the subject's jaw. One could be sure that it would provide effective feedback whenever a sleepy head nodded.[32]

The most concentrated line of scientific investigation in Germany associated with aviation psychology and physiology was the study of the condition known as anoxia, or oxygen deficit. The symptoms of anoxia were recognized to be fatigue-like, so this research too must count as research into fatigue. Studies of anoxia take us back to the low-pressure chambers that were proudly displayed to Wing Commander Livingston on his fact-finding mission to Germany in 1937. As a research topic, the effects of oxygen deficit outnumbered all the others pursued by German specialists.[33] Once again the difference with the Cambridge experiments is clear. A low-pressure chamber is not a cockpit.[34] I do, however, want to take a look at another German experiment on fatigue and oxygen metabolism, although it is one that does not depend on a low-pressure chamber. This work took place at an institute that can be seen as an approximate German equivalent of the Applied Psychology Unit in Cambridge. The institute in question was one of the targets of Russell Davis's immediate postwar visit to Germany. In his intelligence report, Russell Davis declared: "Virtually the only institute which before the war carried out research in applied psychology and which retained its integrity at the end of the war was the Kaiser-Wilhelm-Institut für Arbeitsphysiologie at Dortmund."[35] In the event, only one person at the institute appears to have been interviewed: a certain Dr. Graf. Why might Russell Davis have been interested in Graf—and how close was the comparison between Dortmund and Cambridge?

Pilot Fatigue at the Front

In 1942 a paper with the title "Versuche über die Wirkung von Sauerstoffatmung bei normalen Druck auf die Leistungsfähigkeit"—experiments on the effects on performance of breathing oxygen at normal pressure—appeared in the journal *Luftfahrtmedizin*.[36] The authors were Gunther Lehmann (1897–1974) and Otto Graf (1893–1962).[37] Graf was a pupil of Kraepelin's and was in charge of the psychological section of the institute. He had carried out extensive research on both industrial fatigue and the effects of a range of stimulants and sedatives, including alcohol and Pervitin.[38] Lehmann was the director of the Dortmund institute, a position that he had taken over from Edgar Atzler after Atzler's death in 1938. During WWII, Lehmann had the rank of

Stabsarzt, captain, in the Medical Corps of the German Air Force.[39] Lehmann had been intrigued by what he heard from pilots when he was, as he put it, at the front. Pilots told him that they had started using their oxygen-breathing apparatus, which was meant for high-altitude flying, almost as soon as they got their aircraft off the ground. They said the oxygen helped them to overcome fatigue.

In their 1942 paper, Lehmann and Graf described the fatigue in rather bland terms by saying that the pilots used oxygen when they were not feeling "completely fresh" (*nicht vollkommen frisch*).[40] Lehmann later revealed that the situation was in fact considerably more serious. In a postwar account of this work written for, and translated by, the American occupation authorities in Germany, the pilots were described as "tired and flight weary." In that translated account, Lehmann explained the circumstances as follows:

> At the beginning of the War pilots in general disliked the use of oxygen masks and usually donned them only at altitudes higher than those required by regulations. However, their attitude changed during the summer of 1940 when fighter units began to go on frequent exhausting missions.[41]

The date is significant. The summer of 1940 was when the German Air Force was attempting to gain air supremacy over the English Channel and south coast of the British Isles.[42] This was the front line to which Lehmann had referred and the circumstances were precisely those that—on the other side of the conflict—gave urgency to the Cambridge Cockpit experiments. The pilots on both sides of the battle were in a state approaching exhaustion. Lehmann and Graf submitted their paper for publication in May 1941, remarkably soon after this unofficial use of the oxygen mask had made its appearance.

Why did Lehmann and Graf think that an experiment was called for? The problem that struck them was that the pilot's report fell outside current physiological theory. The uptake of oxygen in the hemoglobin in the bloodstream was understood as necessary for recovery after intense muscular exertion, but the pilots, though exhausted, were not tired as a consequence of muscular exertion. Breathing oxygen should have been irrelevant. Were the pilots deceiving themselves and the effect was "psychological" in the dismissive and derogatory sense of the word—that is, it was merely apparent? Or was current physiological theory incomplete? The baseline of understanding was work by A. V. Hill, originally from the Physiological Laboratory in Cambridge.[43] Lehmann and Graf suspected that the gross peripheral effects in and around the muscle itself, as studied by Hill, masked other, more subtle effects. Perhaps oxygen tension in the blood had an effect on the working of the cerebral cortex. The higher and conscious functions that are subserved by the cerebral cortex are the first ones to be affected when oxygen tension is low. Perhaps the converse holds, and the cortex would prove immediately responsive to increased oxygen tension at normal air pressure. In previous experiments, Anthony, along with H. F. Rein of Göttingen, had shown that blood-oxygen tension had an effect on a number of internal organs, so here was a chance for Lehmann and Graf to show that the all-important cortex should be added to that list.[44]

Two sorts of test were devised to investigate this possibility: (i) a bicycle-ergonometer test and (ii) a repetitive, Kraepelin-style test of intellectual processes. During these tests, the experimental subject had to wear an oxygen mask and would sometimes be breathing pure oxygen, sometimes normal air, and sometimes oxygen-depleted air. To avoid the effects of suggestion, the subjects were not told what they were breathing. Lehmann and Graf admitted that the subjects could hear (though not see) when the supply was switched from one source to another, but they were confident that an examination of the work curve permitted them to rule out any effect of suggestion based on stimulation from this source. Any such stimulation, they argued, would have produced an immediate and observable "spurt" in the performance at the beginning of the work session. There would be an observable expression of what Kraepelin called an *Anfangsantrieb*, but the shape of the work curve spoke against this possibility.

What did Lehmann and Graf find? First, it was established that, for subjects already laboring under a moderate but uncomfortable state of fatigue, pedaling the bicycle apparatus seemed easier when breathing pure oxygen than when breathing normal air. It seemed hardest when breathing oxygen-depleted air. When breathing pure oxygen, subjects pedaled faster and with less subjectively felt effort. Given that the muscular demands of the task were deemed not so great that oxygen depletion of the hemoglobin was at issue, Lehmann and Graf concluded that, physiologically, the task was not made objectively easier by the oxygen. The effect must be on the subject's willingness to confront the fatigue. Oxygen, they concluded, increased willpower. Since the will was often assumed to have its seat in the cortex, this conclusion fitted Lehmann and Graf's hypothesis.

The second set of tests required the subjects to perform mental arithmetic and to rapidly identify target letters surrounded by sets of irrelevant letters. Again, breathing pure oxygen improved the speed and accuracy of the subject's performance—at least in most of the tests. Here, though, Lehmann and Graf encountered an anomaly that led them to complicate their interpretation. Although there was a general improvement, some subjects showed no improvement when breathing pure oxygen, and some even turned in worse performances. Lehmann and Graf introduced two other variables to explain this finding. They argued that well-practiced mental activities—activities that had become automatic—were relatively immune to improvement by breathing pure oxygen. They also argued that subjects who were already excited and anxious about the tests they were to perform were made too excited and too anxious by breathing pure oxygen. Oxygen was a stimulant, and these subjects had become overstimulated. For them, breathing an oxygen-depleted mixture would have a soothing effect. They expressed their results by saying that imbibing oxygen was only advantageous while the performance was still responsive to the exercise of will. The oxygen augmented the conscious determination to persevere with the task.

Here is how Lehmann and Graf sum up their results in the final section of their 1942 paper in *Luftfahrtmedizin*:

> *Die erhobenen Befunde sprechen dafür, daß die O2-Atmung vor allem die Bereitschaft zu Willensantrieben erhöht und daß die beobachteten Leistungssteigerungen durch eine Häufung und Steigerung solcher Antriebe zustande kommen. Wir verstehen dabei unter „Antrieben" im Sinne der Kraepelinschen Schule die verhältnismäßig rasch vorübergehenden positiven Schwankungen in der Triebkraft der Arbeitswillens, die auch subjektiv beobachtet werden können als Momente erhöhter Kraftanspannung.*[45]

> The findings indicate that breathing O2 mainly increases the readiness to exercise the willpower and that the observed improvements in performance arise from the increase and buildup of this drive. By "drive" we follow Kraepelin and his school and refer to the (sometimes rapidly fluctuating) driving force of the will to work which is subjectively experienced as a heightened state of tension.

The fatigued German pilots who had talked with Lehmann had not been deluding themselves when they reached for their oxygen masks at the first opportunity. In Lehmann and Graf's terms, they were inhaling willpower.

The Dortmund *Fahrgerät*

This was the point at which the 1942 publication stopped, but it was not the end of Lehmann and Graf's wartime research into oxygen and fatigue. In fact, their research was about to get more interesting and move closer to the Cambridge position. Lehmann and Graf tried to bring about a convergence between Kraepelin's style of work and that of the Psychotechnik tradition. Despite losing an arm in the Great War, Graf had a reputation for tinkering with apparatus. In the early 1930s, he had devised an ingenious apparatus that

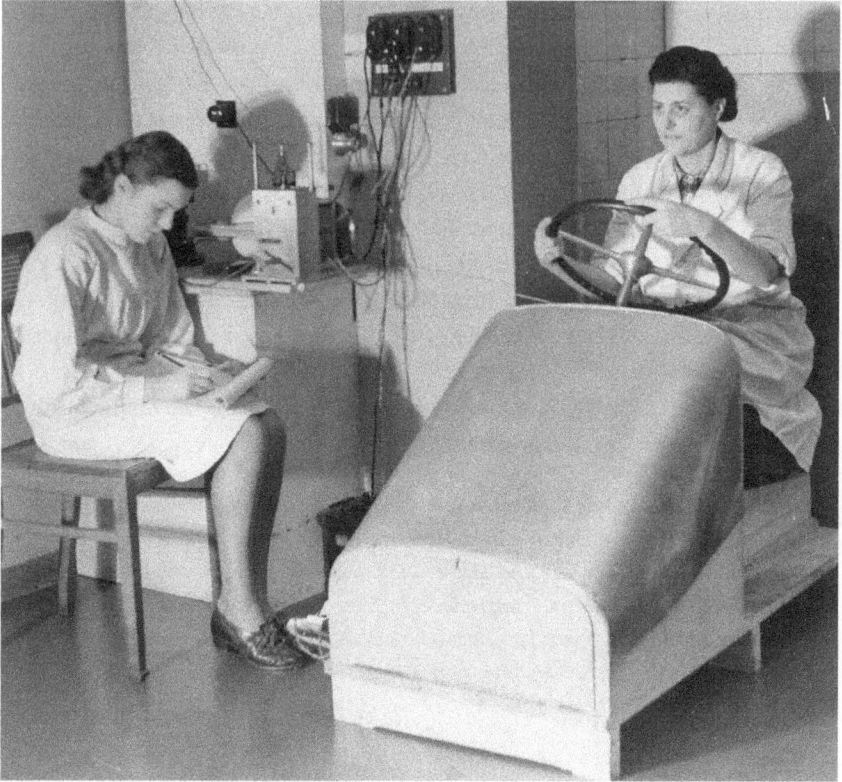

FIGURE 7.2. The Dortmund *Fahrgerät*. This device was built by Otto Graf at the Kaiser-Wilhelm-Institut für Arbeitsphysiologie in Dortmund, as a follow-up to earlier work in the institute on pilot fatigue and oxygen intake. This research could have led to the commissioning of a Cockpit-like study of fatigue but did not do so. Preconditions that were present in Cambridge were lacking in Dortmund. By permission of the Archiv der Max-Planck-Gesellschaft, Berlin. Call numbers VI. Abt., Rep. 1, KWIArbietsphysiol II/37.

simulated the demands of driving a motor vehicle. An early version of the apparatus was used in a 1933 paper on the relation between blood alcohol level and the psychological effects of alcohol.[46] A more elaborate version of the apparatus, now looking like the driving cabin of a motor vehicle, was shown in Graf's 1943 paper "Eine Methode zur Untersuchung der pharmakologischen Beeinflussung von Koordinationsleistungen"—a method for studying pharmacological effects on coordination (see fig. 7.2).[47]

The experimental subjects were seated in front of a steering wheel and were also given the use of an accelerator pedal. They were confronted with a screen showing a moving track that had the appearance of a winding road. A marker indicated the position of the car relative to the road. The task was to steer around the bends and keep the car on the road. The speed along the track could be fixed by the experimenter or varied at the discretion of the driver. Instructions could be varied—for example, sometimes demanding speed and sometimes demanding safety. It was possible to test subjects after a night's sleep loss and to combine this condition with doses of caffeine or Pervitin or some other drug. Degrees of light and dark and limitations on track preview could be manipulated. The apparatus included an automatic counter to register the number of times the car left the road, and it also provided a printout of the frequency and extent of the use of the accelerator. The printout provided a vivid picture of how smoothly or roughly the driver handled the vehicle. Lehmann and Graf now tested their theory about the advantageous cortical effects of breathing pure oxygen, only this time they used the car-driving apparatus that made the task look much closer to the one confronting the pilots who had talked to Lehmann. This aspect of the work, though carried out during the war and following the plan of the paper in *Luftfahrtmedizin*, was not published until after the war.

Disconcertingly, the initial results, in Lehmann's words, seemed "contradictory and confusing." The first experiments with the motorcar apparatus "revealed an increase in errors when oxygen was inhaled and a decrease when the mixture containing 12.5 percent oxygen was inhaled."[48] Was oxygen now apparently delivering the opposite of the expected stimulation? The problem, they concluded, again arose from the use of inexperienced and nervous subjects. Given more practice, the effect of the oxygen reverted to being a help rather than a hindrance. For the well-practiced but fatigued subject, breathing pure oxygen restored performance, while breathing oxygen-depleted air made the driving less efficient. As with the previous experiments, the central psychological fact here, argued Lehmann and Graf, was that of overexcitement. Oxygen is indeed a stimulant, so an already excited subject will become overexcited. Once the need to avoid overstressed subjects had been realized, the driving experiment proved a success. By switching back and forth at ten-minute intervals between pure oxygen, ordinary air, and oxygen-depleted air, Lehmann and Graf were able to show that they could reliably manipulate the performance of the driver. Errors went up or down in a way that demonstrably depended on the oxygen tension.

One of the ideas that Lehmann and Graf used to describe the results of the driving experiment was that fatigue and drugs caused *Blockierung*, or mental blocking, though Bills's name was not mentioned nor was his work cited.[49] Just as Bartlett treated Jacksonian ideas as a resource to be developed, so Lehmann and Graf did the same with Kraepelin's ideas. Graf explicitly likened his driving apparatus to that used "in der Psychotechnik zur Kraftfahrerprüfung"— that is, in the psychotechnical testing and selection of drivers.[50] But if the driving tests were in the tradition of German psychotechnics, they were also part of an attempt to improve what Graf called the "old, reliable methods introduced by Kraepelin," such as the tests of mental arithmetic.[51] Despite their alleged reliability, Graf increasingly saw these as artificial because of their simplicity and repetitive character—which was precisely the complaint against them that Bartlett had made in his Ferrier Lecture. Graf continued to use them but, like Bartlett, felt the need to introduce more realistic work tests—hence the motorcar apparatus. Some of Graf's rhetoric against artificiality was almost Bartlettian in tone. It was in this spirit that, throughout the war, Graf continued work using his simulated driving tests to explore the effects and interactions of stimulants such as caffeine and Pervitin and mixtures of caffeine and cardiazol.

In rationalizing their results, Lehmann and Graf sometimes used terminology that resonated with Jacksonian talk of "levels." They differentiated both anatomically and functionally between the cortex and the subcortex. Caffeine, conjectured, Graf, operated on the cortex, while Pervitin and cardiazol operated on the subcortex and brain stem. The cortex was the seat of higher mental processes such as concentrating one's attention and making conscious decisions. Routine patterns of motor behavior, by contrast, were described as subcortical. The *subcorticalen Zentren* contain preformed patterns of movement: *präformierte Bewegungsabläufe*. There is clearly some convergence with the terminology of Jacksonian levels, though its significance is difficult to assess. A lot depends on precisely how the relation between these two functional entities, the cortex and the subcortex, was conceived, and this point remained unclear.

Lehmann and Graf said that an increase in oxygen tension increased the number of what they called "cortical impulses." But what are "cortical impulses"? When discussing the tremor that attends the aiming of a handgun and the oscillations present in other forms of skilled movement, they describe it as

> *also einer offenbar störenden Durchkreuzung des wohl vor allem in* subcorticalen *Zentren präformierten Bewegungsablaufes durch unzweckmäßige Impulse vom Oberbewußtsein, also von* corticalen *Zentren her.*[52]

> a manifestly disturbing frustration of the preformed pattern of movements primarily located in subcortical centers, that is the result of an inappropriate impulse coming from the cortical centers, as a result of too great a self-consciousness.

Do these inappropriate impulses act as unspecific sources of arousal or in virtue of the information they convey? In the German wartime literature, it is difficult to tell.

If we look at the way Graf rationalizes his experimental results, there are hints of divergence as well as convergence between the neo-Kraepelin schema and the neo-Jacksonian schema. Graf spoke of a calculus of postulated excitations and inhibitions. Some substances act as stimulants, others as inhibitors. When two substances are involved in an experiment, for example caffeine and cardiazol, their combined effect depends on whether the minus signs of the inhibitions outnumber the plus signs of the excitations. Too great an accumulation of negative effects, and the end result was a debilitating fatigue state with its inevitable lowering of performance and work output. (This was Anthony's inhibiting "fatigue process.") Thus, said Graf in his (translated) postwar contribution to *German Aviation Medicine*:

> Work is a complex entity consisting of elements. The decisive points are the proportion in which these elements participate in building up the efficiency, the importance they have within the entire efficiency, and their mutual interaction.[53]

Efficiency, he said, was

> only a general integral, which is differentiated into the individual component functions the sum of which forms the complex entity "efficiency." The integral of efficiency may be equal to zero if a number of antagonistic functions compensate each other.[54]

It remained unclear how the occasional references to higher and lower functional levels were to be coordinated with this presumed additive calculus. Graf argued that when cardiazol is mixed with caffeine, the cardiazol, though acting on the brain stem, can have the effect of calming the cortex. Thus "the cortical effect is weakened by sub-cortical processes, and cardiazol which by itself has scarcely any effect on the higher psychic processes, obtains such an effect when combined with caffeine."[55] Here Graf was rationalizing his findings by saying that the lower level is inhibiting the higher level—the opposite of what is supposed in the Jacksonian tradition.

If the theoretical interpretations offered by Lehmann and Graf were still evolving, the potential of their experimental methods for the study of fatigue is obvious. If the *Fliegerdrehkammer* was too simple to test flying skill, the *Fahrgerät*, with its steering wheel and accelerator, could have been used to pose more complex problems and register more complex responses. Graf's apparatus was much closer to the Cambridge Cockpit than Everling's, but a similar qualification needs to be made about the use to which it was put. Suppose the military or civil authorities in Germany had asked Lehmann and Graf to pursue the problem of fatigue by upgrading the Dortmund motorcar and building a realistic Dortmund Cockpit. There is every reason to believe that they would have grasped the opportunity. This point becomes apparent if we look into the institutional and administrative background of the small and limited experiments that I have just described.

Arbeitsphysiologie and the German Air Ministry

There was no German equivalent of the Flying Personnel Research Committee. How, then, was German research organized and commissioned? In an account written immediately after WWII, Hubertus Strughold declared that the initiative for a research project usually came from the directors of the various leading research institutes. According to Strughold, there would be central guidance, and sometimes the direct commissioning of research by the *Chef der Sanitätswesens der Luftwaffe*, but research groups, he said, were both permitted and encouraged to decide their own research priorities within a general framework of economic planning decided by government ministries.[56]

Strughold's account is broadly consistent with the picture drawn by later, independent historical researchers. It is also consistent with the material in the archives of the Max-Planck-Gesellschaft and the accounts given by scholars who have made a special study of the Kaiser-Wilhelm-Institut für Arbeitsphysiologie. The Dortmund institute already had close connections with the army and Navy.[57] In the lead up to the war in 1939, however, efforts were made by its then-director, Atzler, to promote links with the German Air Force. Atzler's effort was encouraged at the highest administrative levels in the Kaiser-Wilhelm-Gesellschaft. The initiative was indeed coming from the sources that Strughold identified.

In May 1935, there had been a flurry of correspondence between the deputy director of the Kaiser-Wilhelm-Gesellschaft, Max von Cranach, and the German Air Ministry. Von Cranach (1885–1945) was trying to set up a meeting between Atzler and the state secretary for aviation, General Erhard Milch (1892–1972). The Dortmund institute, von Cranach had told Milch, was at the disposal of the Air Ministry.[58] Von Cranach also recommended that, prior to meeting Milch, Atzler should pay a visit to Dr. Erich Hippke (1888–1969), who was the head of the Department of Medicine at the Air Ministry.[59] In the event, Atzler never reached Milch, but on May 27, he reported back to von Cranach that the meeting with Hippke had gone well. Hippke, he said, had been honest with him. Hippke had established his own research institute in aviation medicine and had been frank in saying that he wanted to ensure that the resources from the Air Ministry should go to this institute. Nevertheless, Hippke recognized that the Dortmund institute could perform important ancillary work. "Ich habe den Eindruck, daß es gelingen wird, im nächsten Jahr auch von Luftfahrtministerium einen Betrag zu erlangen," said Atzler hopefully—"I have the impression that over the coming year money from the Ministry will be forthcoming."[60]

The hope did not last long. In a letter of September 30, 1935, Atzler informed von Cranach that he now realized that he had been fobbed off by false promises. It was clear, said Atzler bitterly, that Hippke had no intention of bringing in the Dortmund institute on a collaborative basis. Atzler also reported that as soon as he realized that Strughold was the head of the Medical Research Institute of the Ministry of Aviation (that is, the institute that Hippke favored), he had approached Strughold with an invitation to join the editorial board of the journal *Arbeitsphysiologie*. After a long delay, he had just heard from Strughold, and the invitation had been declined. So both lines of approach, he reported to von Cranach, had failed.[61] The Institut für Arbeitsphysiologie was in danger of being frozen out of funding by other research institutes who were closer to the Air Ministry in Berlin.[62]

There is no evidence that this situation changed radically during the war and after Lehmann had taken over from Atzler. There were some moves in the desired direction, but they were not great enough to generate any marked shift in the research program of the Dortmund institute. In 1937, Strughold's name finally appeared amongst the editorial board on the title page of *Arbeitsphysiologie*.[63] Another sign of rapprochement was that the oxygen-breathing experiments of Lehmann and Graf were published in *Luftfahrtmedizin*, a journal edited by Strughold. But who financed the oxygen-breathing experiments? Perhaps it was one of the crumbs from the table of the Air Ministry for which Atzler and von Cranach had hungered, or perhaps it was financed in-house, or perhaps it was financed by the DFG, the Deutsche Forschungsgemeinschaft, as suggested in remarks by the historian Ernst Klee in his *Deutsche Medizin im Dritten Reich*. Klee noted that relevant documentation that might solve this problem has disappeared.[64] Whatever the source of finance with regard to that particular piece of research, the Dortmund institute remained marginal. This fact explains why the study of fatigue was of so limited a character: the institute was too far from the center of power.

If an explanation in terms of marginality is correct and can account for the small scale of the Dortmund research in fatigue, it nevertheless raises another problem. Why did the other, apparently more favored, institutes also do so little work on fatigue? The minimal character of any direct attack on fatigue is again evident in Strughold's statistical breakdown of the wartime research published (after the war) in *German Aviation Medicine*.[65] Why was the subject of fatigue approached, if at all, indirectly—for example, via one form or another of the problem of oxygen deficit? Lehmann and Graf's paper was just one small experiment using a very small number of subjects. It stands in marked contrast to the dozens of experiments devoted to high-altitude oxygen deficit. Why this overall discrepancy? And if Cambridge psychologists could generate new ideas about fatigue, why didn't their counterparts do the same in Germany?

Part of the answer may be that the Dortmund institute was not alone in finding itself distant from the center of power. Strughold listed no fewer than six distinct research centers that were doing work relevant to aviation physiology.[66] He also identified two overarching but differing institutions that guided research, namely, the Lilienthalgesellschaft and the Deutsche Akademie der Luftfahrtforschung. In his immediate postwar report from Germany to the British Intelligence Objectives Subcommittee (BIOS), Russell Davis corroborated this picture when he reported, "No research into design was sponsored by the central laboratories. There was no evidence that the service departments concerned with design (e.g., the 'Technisches Amt der Luftwaffe') sought the advice of psychologists, physiologists, or doctors."[67] The decentralized structure of the institutions supporting psychologists, physiologists, or doctors might explain why even those persons who seemed to be at the center of things were, on closer inspection, not as central as they thought, or as central as they might appear to an outsider.[68]

Immediately after the end of hostilities in 1945, the Royal Air Force Institute of Aviation Medicine at Farnborough dispatched Squadron Leader E. A. G. Goldie and Squadron Leader J. C. Gilson to Germany to report on the activities of the Physiological Institute in Göttingen.[69] Goldie and Gilson reported back to the Flying Personnel Research Committee on June 15, 1945.[70] They expressed surprise at what they found. The Luftwaffe, they said, clearly had access to a lavishly equipped laboratory, comparable to any facilities available in Britain or America, and yet it was "surprising to find the impression that this Institute had not been at all in close touch with operational and service questions." The Flying Personnel Research Committee, by contrast, had been created precisely with the goal of ensuring that, in Britain, psychological and physiological researchers *would* be kept in close touch with operational and service questions. Goldie and Gilson reported that the Göttingen institute appeared to have been "largely kept on a programme of pure physiological research." Part of the problem, they surmised, was interdepartmental secrecy—an arrangement that discouraged cooperation. When Goldie and Gilson disclosed to Strughold and Rein some performance data about a prototype German rocket-propelled aircraft, the vertical takeoff Bachem Ba 349 *Natter* (Viper), they showed "quite genuine amazement." "It is unthinkable in our system," Goldie and Gibson concluded, with evident satisfaction, that "scientists of the status of Strughold and Rein would have been so out of touch with this development which was apparently presenting considerable physiological problems."[71]

I began this chapter by asking whether there was a German Cockpit. I have returned a negative answer and have explained why a negative answer is plausible. The study of fatigue, as a psychological phenomenon, had been abandoned for richer pastures in physiology. Researchers were scattered across a range of quasi-independent institutes. Little attempt was made to pressure them to come up with new ideas of direct relevance to the operational problems created by pilot fatigue. Crucially, there was a lack of central organization and direction that might have mobilized a more consistent engagement with the nature of fatigue. The result, whether intended or unintended, was a concentration on the extreme ends of the scale of fatigue-like effects. Fatigue was implicitly defined, either as serious pathology and hence a medical problem, or as a problem of high-altitude physiology. Great weight was given to physiologically identifiable patterns of causation, and relatively little to tracing out revealing psychological and behavioral connections.

The Priority of Practice

Another feature of the German scene deserves notice. The Jacksonian tradition was not as salient in the science of German-speaking continental Europe as it was in Britain.[72] Concepts of fatigue and intoxication were typically framed in terms of inhibition and excitation, rather than in terms of levels of control, disinhibition, and release. The analysis of experimental results followed the pattern of Kraepelin's original account of the work curve. Whether the experiment involved lifting weights in the finger ergonometer, or the mental fatigue of crossing out letters on printed pages, or performing vast numbers of simple arithmetical calculations, the same interpretive pattern was always used. An empirical work curve was drawn and interpreted as an interaction of a number of psychological factors: for example, motivating effects, practice effects, and fatigue effects, with warming-up effects and distraction effects to be added where it seemed appropriate. Kraepelin had framed the problem in this way in his classic 1902 paper "Die Arbeitscurve":

> *Unsere bisherigen Betrachtungen haben uns gezeigt, das die Arbeitscurve eine recht verwickelte Zusammensetzung aufweist. Uebung und Ermüdung, Gewöhnung, Anregung und Antrieb in wechselnder Größe, dazu Uebungsverlust und Erholung wirken mit und gegeneinander, um alle die mannigfaltigen*

Gestaltungen der Arbeitscurve zu erzeugen, die uns bei der Untersuchung ver-
schiedener Personen und unter verschiedenen Bedingungen begegnen.[73]

Our analysis up to this point has shown us that the work curve involves complicated interconnections. Practice and fatigue, adaptation, stimulation, and drive in changing amounts, as well as loss of practice and recovery all work with, and against, each other and all contribute to the manifold configurations of the work curve that we have encountered in our studies of a range of persons under a range of conditions.

Much here was well-observed about the different factors that could be cited in explanation of the level of the work curve at any given moment. There was so much drive, so much willingness, so much practice, but also so much monotony, boredom, disinclination, distraction, and fatigue. All contributed their respective plus or minus to the resultant, but it was impossible to assign any numerical values. There were too few equations and too many unknowns. This underdetermination threatened to bring Durig, Simonson, and Anthony to a halt, and the same problem was still evident in the theoretical ideas used by Graf and Lehmann. Bartlett spotted one of the few ways in which it might be possible to cut through all the conceptual indeterminacy. Fatigue had been assumed to be a negative factor and to reduce the output of work. But what if the usual signs and symptoms of fatigue were sometimes accompanied by more work, not less? This fact, if established, might point to a nervous system with an architecture of levels and help to simplify the confusing data.

The reason for scanning the German scene for Jacksonian formulations is that it could help locate experimental work based on assumptions that were comparable to those underlying the Cambridge attack on fatigue. Of course, the mere presence of such verbal formulations does not prove that those who speak in this way necessarily put their words into action in the same way when it comes to experimentation. The acceptance of a Jacksonian theory did not itself explain the construction of the Cambridge Cockpit; rather, it rationalized it. Conversely, adherence to Kraepelin's ideas would not itself preclude the construction of a German Cockpit. Something must indeed explain the fact of construction, or nonconstruction, of the Cockpit, but that "something" is not the mere espousal or denial of a particular body of theoretical ideas. We stay closer to the historical facts if we take a hint from Bartlett and treat theoretical accounts, whether from Dortmund or Cambridge, as having the same status as the folk stories discussed in *Psychology and Primitive Culture*. An ethnological stance would suggest that scientific theories were accounts that scientists gave themselves about their own shared experimental practices; they are the "autobiography" of an experimental institution.

WAS THERE A GERMAN COCKPIT? 151

From the outset, my argument has operated within these broad Bartlet-tian guidelines. The construction of the Cambridge Cockpit was prompted by following a template provided by previous *experimental practice*—hence my comparison of the exotic Spitfire with the exotic folk story. But the similar-ity between these cases would only be salient to the appropriately socialized mind of an appropriately positioned person. Conversely, in locations where the prior experimental practice was based on the protracted repetition of highly simplified responses, it would be correspondingly difficult to take the step to the complex Cockpit experiments. This was essentially Bartlett's own argument in his Ferrier Lecture. It was Kraepelin's paradigms of experimental practice that prompted Bartlett to say that Kraepelin might have done more harm than good. Local experimental practice made it difficult to build a Ger-man Cockpit—I say difficult, but not impossible, because Graf's motorcar experiments were analogous in some respects to those envisaged by Craik and Bartlett.

Graf's use of a driving-skill test, which had been going on since 1933, makes sense within an ethnological perspective as an example of cultural con-tact. Graf was drawing upon the experimental practices of the psychotechnic tradition—and doing so despite the fact that ideologically it was out of favor. His use of exemplars drawn from psychotechnics was evident when he said:

Diesen unseren Anforderungen schien am besten eine Methode zu entsprechen, wie sie in der Psychotechnik zur Kraftfahrerprüfung verwendet wird und wie wir sie selbst schon in ähnlicher Form verwendet haben. Das Prinzip besteht darin, daß die Vp. mittels einer Lenkvorrichtung einer bestimmte Kurve so nachzu-fahren hat, daß möglichst wenig abweichungen von dem vorgeschriebenen Wege auftreten. Ich verweise hinsichtlich der verschiedenen Modifikationen solcher Prüfungen auf die Sonderuntersuchung von Hallgebauer *und beschreibe hier nur meine Methode, wobei ich betonen möchte, daß ich das Ideale einer sol-chen Versuchsanordnung in dem Kraftfahrprüfungsapparat von* Bautze *nach* Hallgebauer *gesehen hätte, der jedoch in Anbetracht der hohen Kosten nicht in Frage kam.*[74]

Our requirements are best met by the methods already used in the psychotech-nic tests of motor vehicle drivers. I have already used similar tests myself. The basic principle is that the subject must use a steering device to follow a prede-termined curved route with as few deviations as possible. I indicate the vari-ous modifications I have introduced into the apparatus used in *Hallgebauer's* studies but will only describe here my own methods. I am aware that the ideal experimental arrangement would be to use the driver-testing apparatus de-veloped by *Bautze* and *Hallgebauer,* but that is out of the question due to cost.

As well as mentioning the names of previous workers in the field of psychotechnics, this passage shows Graf engaged at the level of steering wheels, empirical measurements, and costs: that is, the stuff of apparatus building. The achievements of those working in the psychotechnic tradition (despite it being out of fashion and suspect) will have been more visible to those, such as Graf and Lehmann, working in the applied field of *Arbeitsphysiologie* than it was to colleagues and competitors with a more fashionable academic, ideological, or military orientation.

Occasional Jacksonians

Kraepelin's approach to work and fatigue did not go wholly unchallenged in Germany. There were German Jacksonians, and defenders of Jackson-like models of levels—for example, Oswald Schmiederberg (1838–1921) and Otfrid Foerster (1873–1941)—but they did not have a significant following and did not work in fields close to aviation physiology.[75] Sometimes, though, it can be difficult to decide whether a position should be seen as affirming an approach derived from Jackson or one derived from Kraepelin.[76] This is a problem characteristic of analyses based on the historical actor's affirmation, or apparent affirmation, of merely theoretical ideas. Consider the following statement by the previously mentioned H. F. Rein of Göttingen. In an address to mark the 142nd anniversary of the *militärärzrlichen Akademie* in Berlin in 1938, Rein had described the contributions of physiological research to modern military problems, particularly those problems generated by the air force. He said:

> *Am empfindlichsten gegen Sauerstoffmangel sind die höchsten Zentralgebiete, die des Großhirnes. Es ist dem Physiologen bekannt, daß einer Reduktion der Funktionsfähigkeit dieser gebiete tieferen Hirnabschnitten eine gesteigerte Herrschaft über die gesamte Körpermotorik einräumt.*[77]

> The highest areas of the central nervous system, the cortex, are those that are most sensitive to oxygen deficit. It is well known to physiologists that a reduction in the functional capacity of those areas makes way for the domination of the entire bodily motor system by the lower levels in the brain.

If the *Herrschaft*, or "domination," is understood to have been the result of a process of release, and if *Abschnitt* is rendered as "level," then this picture looks like the Jacksonian one—but the validity of that rendering depends on things that are not made explicit in the quotation or the lecture. Perhaps what is here rendered as "domination" is not the result of the release of lower levels but merely victory in a struggle between two areas (*Gebiete*) that are functionally on the same level, even if they are designated by the words "higher" and "lower" on anatomical grounds. The trouble is that Rein never puts the allegedly well-known idea to work in the subsequent discussion. The Jacksonian reading is not reinforced by the body of the lecture. Fatigue, as such, was not even identified as a central research problem. For Rein, the two main problems facing aviation physiologists were acceleration and altitude.[78]

The problem of identifying German exponents of a Jacksonian approach can be illustrated from another angle. Consider the work of Heinrich Lottig (1900–1941), who was the head of the Institut für Luftfahrtmedizin in Hamburg. His main concern was the effects of low air pressure.[79] Lottig used a low-pressure chamber to show how the ability to perform cognitive tasks was diminished as air pressure was reduced. He asked subjects to write down the sequence of numerals in descending order starting from 1000, 999, 998, and so on. This exercise was known as the *Lottigsche Zahlentest*.[80] In chapter 1, I described how Squadron Leader Livingston met Lottig during his prewar visit to Germany and was impressed by these experiments on young volunteers who sought to stave off the deterioration of their performance. Livingston understood that Lottig saw the *Zahlentest* as a test of character and will.[81] This perception was confirmed by a publication of 1939 where Lottig presented his work as a means of investigating the relation between the personality and the vegetative nervous system. Ultimately, Lottig said, mind and body form a unity, although, like a number of other German psychologists, he spoke of a person's character as structured into different *Schichten*, or levels.[82] After reviewing an extensive range of experimental work by aviation physiologists, Lottig said:

> *Die Demaskierung tieferer Schichten des Charakters im Unterdruck im Sinne der Enthemmung und des Verschwindens unechter oder unangemessener charakterlicher Züge erinnert in vieler Beziehung an ähnliche Verhältnisse im Alkoholrausch oder bei der Hyperventilation.*[83]

> This unmasking of the deeper layers of the character at low air-pressures operates by disinhibition and the disappearance of inauthentic and inappropriate character traits and reminds us of similar relations in alcohol intoxication or hyperventilation.

The reference to hyperventilation refers back to a previous claim by Lottig that hyperventilation results in stirring up and loosening up the mental depths: "sie wirkt auch aufwühlend und auflockernd auf seelische Tiefen."[84] Alcoholic intoxication was assumed to work by the same principles. Lottig's account undoubtedly shared important elements with Jackson's theory, and a hasty reading might suggest that the two were identical—apart, that is, from the puzzling reference to the disappearance of *inauthentic* character traits. For Lottig, the authentic self was located as much in the lower as the upper strata of the body and mind. Imposing the stress of lower air pressure may result in a form of "release," but Lottig inverted the Jacksonian connotations of what is higher and what is lower.[85]

Sometimes, however, clear and authentic Jacksonian images do surface in the German literature. One initially surprising example is provided by Kraepelin himself. Kraepelin was fully prepared to utilize Jacksonian metaphors of levels, inhibition, and release, if and when it suited his purpose. As Kraepelin scholar and editor Eric Engstrom has shown, Kraepelin had strong right-wing political opinions. He loudly declared his fear of the revolutionary mobs that, he was sure, would emerge if ever the social hierarchy in Germany were successfully challenged by the political left. In a lengthy article in the *Süddeutsche Monatshefte* in April 1919, Kraepelin spoke of the dangers of mass movements of the left and how, in such movements, the individual will is dominated by ancient and violent herd instincts:

> *Bei jeder Waffenbewegung, die sehr rasch der Hereschaft ihrer Führer zu entgleiten pflegt, kommt es somit zu einer Zurückdrängung der höheren Bestandes- und Willensleistungen zugunsten unklarer Triebregungen, ganz entsprechend den hysterischen Stürmen, in denen die zielbewußte Persönlichkeit durch die unbeherrschten Mächte früherer Entwicklungsstufen überwältigt wird.*[86]

> Thus in every mass movement, which in general tends very quickly to slip the reins of its leaders, a suppression of the faculties of higher reason and volition occurs in favor of darker instinctual urges. This corresponds directly to hysterical tempests in which the resolute personality is overwhelmed by uncontrolled forces from earlier stages of development.[87]

Kraepelin was responding to the left-wing uprising in Munich after the armistice in 1918. There were outrages by both left and right, but it was the paramilitary counterrevolutionaries, supported by Kraepelin, who could truly be said to have unleashed brutality and bestiality.[88] For the present argument, however, the important point is that Kraepelin's resort to Jacksonian ideas was context-bound. The image of evolutionary levels (where the higher levels inhibit the lower levels) was clearly lodged in his mind and enthusiastically employed in his politics, but it was kept out of sight in other contexts. Was Kraepelin guilty of logical inconsistency? Or should his expedient resort to Jacksonian ideas be seen as characteristic of all practical thought? The second option offers the more realistic analysis. Seen in this way, the episode can be assimilated to the principle that society explains the stories that its members tell, rather than the stories explaining the society.

ALLIED INTELLIGENCE VISITS TO POSTWAR GERMANY

Returning now to WWII, I have already mentioned Russell Davis's brief foray into intelligence work. The British effort in this field was small compared to the American effort. When hostilities in Europe came to an end with the unconditional surrender of Germany on May 7, 1945, Lt. Colonel Paul Morris Fitts immediately joined the large number of other American investigators who had already descended on the occupied regions of Germany in search of scientific and commercial intelligence.[89] After beginning his career in the academic world, Fitts had been appointed to the Aviation Psychology Program during the war. He was to become the Chief of the Psychology Branch of the Aero Medical Laboratory within the USAF Air Material Command. Fitts's mission was to find out what contribution psychologists had made to the German war effort, particularly in the realm of equipment design.[90] His itinerary involved visits to government research institutions as well as to universities in Frankfurt, Heidelberg, Würzburg, Munich, Jena, Halle, Göttingen, and Marburg. What did Fitts find in Germany?

The most important German research that Fitts reported back to the American authorities came from the Medizinisches Institut in Garmisch-Partenkirchen. This institute had originally been based in Munich as part of the Luftfahrt-Forschungsanstalt, but in 1944, for reasons of safety, it had been moved out of Munich into the famous mountain-skiing-and-tourist region. Led by Ulrich Henschke, a medical researcher, this multidisciplinary group worked on a wide range of problems that included new control systems for fighter aircraft, remote guidance systems for gliding bombs, the use of high-powered telescopes for use by aircrew, new forms of gunsight, the testing of performance-enhancing drugs, tests to select aptitude in instrument flying, the development of electromechanical prostheses for victims of serious injury, as well as studies devoted to fatigue.[91]

It is unclear whether Fitts actually met Henschke and his team in Garmisch-Partenkirchen or whether he merely transmitted reports of their work that had already been written up at the behest of the Air Technical Intelligence Headquarters of the new, Allied occupation authorities. (Bavaria had been subject to military occupation for some time before the final capitulation.)[92] In June 1945, Henschke drew up a three-part report dealing with the history of the institute, its past and current research, and the biographies and publications of his co-workers.[93] He also provided the Americans with a number of progress reports that he had previously submitted to the German authorities financing the institute. The Americans were particularly interested in experiments, originally reported to the German authorities in 1944, exploring the idea that a fighter pilot could respond more effectively if he steered the aircraft by two hand-levers rather than rudder pedals. The action of the pilot in rapid, twisting maneuvers would then resemble that of a skier descending a slalom course.[94]

What did the Garmisch-Partenkirchen documents reveal about fatigue? This particular problem was the concern of Henschke's colleague Dr. Fritz Hollwich.[95] It is significant that in the original documentation, the German word used to designate his area of wartime research was *Abgeflogensein*. This label had gained currency during WWI. It referred to a state that fell under the general rubric of "fatigue," but it designated extreme and pathological states of exhaustion caused by the stress and danger of flying. While there was general agreement about the extreme nature of the condition, there was disagreement about the other connotations of the word, in particular about whether it was a matter of physiology or psychiatry. Was it mental or physical in nature?

An RAF officer, commenting on the German research, submitted an intelligence report in which he noted that younger German medical officers treated *Abgeflogensein* as a state having a physiological basis, while more senior medical staff believed it was a psychological state of neurosis.[96] Hollwich's position was aligned to that of the younger group. He explained that his special concern was with tissue damage of the kind that might be caused by the stress of repeated high-speed maneuvers such as the banking and diving of an aircraft. This description suggests that he was working on the particular stresses that were experienced by the crews of dive-bombers such as the Junkers 87. He was pursuing an investigation based on the hypothesis that pilot fatigue of this kind was associated with changes of permeability in the system of blood capillaries. Hollwich was not alone in thinking along these lines.[97] Weaknesses of the capillary system, he said, might also be implicated in "Föhn sickness." The reference was to the long-standing suspicion, held in medical circles in central Europe, that the season of the notorious Föhn winds was an inauspicious time to conduct surgery.[98]

The Flying Personnel Research Committee was already aware of the German's physiological approach. There was a series of reports to the committee, report nos. 321 to 321(d), of July 1941, based on extracts from the German scientific literature, which noted the belief that, when anoxia was combined with the effects of acceleration, there was "an increased tendency to 'Kapillarisierung' of the muscles."[99] FPRC report no. 384, of the same year, *Medical Services in the German Air Force*, based on information from the captured crew of a long-range bomber, made it known to the committee that the German authorities paid particular attention to "the state of the blood vessels and to blood pressure" and "believed that many patients suffer from distension of the blood vessels."[100] For Hollwich, fatigue was interesting when it could be detected by a microscope. This approach was scientifically legitimate, but it was very different from the approach adopted by the Cambridge psychologists.[101]

Apart from physiologists and medical researchers, there was one active psychologist in Henschke's group. In the reports on the institute's work submitted to the occupying forces, Siegfried Gerathewohl explained that he had been involved in pilot aptitude testing since 1936. Despite being dismissed by Göring, he was still working on ways of improving such procedures. He was pursuing experimental work that might reveal an aptitude, or lack of aptitude, for night flying in which the pilot depended on instruments.[102] The problem with night flying was that the instruments would say one thing about the behavior of the aircraft, while subjective sensations and feelings would say something very different. Some people were better than others at what German researchers called "objectification"—that is, following their instruments. By using revolving chairs and other pieces of apparatus, Gerathewohl was seeking to filter out those who would make superior pilots by virtue of their superior powers of objectification.

Gerathewohl also told the Allied authorities that he was working on a general monograph on aviation psychology. But it was not until 1953 that the book finally appeared as *Die Psychologie des Menschen im Fleugzeug.*[103] The reason for the delay was that Gerathewohl, along with Henschke, and a number of other workers in the field, had transferred from Germany to the United States. They worked in association with the military authorities on aeronautical and aerospace projects. Gerathewohl had taken the opportunity to bring the psychology in his book up to date and describe postwar American and British developments, including his own contributions.[104]

Gerathewohl had little to say about fatigue in his 1953 monograph.[105] One theme that did emerge, however, was that in his opinion it was unjust to accuse German psychologists of lack of scientific objectivity in their wartime work. Gerathewohl quietly made the case for the originality and leading role of wartime aviation psychology in Germany. He missed no opportunity to bring the wartime German literature into the postwar conversation. The significance of this point for my argument is that if German psychologists had ever performed experiments that had the striking character of those done on fatigue in Cambridge, then Gerathewohl would have drawn attention to it. It is safe to conclude that there was no Garmisch-Partenkirchen Cockpit.

THE RED TELEPHONE

The situation in which Craik and Bartlett worked contrasted at almost every point with the situation of aviation physiologists and psychologists in Germany. There was a long history of Jacksonian thinking amongst British physiologists and psychologists. Amongst Cambridge experimental psychologists, as we have seen, this history went back to the work of Rivers and Head. The Cambridge group was also in the fortunate position of having its leading member, Bartlett, on the administratively important governmental and military committee that made the decisions about what research needed to be done and who was to do it.[106] Cambridge psychologists were therefore close to the center of power and were trusted with its secrets, in a way that was not true for Lehman and Graf, or even Strughold and Rein—and, in Britain, there was undoubtedly a center of power to be close to. In summary: Cambridge psychologists could draw upon personal connections, available resources, high-level encouragement, a degree of autonomy and lack of competition, as well as a very particular background of previous work and accomplishments in the field of psychological research. They therefore had a capacity for action that was wholly different to that of their German counterparts. These circumstances also contrived to give the resulting research on fatigue in Britain a very particular scientific orientation. It is hardly surprising that British and German psychologists should then construct very different responses to the problem of pilot fatigue, one group building a Cockpit while the other didn't.

On a lighter note (but it epitomizes the point I want to make), Richard Gregory recalled that Bartlett had always been rather proud to have on his office desk in Cambridge a special telephone—a red telephone—that had been installed during the war by the Post Office.[107] Such devices "scrambled" the electrical signal and prevented electronic eavesdropping. Bartlett was rather disappointed, added Gregory, when, after the war, the device was eventually removed. This accoutrement provided Bartlett with a privileged and secure link to persons of influence in Whitehall and other leading decision-makers in science. The red telephone was a potent symbol. It showed that Bartlett was truly part of what (with greater or lesser enthusiasm) the British refer to as "the Establishment."[108]

Was There an American Cockpit?

Was there a Craik-style fatigue machine at the Naval Research Laboratory in Washington or at Randolph Field in Texas? These would have been plausible locations at which to find an American counterpart to the Cambridge Cockpit, just as the Technische Hochschule in Charlottenburg, the Kaiser-Wilhelm-Institut in Dortmund, or the Medizinisches Institut in Garmisch-Partenkirchen, were plausible locations for a German counterpart. I have already concluded that there was no German Cockpit. I now argue that the same negative answer must be returned in the case of the American war effort. I have found no American equivalent of the Spitfire that Craik and Bartlett dragged into the Psychological Laboratory. Perhaps the Cambridge Cockpit was unique. After briefly outlining the empirical case for Cambridge exceptionalism, I shall turn from negative argumentation to positive argumentation. I shall examine how some of the leading American psychologists chose to address the nature of fatigue in general and pilot fatigue in particular. I shall also point to the ways in which the American approaches and the British Cambridge approach differed from one another and, in more than one respect, were on a collision course.

The Case for Cambridge Exceptionalism

No one who searches for an elusive artefact or document and fails to find it can be sure that they have looked everywhere. Proving a negative is notoriously difficult. The scope of psychological work in wartime America was huge and so, therefore, is the corresponding scope for oversight. But there are considerations that pull in the opposite direction and support the negative conclusion that there was no American equivalent to the Cambridge Cockpit. The professional psychological journals were full of reports of work done to support the war effort after America was drawn into the conflict.[1] Historians of psychology also know that in the years following World War II, numerous reports were drawn up cataloguing the psychological work that had been done for the military and for government agencies such as the Office of Scientific Research and Development.[2] If there had been an American Cockpit, it would, surely, have been described somewhere in this postwar audit.

In 1946, a team under Lieutenant Commander J. W. Macmillan of the United States Navy Reserve was responsible for a detailed survey of work on *The Role of Fatigue in Pilot Performance*.[3] The survey, which was not for publication, was commissioned by the National Research Council Committee on Selection and Training of Aircraft Pilots. Macmillan was well aware of the Cambridge Cockpit experiments and discussed them at length. The reason he gave for paying attention to the Cambridge work is revealing: "These studies have been reported in some detail," he said, "because they are the only ones of their kind known to the present authors."[4] One man's opinion regarding the absence of an American equivalent is not proof, but other authoritative sources corroborated Macmillan's conclusion. Consider, for example, the postwar research reports of the Army Air Forces Aviation Psychology Program. These reports would have been likely to have mentioned an American Cockpit, if such a thing existed. This series of bulky volumes, known as the "Blue Books," included three that are particularly relevant to the search for an American Cockpit: These are (i) the synoptic *Report 1*, by J. C. Flanagan;[5] (ii) *Report 4*, by A. W. Melton,[6] on apparatus tests such as the complex coordination test designed for aircrew selection; and (iii) *Report 19*, by P. M. Fitts, on the psychologists' contribution to equipment development.[7] Immediately after the war, Fitts had visited Cambridge and, in his preface, mentioned lengthy conversations with Bartlett. Of these three reports, it was only the contributors to Fitts's volume who discussed the Cambridge Cockpit, and significantly, they made no mention of any American equivalent. If determined postwar audits of this kind made no mention of an American Cockpit, the probability must be that no program of fatigue research was carried out that was comparable to that in Cambridge. The Macmillan report seems to have been correct. Fatigue was certainly a problem confronting American psychologists, but they conceptualized and investigated it in a different way than the Cambridge psychologists did. What, then, were psychologists in the USA doing that related to skill fatigue?

LARGESSE AND LEARNING THEORY

Writing in the *Annual Review of Psychology* in 1961, the distinguished psychologists Edward and Ina Bilodeau recalled the postwar years in the following words:

> Extensive programs of motor-skills learning began after World War II. There were two direct causes. One was Hull's book in 1943 . . . The other cause was the familiarity with apparatus engendered by developing a battery of hardware to select aircrews for the United States Air Force. . . . In 1949, for example, the Human Resources Research Center (later, Air Force Personnel and Training Research Center) opened and supported an extraordinary, active program of in-service and contract research that in its first five years set the field ahead by more than 20.[8]

The phrase "battery of hardware" refers to apparatus-based tests for selecting aircrew of the kind that Melton described and evaluated in *Report 4*. I shall say more about Hull's book in a moment, but it is worth putting some numbers to the Bilodeaus' statement about the level of the support of which they and their colleagues were beneficiaries. In an address to the first International Symposium on Military Psychology in Brussels in July 1957, Melton gave two pieces of financial data. The first referred to recent budgets for contract research and development: that is, contracts between academic institutions and the government or military. These, he said, had been running at $5,000,000 per year for the past few years. The in-house military spending on psychological research was even greater, and Melton put it at $18,000,000 per year.[9]

How did American psychologists react to this largesse? The answer is that the practical focus of the wartime work had left them frustrated, and they wanted to get back to fundamentals. In 1947, at the end of *Report 4*, Melton had noted "the many fundamental purely psychological problems basic to test constructions which need assiduous investigation."[10] Robert Gagné, who had contributed to the wartime work that Melton reported, complained about the lack of adequate theory. This lack, he said, "creates a void in this area of human behavioral knowledge."[11] How was the study of fatigue to be rescued from that void? The answer, for many psychologists, was that fatigue was identified with inhibition, and inhibition was integrated into a rigorously articulated theoretical framework known as "learning theory." Learning theory included both animal learning and the human learning of motor skills. It was the primary field for the application and elaboration of behaviorist principles. This is where Clark Hull (1884–1952) came into the story. The reference made by Edward and Ina Bilodeau, in the above quotation, to "Hull's book in 1943" was to Hull's impressive *Principles of Behavior*.[12]

Hull's book was a late and highly refined expression of the behaviorist tradition. In retrospect, it might be seen as the apogee of the behaviorist project. To help confront some of the complexities of Hull's theory, it may be useful to trace his basic ideas back to some of the earlier and simpler statements of behaviorism. J. B. Watson's 1919 classic, *Psychology from the Standpoint of a Behaviorist*, set the tone. The standpoint was one of avoiding engagement with the mysterious, private theatre of mental happenings. Psychology was not the science of the "mind," but nor was it a science based on speculation about hidden synaptic connections in the brain. In our everyday interactions, we know nothing of either of these things. We understand one another by observing outer behavior. Why not build a science on that basis alone? The process of "thinking," for example, might be understood as no more than silent, inner speech. When this general orientation was applied to the phenomenon of fatigue, Watson was led to declare that fatigue was simply not a serviceable concept. He urged psychology students not to waste their time reading general discussions of fatigue. What he called the "helpless position" of the study of fatigue was that the very idea of fatigue was stranded in the intermediate realm between the mental and the physical. The concept of fatigue was irreparably tainted with metaphysics and should be left to philosophers.[13]

Hull followed Watson's advice about fatigue when he formulated his own more complicated and sophisticated form of behaviorism. Psychology, argued Hull, must become a "system"—not, of course, a metaphysical system, but a scientific system. Historically, the distant guiding light was Newton's *Principia*, and Hull wanted psychology to meet the highest standards of scientific rigor as they were understood in America in the 1930s and 1940s. He wanted psychology to be formulated so that all the postulates and deductions would be made fully explicit. Hull wanted behavioral psychology to be a "molar" rather than a "sub-molar" neurological enterprise, though he was not opposed to making a few expedient links. In Hull's settlement between psychology and physiology, theoretical concepts having the status of "intervening variables" were permitted, but only if they were "operationally defined"—that is, defined in terms of observable quantities and measurable behavior.[14]

In operational terms, when conducting laboratory experiments on (say) learning in rats, the "strength" of a learned habit of responding, such as turning right in a maze, might be identified with the number of times the response had been rewarded. The number of hours of food deprivation might operationally define the strength of a "drive." The persistence of food-seeking behavior would retrospectively define the readiness to respond, or what Hull called the "reaction potential." Following Watson, Hull assumed that the concept of "fatigue" was compromised by subjectivity, so it could not be used as an intervening variable. Psychologists, he held, needed to replace talk of fatigue by something more scientific. Philosophers might have sneered at Hull for taking physics as his ideal, but in writing the *Principles of Behavior*, he was both bold and ingenious. Even psychologists who disagreed with Hull knew that he was to be taken seriously.[15]

Referring to the period from 1945 to 1960, Bilodeau and Bilodeau noted:

> The recent history of motor skills learning is remarkable for the intensive ten to fifteen years of research devoted to the decrement effects of practice and their dissipation with rest. . . . As to so much other research, Hull provided the impetus, and a theoretical framework, to the boom in decrement research.[16]

The word "fatigue" was unacceptable in polite behaviorist company, but the study of work decrement and its dissipation with rest was considered acceptable. In reality, this line of research was just the study of fatigue, or some aspects of fatigue, under another name. But however it might be interpreted, Hull and his co-workers assembled a mass of experimental data, some from animal learning, some from human motor-skill learning, and gave it order and coherence—at least for a while.

Fatigue as a "Drive"

Hull's starting point was simple. He said that every response generated a slight disinclination to repeat that response. That disinclination was identified as "inhibition" in order to indicate that it worked against the "excitation" or "reaction potential" assumed necessary if a stimulus were to evoke any response. The negative motivation attached to a response was labeled "reactive inhibition" and symbolized by I_R. Hull did not equate I_R with fatigue, but he admitted (somewhat archly) that there was a connection:

> The reaction decrement which we have attributed to reactive inhibition bears a striking resemblance to the decrements which are ordinarily attributed to "fatigue."[17]

Of course, the resemblance was no accident. Hull could have said that every response is accompanied by, and generated, a quantum of fatigue. The symbol I_R can be seen as an abstract representation of the lactic acid that can bring a muscle to a standstill, but which then dissipates with time.[18] Although there was no entry under "fatigue" in the index of the *Principles*, fatigue clearly lay at the very heart of Hull's thinking. In conformity with traditional thinking about fatigue, Hull specified that the quantity of I_R increased with the number of times (n) a particular response was made. The amount of inhibition generated also increased as the effort involved in making the response increased, that is, as the work (w) increased. Hull also postulated that reactive inhibition decayed rapidly and exponentially, so that after a short time (t) the amount of remaining inhibition became very small. Appropriating Hull's own mathematical idiom but still expressing these points in a wholly general way, reactive inhibition was some mathematical function f of the variables n, w, and t. Thus:

$$I_R = f(n, w, t).$$

Hull located reactive inhibition (I_R) within the broader framework of learning and behavior. He defined learning as habit formation, and behavior in general was conceived as an interaction of drives and habits. Habit strength, he said, naturally had a limit. It had a maximum value that is approached asymptotically.[19] A similar point holds for the "learning curve." Think here of a test of learning in which a hungry laboratory rat demonstrates successful learning by running quickly to the food box at the end of a maze. Learning was represented by a monotonic curve that approached a limit where performance is perfect. Hull acknowledged that, in ordinary usage, "habits" are simply well-worn modes of action, but he insisted that, in the context of learning theory, a habit is "a *persisting state of the organism* . . . which is a necessary, but not a sufficient, condition for the evocation of the action in question."[20] The emphasis was Hull's own, and the point was important.

The persisting state, the learned habit (H), is an enduring association between a stimulus (S) and a response (R) that is strengthened by "reinforcement." Reinforcement was the reduction of a basic "drive" (D) such as hunger, thirst, or sex. When drives are high, they impel an organism into action until the drive is satisfied, and the drive level thus reduced. An organism learns how to satisfy drives because the patterns of stimulus and response that lead to the reduction of drives become increasingly strong habits—up to their natural limit. A response R is then said to have become "conditioned" to a stimulus S, by S and R being conjoined in the presence of drive reduction.[21] Hull noted that I_R was a drive, so any reduction in I_R could serve to reinforce an S-R connection. In lay terms, fatigue acts as a drive because it is nice when it stops and we can have a rest. Fatigue is akin to hunger and pain or tissue damage, and Hull himself made precisely these connections.[22]

Fatigue as a Habit

Coming back to Hull's mathematical system, these points were expressed by saying that a response has to be evoked by an "excitatory potential" that is symbolized by ${_sE_R}$. This symbol was defined as the product of the strength of the drive (D) and the strength of the habit, symbolized by ${_sH_R}$. Thus, the excitatory potential is given by the equation:

$$_sE_R = D \times {_sH_R}.$$

The rationale for the multiplication sign is that if either drive is zero or habit strength is zero, then the excitation will be zero, so there will be no actions that are based on learned tendencies. More generally, the multiplicative relation means that a strong drive will activate a weak habit, and a strong habit will find expression even if the drive is weak.

At this point Hull introduced a second concept of inhibition into his account. As well as "reactive inhibition," he concluded that there must be a phenomenon called "conditioned inhibition." Conditioned inhibition has the status of a habit and was symbolized by ${_sI_R}$. Fatigue now operated as both a drive-state and as a habit. This dualism became part of Hull's recasting of the commonsense concept of fatigue. There were now two sorts of fatigue performing different tasks and working in different ways: "Whereas ${_sI_R}$ involves the whole neural receptor-effector mechanism of habit," explained Hull, "I_R involves only (or mainly) the effector portion of this mechanism."[23]

This complicated dualism was designed to meet the results of experimentation. The simple postulation of I_R and its dissipation could be used to explain (i) spontaneous recovery from the extinction of a conditioned response; (ii) the general result that distributed trials are usually better for learning than massed trials; and (iii) a phenomenon known as "reminiscence," wherein learning is found to improve, not merely during but *after* the end of training and reinforcement.[24] But experiments had shown that "spontaneous recovery" from experimental extinction was rarely wholly complete. The reactive inhibition dissipated with time, so what was the cause and nature of the residual inhibition? To explain the residual inhibition, Hull assumed that a component of inhibition must have become conditioned to the circumstances that produced the original reactive inhibition I_R. This component was the conditioned inhibition ${_sI_R}$. He justified the conclusion as follows:

> In accordance with the "law of reinforcement" . . . this cessation of activity would be conditioned to any afferent stimulus impulses, or stimulus traces,

which chanced to be present at the time the need decrement occurred. Consequently there would arise the somewhat paradoxical phenomenon of a negative habit, i.e., a habit of *not* doing something.[25]

While reactive inhibition decays, conditioned inhibition becomes a permanent fixture. The total inhibition at any point in time would now be the sum of the reactive inhibition (I_R) and the conditioned inhibition $(_sI_R)$. The positive excitatory potential $(_sE_R)$, available to generate a fatiguing response, would be diminished by the sum of these two negative quantities. The remaining, *effective*, excitation (designated by the bar over the symbol) is then given by the equations:

$$_s\bar{E}_R = {}_sE_R - (I_R + {}_sI_R)$$
$$= (D \times {}_sH_R) - (I_R + {}_sI_R)$$

These equations, and their interpretation, proved to be an enduring source of both inspiration and trouble for American psychologists—and even Hull sometimes had his doubts.[26]

A Somewhat Paradoxical Phenomenon

How can not-responding be a response? What was being conditioned to what? The puzzlement over the equations was expressed in a number of ways. First, notice that in Hull's equation, I_R and $_sI_R$ are simply added together. Wasn't Hull here abandoning the idea, implicit in his system, that drives and habit strengths only made sense when related multiplicatively? Secondly, in defining effective excitation, it was plausible that I_R could be subtracted from $_sE_R$, but it was not obvious that $_sI_R$ could be subtracted from $_sE_R$ in a meaningful way. Both I_R and $_sE_R$ were motivational factors. Operations of addition and subtraction make sense when, as in this case, they concern factors of a like kind. But conditioned inhibition, $_sI_R$, wasn't a source of motivation: it was a habit. Wasn't subtracting a habit strength from an excitation like trying to subtract chalk from cheese?

The single most striking criticism of Hull's equations was formulated in 1954 in an important paper entitled "The S-R Reinforcement Theory of Extinction," by Gleitman, Nachmias, and Neisser.[27] Hull's theory, they said, was the most highly developed of current learning theories, and they admitted that it offered elegant explanations of a range of experimental results. Nevertheless, it was beset by deep problems. Paradoxical consequences could be derived from the theory. It logically implied that there is no learned act which can be performed for any length of time. No special experiments were needed to see that this outcome was unacceptable. They pointed out that daily life "is full of countless activities which we perform again and again with no sign of decrement. We turn door knobs, say 'how do you do' sit down on chairs, and recline on beds, and have done so since childhood."[28] If Hull were right, all these activities would be impossible. Necessarily, said his critics, given Hull's postulates,

> there is no learned act which can be performed for any length of time; its very repetition—regardless of reinforcement—must lead to its eventual elimination.[29]

The problem was this: When a habit is repeatedly reinforced, the strength of the excitatory potential $_sE_R$ approaches its asymptote. At this point, no further reinforcements can add to the habit strength. But there is no similar limit on the growth of inhibition. In postulate 8 of the *Principles*, Hull specified that the net inhibition "generated by a sequence of reaction evocations is a simple linear increasing function of the number of evocations."[30] From the point at which the conditioned inhibition matches the maximum level of excitation, further evocation of the response will simply lead to more inhibition and, ultimately, to extinction. The learning curve will rise to a maximum and will then turn downward until it reaches its baseline—and it will never come up again.

Can the System Be Repaired?

Others had previously noticed the problem that repeated responses posed for the learning curve.[31] McGeoch and Irions suggested that perhaps reactive inhibition should be subtracted from the habit strength rather than the reaction potential. Gleitman, Nachmias, and Neisser pointed out that such subtraction would be inconsistent with Hull's commitment to the relative indestructibility of learned habits and the merely "masking" role he accorded to the sources of inhibition. The eminent psychologist Ernest Hilgard expressed the mounting concern over the equations when he said that the "systematic considerations here are actually very complex and are treated rather sketchily by Hull." Hilgard focused attention on the component of fatigue represented by I_R, the reactive inhibition. Again using D as the symbol for drive, he noted:

> Hull did not follow up all the implications of treating I_R as a drive. He did not subtract it from D, which as a matter of course his system requires; he did not use it in conjunction with D as a multiplier in the determination of reaction potential.[32]

Hilgard seemed to be saying that Hull's equation should be replaced by the following:

$$_s\bar{E}_R = [(D - I_R) \times {_sH_R}] - (I_R \times {_sI_R}).$$

This equation is the symbolic version of Hilgard's criticism that was given by A. R. Jensen in his paper "On the Reformulation of Inhibition in Hull's System."[33] The suggested reformulation avoided the subtraction of incommensurables—the chalk-and-cheese problem—but it lost the capacity to explain incomplete spontaneous recovery.

Jensen analyzed a number of other influential, theoretical criticisms of Hull's equations and proceeded to cast them into symbolic form.[34] Overall, he examined some seven alternative or modified equations as well as assessing the numerous interpretations of the particular mathematical expressions of which they were composed. Jensen argued that the modified equations all failed to resolve the problems posed by the original equation. The new equations either introduced new obscurities, or repeated the same mistake, or led to new but false predictions, or (as in the case of the suggestion mentioned above) failed to do justice to the facts that Hull was able to explain by his original version of the equation.

Jensen pointed out that, formally, using Hull's original equation the Gleitman objection could be overcome by increasing the value of the drive. Extinction occurs when $_sI_R = D \times {_sH_R}$. Jensen argued that, if it is assumed that both $_sI_R$ and $_sH_R$ are close to their asymptote, then a response can be created by increasing D. The value of the habit strength multiplied by the new and greater drive strength will then be greater than the asymptotic value of $_sI_R$. Formally, this argument is correct, but is it psychologically meaningful? Empirically, Jensen admitted that subjects had been known "to practice the pursuit task day after day for months, long after having reached an asymptote for time on target, yet they show no loss of the skill."[35] Did this observation mean that, to accomplish this increasingly routinized skill, the drive strength of the subject had to be ramped up every time? If so, this point would imply that turning doorknobs or saying a casual "good morning" were high-drive phenomena. Surely the opposite is the case. Perhaps different principles are needed to explain such taken-for-granted aspects of behavior. Taken-for-granted behavior continues to consume some mental capacity, but high drive is precisely what is not needed. After this detour over the role of drive, Jensen finally concluded that all the attempts to save Hull's equations had proven futile. "The very building blocks of the theory" he said, "are inadequate."[36]

Bilodeau and Bilodeau, in the 1961 *Annual Review of Psychology*, praised Jensen's paper and acknowledged the depth of the problem. Different experimenters now seemed to have their own understanding of what counted as conditioned inhibition or what surrogate should be adopted. Basic concepts and interpretations were fragmenting. Even the concept of reactive inhibition (I_R) was falling apart as it was identified variously as "stimulus inhibition," "neural satiation," or "central inhibition" in both experimental and theoretical reports.

> All this makes it impossible to know if we have verified Hullian or neo-Hullian thinking, particularly since there are all shades of meaning to the same term; we greatly need a separation of I_{R1}, I_{R2}, . . . I_{Rn} and a thorough theoretical reworking.[37]

The *Principles of Behavior* was perhaps the most influential American-based, resource for understanding the nature of fatigue (or for understanding what used to be called "fatigue"). But the unifying analysis that Hull seemed to offer had become a matter of dispute rather than consensus. The study of work decrement had descended into conceptual chaos—the "helpless" condition bemoaned by Watson and from which the rigor of the *Principles* was meant to rescue the discipline. Almost the only ray of light discerned by the two Bilodeaus in their extensive review of publications in the field was the work of the aviation psychologist Jack Ashton Adams.

"A Fatigue Theory of Our Own"

Jack Adams (1922–2010) served with distinction in the US infantry during WWII. After the war, he trained as a psychologist and from 1951 to 1957 worked at the US Air Force Personnel Training and Research Center at Tyndall Air Force Base. Later, at the University of Illinois, he was director of the Aviation Psychology Laboratory.[38] Adams was an expert on training schedules, the predictive power of tests, and the use of aircraft simulators as training devices, but his work ranged widely over the topics of tracking, vigilance, learning, and motor skills. His style was forthright, as was his demand for conceptual clarity and testability. Bad definitions meant bad science. He wanted "precise deductions concerned with quantitative performance functions."[39] He had reservation about the appeal to information theory and comparisons between humans and servomechanisms. These fashionable ideas, he said, rested on mere analogy, and analogies were unscientific.[40]

Adams took the view that Hull's overall account of inhibition was untenable. The concept of conditioned inhibition $({}_{S}I_{R})$ gave the wrong performance predictions.[41] Adams could find no clear evidence that such a thing existed. Nevertheless, he accepted that the simpler concept of reactive inhibition (I_R) could, and should, be salvaged: "The final death rattle" of Hull's theory, he declared, "is yet to be heard." The lingering tendency to use Hull's ideas, said Adams, deserved sympathy:

because fatigue (if this colloquial term will be allowed for its expository rather than scientific usefulness) is a state variable that stands as one of the most impressive determiners of performance, and Hull's work inhibition postulates have been the only systematic attempt in psychology to order work-rest phenomena.[42]

Adams proposed to his fellow specialists in motor skills that we should "marshal our findings for work and rest (which are considerable) for a fatigue theory of our own."[43] What would such a theory look like? In their literature review, Bilodeau and Bilodeau had singled out for praise what they called "an ingenious experiment" by Adams.[44] They were referring to Adams's 1955 paper: "A Source of Decrement in Psychomotor Performance."[45] This paper can provide us with a pointer to what Adams meant by a "fatigue theory of our own."

"An Ingenious Experiment"

The experiment in question involved the use of the Pursuit Rotor—the "indestructible pursuit rotor," as Adams called it, on account of its popularity and prominence as a test used for pilot selection.[46] In Adams's experiment, over three hundred subjects were drawn from the Lackland Air Force Base in Texas. As described previously, the subject had to hold a stylus in contact with a small target moving in a circular trajectory—like a coin on a gramophone turntable. Time-on-target was the measure of success. Performance on the Pursuit Rotor was known to decline rapidly with time, and in Adams's eyes, this characteristic made the Rotor a simple but effective fatigue machine. Obviously, the responses of the subject involved more than moving the arm and hand. Time-on-target, he said, was actually a composite score because several component responses were involved. He needed to take account of the reactive inhibition of each of these components in order to understand their contributions to the total fatigue decrement.

The component that Adams chose to isolate for study was referred to simply as the visual response. Visual responses were implicated because, to keep the stylus on the target, it was necessary to watch for any divergence between stylus and target. The divergence was an error signal that would provide the stimulus to move the stylus back into contact with the target. The question was: What did the continuous repetition of this visual response contribute to the fatigue? Adams's idea was:

> Since this visual response of pursuit and discrimination is critical to successful performance of the task, inhibition of it through repetition should be reflected as decrement in the measured time-on-target score.[47]

The ingenuity of the experiment lay in isolating the fatigue of the visual-response component from the fatigue of the motor-response components.

The essence of Adams's experimental design was that a control group (B) would practice on the Pursuit Rotor, then take a fifteen-minute break, then resume their task on the Rotor. Before the break, any fatigue of the various contributory responses, such as the visual response, would generate a performance decrement (although its presence might be obscured by improvements in performance due to practice effects). But the fatigue would dissipate during the rest, and this fact would betray itself by the spontaneous improvement in performance immediately after the rest. The demands made on the experimental group (A) were different. They also had two practice runs on the Rotor, but, instead of a rest break, the interval between them was occupied with the task of watching another person perform on the Rotor. In Adams's terms, this arrangement meant that their "visual responses" were *not* given a rest. The process of looking at someone else perform the same task would separate out the fatigue in the visual response from the fatigue in the other components of the task. The aim was to see whether, when the task was resumed, the response of the experimental group (A) was worse than that of the control group (B). If it was worse, the blame could be laid at the door of the reactive inhibition, I_R, built up by the continued demands of making the extra visual responses of watching someone else at work on the Rotor.

Adams assumed that every attempt to correct a tracking error shows that a prior visual response has been made. To ensure that corresponding visual responses were being made by the onlookers in group (A), Adams required them to press down a button as long as the person they were observing was successful in keeping the stylus on the target. Then he had to allow for the fatigue contributed by those acts of button pressing, and so on. All of these complications were scrupulously addressed by the addition of further controls. Adams could then home in on the visual response whose contribution he wanted to study. A statistical analysis of the scores revealed that "the activity of group A in the interpolated interval clearly served to depress performance below the rest condition."[48] In Adams's terminology, these findings mean that "repetition of the visual response . . . results in the generation of reactive inhibition."[49] Thus, Adam's retention of Hull's concept of reactive inhibition (I_R) appears to be vindicated—at least, with regard to visual responses.

Adam's paper was published in 1955, but more than a decade later it still represented his position. In his 1967 book *Human Memory*, he reiterated his stance. After reaffirming his rejection of conditioned inhibition $({}_SI_R)$ he argued:

This left I_R as the viable element of Hull's work inhibition, and it was a formal way of stating much of what the layman means by fatigue . . . Although undoubtedly wrong in some of its details, there is no doubt that psychology needs a concept of temporary work inhibition in its theoretical arsenal.[50]

Adams admitted that the concept of work inhibition—that is, fatigue—was now in a "region of empirical limbo without theoretical explanation."[51] But he still took for granted that fatigue was a function of the work (w) involved in making a response, the number (n) of responses, and the time (t) between responses. These assumptions might be plausible when applied to muscular fatigue, but could they be generalized to the perceptual system? How did Adam's stance differ from Kraepelin's old idea that fatigue, whether muscular or mental, is best studied by making the subject endlessly repeat a simple response? The answer is that it doesn't differ. Adams was making the very assumptions that Bartlett challenged in the 1940s when planning the Cockpit experiments. The study of fatigue had gone around in a circle.

Broadbent's Reaction

If Adams's experiment earned praise on one side of the Atlantic, it raised a smile on the other side. In a parody of Milton's famous line, "They also serve who only stand and wait," Broadbent joked that Adams had demonstrated that "They also show fatigue who only sit and watch."[52] Broadbent did not doubt Adams's data, but he doubted Adams's methodology and conclusion. Adams had failed to take into account that other theories could yield the same prediction as the theory he had singled out for testing. Adams's theory was only supported by the experiment if it were granted that these other theories were also equally supported—and these other theories might have more promise than the idea of "visual responses" that prompt a temporary increase of reactive inhibition (I_R). Perhaps, argued Broadbent, the fatigue decrement in the Rotor performance was located in perceptual rather than response processes. Broadbent drew attention to the monotonous and repetitive character of the Rotor test. Perhaps the assumed accumulation of reactive inhibition was really the increasing probability of a filter switch to some alternative source of stimulation.[53]

Broadbent was here deploying all the arguments that he had previously used against Pavlov, and in doing so, he was also deepening and extending Bartlett's wartime arguments that had been directed against Kraepelin. But while a Filter Theory could make sense of Adam's experimental findings, for Adams, talk of a "filter" was a resort to mere analogy and hence unscientific.

Adams seemed unaware that his own explanatory concepts were also based on a supposed analogy between internal stimulus-response connections and external stimulus-response connections. Adams was modeling a perceptual process on the supposed consequences of a limb movement.

A Failure of Perception and Communication

When Bilodeau and Bilodeau wrote their article on motor skills for the 1961 *Annual Review*, they excluded contributions from Cambridge psychologists on the grounds that they were not relevant. When, in 1964, Adams wrote his own account of the state of motor-skills research for the *Annual Review*, he was in no doubt that Cambridge work was relevant—but only if handled selectively and with care. There were long-standing differences of approach, said Adams. American research dealt with behaviorally large units of behavior, such as tracking performance, while giving less emphasis to constituent mechanisms. He detected a fair amount of agreement amongst British psychologists over the Cambridge approach—in which, he said, the individual subject was assumed to possess:

(i) a number of sensory input channels,
(ii) a short- and long-term store,
(iii) a decision mechanism of limited capacity,
(iv) a number of effector mechanisms, and
(v) a mechanism of temporal expectancy governing the timing of responses.

This summary of the Cambridge approach is highly questionable. First, Adams suggested that a channel is a sensory modality. As used in Cambridge, the word "channel" referred to a source of information whose signals have a common, empirical characteristic.[54] Secondly, where was the perceptual filter? Mention was made of limitations of capacity but not of the filter that protected limited-capacity mechanisms from overload. Thirdly, no mention was made of the concept of "information," whether in its informal or technical usage. Adams made the mistake of asserting that the "keystone of the main British position is a one-channel decision mechanism" that allows the subject "to attend to only one stimulus event . . . at a time."[55] In fact, Cambridge psychologists always allowed that two or more stimuli can be processed at the same time, provided that their combined demands on the information processor did not overtax the limited channel capacity.

Adams was writing as if *Perception and Communication* had never been published. The explanation may lie in an earlier paper by Adams. In a 1962 survey of theories of vigilance, Adams said:

Broadbent's loosely structured views generally have prevented them from being instruments to guide the experiments of laboratory workers in this area,
and it is difficult to see them ever becoming useful unless a basic rephrasing
of their tenets is undertaken to give them the precision that a science asks of
a theory.[56]

Adams struggled to count Broadbent's contribution as science at all. For
Broadbent, by contrast, it was deliberate strategy to begin a scientific investigation with loosely structured ideas and home in on refinements as evidence accumulated. The rationale for the Cambridge stance was that premature precision carried a cost, and the fate of Hull's well-meaning efforts was
a case in point. Adams and Broadbent were involved in a head-on collision
between their respective advocacy of inductive and hypothetico-deductive
approaches. Fortunately for the delicate business of international scientific
cooperation, not everyone in America shared Adams's negative opinion. For
Arthur Melton, Broadbent's book was "an important systematic contribution
to engineering psychology."[57] I shall now move on to look at the subfield to
which Melton referred: namely, engineering psychology.

A Mathematical Cockpit

In October 1945, Franklin Taylor was appointed as the head of the Psychology
Section of the United States Naval Research Laboratory.[58] During the war, he
had worked on fire-control systems for the National Defense Research Council.[59] Given the important maritime role of aircraft carriers, Taylor and his
colleague Henry Birmingham also investigated the activity of pilots. How did
they control their aircraft when engaged in the delicate and dangerous task of
landing on the deck of a carrier ship? In studying this skill, these researchers
used the mathematical techniques of the servo-engineer. And they cited with
approval the publications of Cambridge psychologists such as Craik, Hick,
and Vince.

Taylor and his colleagues argued that it was important to think in terms of
an overall "man-machine system" and ensure that all the "system elements,"
whether human, mechanical, or electronic, were "reduced to the same system relevant terms." Only then could the interaction of all these elements be
understood in a coherent fashion.[60] Relevant descriptive terms, said Taylor,
were "bandwidth," "gain," "noise," and "coding," etc. This was not the usual
language of the psychologist. Adams complained about Taylor's neglect of
"procedural variables" of the kind that were traditionally used to describe
the state of the human subject. Taylor concentrated on "task" rather than on

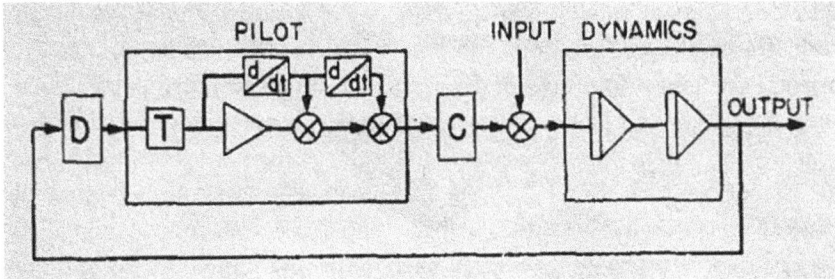

FIGURE 8.1. The pilot-aircraft system "considered mathematically." US Navy psychologists argued that in a "man-machine system," the tasks confronting the "man" and the responses of the "machine" should be described in the same mathematical language. The symbol d/dt represents differentiation. The triangle symbol represents amplification, and the triangle with the double vertical line represents integration. D stands for the instrument display, T stands for the lag due to reaction time, and C stands for the manipulation of the controls. From Taylor 1960, 647.

procedural variables. This approach, said Adams, "does not deserve being elevated to a research philosophy."[61] Taylor, by contrast, was determined to show that his approach was indeed the research philosophy that would carry psychologists into the cybernetic future.[62]

The engineering analogies of Taylor and his colleagues were always accompanied by characteristic "box-and-arrow" or information-flow diagrams, and these call for some explanation. In his influential 1960 paper "Four Basic Ideas in Engineering Psychology," in the *American Psychologist*, Taylor discussed a pilot flying an aircraft in gusty wind conditions. He directed his reader's attention to a diagram which had the uncompromising caption, "The pilot and his aircraft considered mathematically." Taylor's diagram is given in figure 8.1.

The diagram was meant to convey the way in which information on the instrument panel in the cockpit is conveyed to the pilot. The pilot's cognitive and motor responses then modify the position of the control surfaces and the output of the engine. Reading figure 8.1 from left to right, the first box (marked D) represents the display of information on the instrument panel. The next box represents the brain of the pilot. The first item within the pilot box is marked T and represents the reaction time of the pilot—a crucial source of lag. I will come back to the other symbols in the box marked "pilot" in a moment, but note that the simple triangular symbol represents a process of amplification and the circles with a cross represent the combination of information from different sources. The next box (marked C) stands for the cockpit controls that are manipulated by the pilot. The downward-pointing arrow designated "Input" refers to the gusts of wind impinging on the aircraft; the final box represents the dynamic properties of the aircraft as a heavy body subject to a range of forces. Here there are two further triangular-shaped symbols, but on one side of the triangles there is a double rather than a single line. These double lines mean that the triangular symbols now refer not to mere amplification but to the mathematical process of integration. Finally, there is the clearly visible feedback loop that carries information about the resultant behavior of the aircraft back to the instrument display and thus informs the pilot of the need for yet further action.

Taylor required readers of his paper to understand why there are two mathematical symbols for integration in the final box representing the dynamics of the system. He said that the figure showed that the dynamics of an aircraft "consists of two cascaded integrators."[63] This technical formulation was meant to be part of the new language of psychology. Psychologists, said Taylor, needed to wean themselves from their old "anthroponomic" habits.[64] Fortunately, the US military authorities had circulated training manuals meant to bring as wide an audience as possible into easy familiarity with these modes of speech.[65] The references to "integration" referred to the fact that aircraft have high inertia. They are sluggish in response, and they tend to be unstable and oscillate. They "hunt" from side to side, or they overshoot and undershoot a desired height. This is why, in the course of landing, when a pilot levels out above the runway, the aircraft continues to sink. Mathematically, high inertia means that the equations of motion governing the aircraft's behavior are second-order differential equations. Two successive processes of integration are required to solve such equations; hence the reference to two cascaded integrations. For Taylor, as the aircraft moves, it embodies and solves these mathematical equations of motion. It acts out the integrations: indeed, the aircraft *is* an integrator.

Controlling through two integrations is difficult, said Taylor. The pilot has a lot to do and a lot to think about, and that is why (coming back to the box marked "pilot" in fig. 8.1) the reader was presented with further mathematical symbols—in fact, two small square boxes containing the symbols for the mathematical process of differentiation. They represent the point and purpose of the cognitive processes that a pilot must go through in order to cope with the tasks in hand. Consider, for example, the tendency of the aircraft to oscillate. Taylor said:

> To stop this oscillation, the pilot actually has to . . . supply an amplification, two differentiations, and two analogue additions . . . Of course, the pilot does not think of his task in this mathematical way. He knows that he has to estimate

turning rate and acceleration and use these estimates to anticipate what the plane is doing. But he does not think of these processes as single and double differentiation and analogue addition. Yet this is precisely what they are, for were they not the plane would not long continue to fly the desired course.[66]

Notice the words: *this is precisely what they are*. The pilot must anticipate, but anticipation involves registering and responding to rates of change, and mathematically, rates of change are given by the differential coefficient. For this reason, Taylor's representation of the pilot in the diagram contained boxes carrying the symbol d/dt, symbols that signaled differentiation. It was the hard, material engineering truth of the matter that concerned Taylor. He was not playing with metaphors: for him, whatever the subjective experience, the pilot *is* solving differential equations—or failing to solve them.[67] The demands are real and unavoidable and may prove beyond the capacity of the pilot, or beyond the capacity of a fatigued or stressed pilot.

The aim of a good designer, argued Taylor, is to provide an appropriate engineering solution to cover such contingencies. To achieve this aim, the designer must restructure the machinery of display and control in order to make life simple for the pilot. This was the essence of Birmingham and Taylor's proposal in their 1954 interim report to the Naval Research Laboratory in Washington.[68] In the same year, and with very little modification, the text of this report was published as "A Design Philosophy for Man-Machine Control Systems."[69] The demands on the pilot cannot be wished away, but they can be shifted elsewhere. They can be moved from one part of the man-machine system to another, and the flow diagram with its boxes and arrows can then be redrawn. Thus, the process of differentiation can be taken out of the head of the pilot. This relocation can be achieved, said Birmingham and Taylor, by linking mechanical differentiators to the two loci of integration embodied in the system dynamics of the aircraft. In the schematic terms of their diagram, Birmingham and Taylor showed how to tap a signal from the output of each of the two integrators and then, after differentiation, feed the signal back to the display. This strategy was known to servo-engineers as "quickening," and it is represented in figure 8.2. A "quickened" display does not wait for the control signal to take effect on the aircraft and then, and only then, register the output and feed it back to the display. The diagram shows how the effects of the two differentiations intercept the feedback loop without waiting for the output. In this way, the effect of the control signal is anticipated and routed back at an earlier stage. This expedient removes an important element of lag in the system and hence helps to offset some of the lag due to the pilot's reaction time.

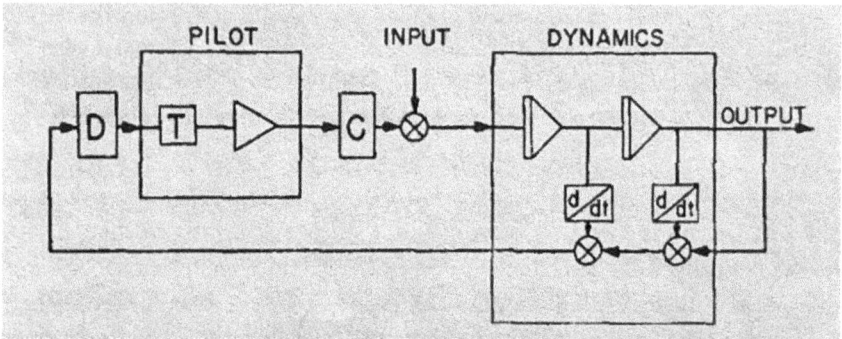

FIGURE 8.2. Quickening of the aircraft control loop. Taylor's design philosophy was to simplify the task confronting the pilot. Using the same symbols and notation as before, it can be seen that some of the complexities evident in the previous figure have been taken out of the box representing the brain of the pilot and located elsewhere in the man-machine system. The new configuration shown here is known as "quickening." All that is now required of the pilot is to act as an amplifier. From Taylor 1960, 647.

The design strategy was to reduce the complexity of the pilot's task and remove sources of stress and fatigue. Inspection of figure 8.2 reveals that, with the new arrangement, the pilot would be merely called upon to respond like a simple proportional amplifier. The more complex cognitive demands have been stripped away and relocated. But engineering solutions always involve weighing benefits against costs. The advantage of quickening is that it reduced lag, but this advantage is purchased at the cost of attenuating contact with the reality that is external to the man-machine system. This strategy is problematic because the signal on the display panel confronting the pilot is now a prediction of the future rather than a measurement of the present. Dangers can lurk in the gap that has now opened up between reality and the representation of reality displayed by the cockpit instruments.

Engineering Simplicity Versus Human Complexity

If aircraft cockpits can be redesigned so that the only thing demanded of pilots is that they behave like amplifiers, then why not replace them with amplifiers? The design philosophy of the Naval Research Laboratory pointed inexorably to the conclusion that pilots should be removed from the cockpit. Taylor and his colleagues admitted that this state of affairs was indeed the engineering ideal, but conceded that, in the current state of technology, humans sometimes still furnished cheaper, lighter, and more flexible engineering solutions. In the meantime, while pilots still had a job to do, how would they react to their attenuated role? How were they expected to comport themselves as they performed their easier but diminished duties? Birmingham and Taylor were quite explicit on the matter: If the demand to think in ways that called for integration and differentiation were removed from the brain of the pilot and displaced into the dynamics of the aircraft, then

> it would be expected that the operator would adjust rapidly to the changed requirements by simplifying his mode of action to a level analogous to simple amplification.[70]

They elaborated this crucial point when they said:

> in order to obtain optimum performance from the control system, it is necessary, not only to design the system so that amplification is all that is required of the operator, but it is also necessary to insure that the operator adopts this, and no other, mode of response.[71]

Here the psychologists and servo-engineers of the US Navy were making some big and potentially dangerous psychological assumptions. The empirical question was: Will humans adopt this, and no other, mode of response? What was the justification that Birmingham and Taylor gave for the expectation that the required adjustment would be rapid and effective? They argued that, while humans were by no means perfectly adaptable and flexible, here the demands on them were being reduced rather than increased, so all would be well. But are these grounds convincing?

To answer this question, let me bring Bartlett (hypothetically) into the discussion of this theme. Bartlett also believed that humans were adaptable and flexible and good at responding to new circumstances, but he had drawn the opposite conclusion to Birmingham and Taylor. Bartlett had long ago concluded that human complexity and creativity would swing into action undiminished by the simplicity of the task. Recall Bartlett's complaint about Ebbinghaus's method of exploring the capacity to remember by feeding the subjects of his experiments with nonsense syllables. Doing so forced the subject to adopt a particular habit and, said Bartlett, stripped the response of the characteristics that made it a matter of interest in the first place. Bartlett directed the same complaint at Kraepelin's simple and repetitive motor and mental tasks. Responses became automatic, and all their potential flexibility and actual variability was obscured from view by being forced into channels that were outside the range of the experimental controls.

In reality, Bartlett never engaged directly with Birmingham and Taylor's work, but if Bartlett was right, their expectations were unfounded. There was no basis for their confidence that pilots in the simplified cockpit would simplify their responses, and the reasons Birmingham and Taylor advanced to support their conclusion were inadequate. Their design philosophy was not based on a correct understanding of the phenomenon of human motor skill, and such an understanding was important even if their goal was to replace humans with machines. Bartlett's insight in both his memory experiments and his Cockpit experiments was that the complexity of a response was determined by the complexity of the responding organism, not the complexity of the stimulus. Perhaps it would be *more* stressful, and *more* fatiguing, for the pilot to be forced to operate within a man-machine system in which only truncated modes of response were available.

Methodologically, Birmingham and Taylor were committing another version of the Ebbinghaus-Kraepelin fallacy. Once again, the error was the focus on the simple and the repetitive. In terms of diplomacy, if nothing else, their design philosophy can hardly have supplied the degree of reassurance and job security that airline pilots, ambitious test-pilots, and would-be astronauts, would have wanted to hear. Pilots have a vested interest in being pilots. Astronauts do not want to be mere passengers: they want something to do, and preferably something complex, important, and interesting. It was not long before these matters raised their heads in a dramatic manner, and Bartlett would not have been surprised. The resulting conflicts provided fascinating sociological material for alert historians of technology. As David Mindell showed in *Digital Apollo*, in the course of planning and executing the Apollo moon landing mission, there was an extended conflict over a design philosophy of the kind advocated by Birmingham and Taylor—and it took place in precisely the area that Bartlett would have predicted: namely, the search for meaning.[72]

Confounded Variables

I now want to look at an argument that crosses the Atlantic in the opposite direction: that is, an argument that moves from America to Britain. If Bartlett's insights provided grounds for doubting the viability of the design philosophy pursued by experts in the US Naval Research Laboratory, these same American psychologists had their own important reservations about some of the assumptions made in the Cambridge Cockpit experiments. Again, the confrontation is hypothetical. No explicit mention was made of the Cambridge Cockpit by Taylor and his colleagues, but there can be no doubt about the relevance of their work when it comes to evaluating this project—and it is equally relevant to assessing the work of Bradford Hill and Williams. Taylor and Birmingham's design philosophy can be used to cast light on the problematic relationship that I have dubbed the Landing-Accident Anomaly.

In the study of any "man-machine system," a measure of system performance will be a measure of the combined effects of its two components, namely, the human component and the machine component. If psychologists are experimenting on such a system in the hope of discovering the properties of the human operator, how do they know which component is responsible for any observed system-performance measure? How are inferences to be drawn about the "man" from overall data about the "man-machine" system as a whole? The component variables are confounded. In 1959, Taylor and Birmingham published a paper designed to show that the depth of this difficulty had not been fully appreciated by engineering psychologists. It was felt, they said,

> that a demonstration might be in order to illustrate just how deceptive it could be to infer the behavioral properties of any one component—human or mechanical—from the performance of the total system.[73]

Their demonstration took the form of confronting an experimental subject with a tracking task to be performed using three different sorts of tracking apparatus. What might explain any differences in the performance of the task? The differences invite the inference that the subjects reacted differently to the different types of tracking apparatus. Perhaps the subject found one of them easier to handle than the other two, or conversely more fatiguing to operate than the others. Taylor and Birmingham's aim was to expose the hazardous character of inferences of this kind. They arranged that, despite any differences in the overall performance, the subject performing the tracking task would be in *exactly* the same state, and would behave in *exactly* the same way, in relation to each of the three different forms of tracking apparatus. These stringent requirements were met by using a robot as the experimental subject—the term "robot" being Taylor and Birmingham's own choice of words. The robot in question was to be an amplifier—just like their ideal pilot in their mathematically ideal cockpit.

Learning and Fatigue as Pure Artefacts

Three types of tracking apparatus were specified as follows. In the first, the system output was proportional to the output of the tracking amplifier. In the second case, it was the velocity of the system output that was proportional to the amplifier output. In the third case, again, the velocity of the system output was proportional to the output of the amplifier, but the gain of the velocity response was increased, and the output of the amplifier was fed back to the input. The three tracking systems are represented in figure 8.3. Once

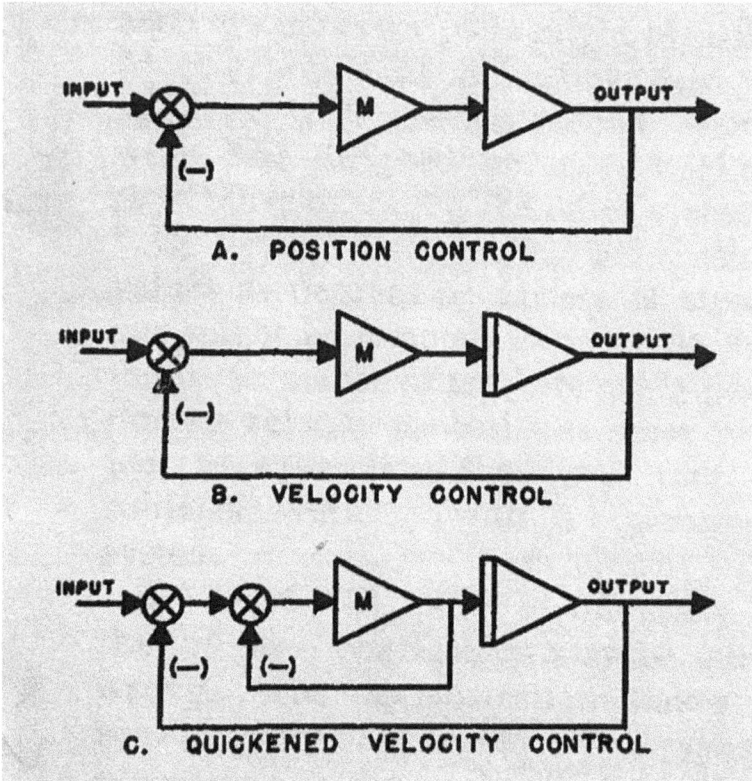

F I G U R E 8.3. Three tracking systems. Three "man-machine" systems are used to demonstrate the problem of identifying the respective contributions of the "man" and the "machine." In each of the three systems (position, velocity, and quickened control), the "man" is replaced by a robot amplifier symbolized by
the triangle marked M. The triangle with the double line again represents a process of integration. From
Taylor and Birmingham 1959, 179.

again, a simple triangle is the symbol for an amplifier, while a triangle with a
double line stands for a process of integration and amplification. The amplifier M, they announced in the caption of their diagram, "represents the robot
man." Using engineer's jargon, the three conditions were labeled, respectively,
"position control," "velocity control," and "quickened velocity control." "The
three dynamic conditions chosen for comparison," Taylor and Birmingham
explained, "were selected because they could all be stabilised by a simple amplifier, as long as there was no time delay present."[74]

The experiment involved feeding a changing signal into the system and
comparing it to the output signal. The signal to be tracked took the form of a
sine-wave input of three cycles per minute. The score for each system was the
integral of the tracking error throughout a thirty-second run. Each system

configuration was run with the same eight, different values of the gain on the robot tracking amplifier. The error score is represented on the ordinal axis; the eight values of the amplifier gain are shown on the abscissa. The gain represented on the graph increases from 1 to 128 in a geometrical series. Figure 8.4 shows the relation between the system error and the amplifier gain for each of the three systems.

How, asked Taylor and Birmingham, would these curves be interpreted if they were understood to have been generated in a psychological laboratory using *human* subjects? They consider two cases. First, suppose that an unsuspecting psychologist who encounters the graph was interested in the process of motor-skill learning. They would imagine that "the values on the abscissa represent successive periods of practice instead of increasing amplifier gain."[75] The conclusion might then be drawn that the subjects learned best when using the velocity system, next best with the position-control system. The absence of learning with the quickened system could be explained

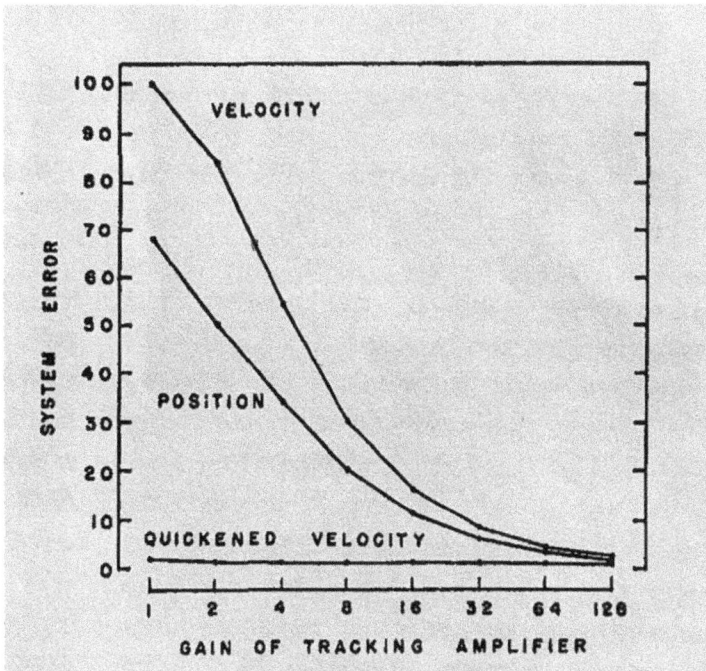

FIGURE 8.4. Performance of the three tracking systems as functions of amplifier gain. If the curves are interpreted as if they were the results of experiments on humans, they would suggest that the subjects behave differently in the three tracking systems. In reality, the "subjects" here are amplifiers that behave in the same way in all three cases. From Taylor and Birmingham 1959, 179.

by the (apparent) fact that that the system was so easy to operate that little or no learning was required.

Taylor and Birmingham next supposed that the unsuspecting psychologist was interested in fatigue. To see how the curves could be seen in this light, they invited the reader to imagine "the curves mirror-imaged through being plotted with the abscissa scale reversed."[76] The curves would then simulate the performance decrement familiar from the study of fatigue or declining vigilance. The plots might then "plausibly be interpreted as indicating that the trackers with the quickened system were less fatigued or more vigilant than those operating the other systems."[77] Taylor and Birmingham's point, of course, was that such an inference would be completely wrong:

> The three very dissimilar performance curves all arose from exactly the same "behavioral" changes in the robot man. The amplifier acted in the same way in each system. It "learned," "fatigued," or altered under "stress" by exactly the same amount in all three dynamic configurations.[78]

Taylor and Birmingham's exercise exposed the danger of overlooking an ever-present logical possibility. The difference in the performance curves might derive wholly or in part from the different ways in which "the systems translate and transform the subject's behavior, rather than to differences in the human behavior itself."[79]

How are such differential translations and transformations generated? Taylor and Birmingham argued that some man-machine complexes will be highly sensitive to small changes in human behavior, while others will not. Some configurations will introduce nonlinearities, and these nonlinearities can even result from combinations of completely linear components. In the present demonstration, said Taylor and Birmingham, "all the electronic elements, including the scoring system, were entirely linear throughout the range employed," and yet the data show that, due to their interaction, the overall system behaves otherwise.[80] Coming back to the problems of laboratory practice, they concluded that

> some of the conflicting results of motor skill experiments on learning, fatigue, vigilance and response to stress are due to the fact that subjects in different experiments were encapsulated in systems with different scale properties.[81]

Taylor and Birmingham were too diplomatic to name individual researchers whose results were compromised in this way—but a possible target for their criticism immediately comes to mind.

The Robot Cockpit and the Cambridge Cockpit

Taylor and Birmingham's abstract talk of a "man-machine system" is oddly similar to Bartlett's references to the "calculating machine" with which he teased his Royal Society audience in his cryptic wartime Ferrier Lecture. This resonance points to a disturbing possibility. Taylor and Birmingham could have illustrated their argument by citing the Cambridge Cockpit experiments and the landing-accident study. Both Bartlett and Bradford Hill used performance measures derived from a man-machine system as a basis for drawing conclusions about one element of the system—the man rather than the machine. Bartlett concluded that the performance decrement of the pilot-cockpit system in his laboratory indicated the fatigue of the pilot. Bradford Hill concluded that the landing-accident statistics generated by a pilot-bomber system indicated the absence of fatigue in the pilot. The robot-tracker demonstration, however, raised the logical possibility that the pilots were in *exactly* the same state of fatigue in both investigations.

Seen in this light, both the eminent investigators reporting to the Flying Personnel Research Committee were committing methodological errors. Their conclusions were inadmissible because, in both cases, variables were confounded. In light of Taylor and Birmingham's argument, it looks as if Bartlett's attempt to revive and clarify the old problem of fatigue had run up against a brick wall. Perhaps Muscio, Watson, and Durig were right in their pessimistic stance toward the concept of fatigue. Perhaps Bartlett was simply wrong to think he could engineer a useful revival of fatigue studies by criticizing Kraepelin along the same lines that he had previously criticized Ebbinghaus. On this analysis, the Cambridge Cockpit fatigue-project seems to have ended in a methodological debacle—though no less a debacle than the Bomber Command landing-accident study.

The Concept of Full Knowledge

The only way to escape this conclusion is to find a loophole in Taylor and Birmingham's analysis, or a way of seeing their insights in a less paralyzing light. The clue is that Taylor and Birmingham themselves admit that sometimes, and in some form, inferences from a system-performance measure to the performance of an element in the system *can* be justified. They went so far as to say:

> When a system measure is used deliberately as an index of human performance with full knowledge of the extent to which the dependent variable is confounded, the practice is certainly justified.[82]

The question is: What is it to have "full knowledge" of the extent to which the dependent variable is confounded? How is such full knowledge possible, and how does it permit an inference that the overall logic of the argument appears to preclude?

A re-examination of Taylor and Birmingham's demonstration with the robot tracker reveals why they had to make a concession of this kind, and why, if properly understood, the concession is wholly consistent with their insights. Recall that, to set up their argument, Taylor and Birmingham needed to assert that the robot amplifier changed in exactly the same way in all three system configurations. They also appeared to know that the behavior of the amplifier was entirely linear throughout the ranges employed. This is the sort of "full knowledge" of which they spoke. But how did they acquire this knowledge? The answer is that they acquired their intimate knowledge of amplifiers from their engineering education and a reliance on the manufacturing practices and standards of the field. Elsewhere, Taylor said precisely as much:

> Most of this information about mechanical elements is already available in engineering or physics texts. . . . The human engineering designer merely has to reach up on the shelf to obtain the physical components with which to structure his systems.[83]

This is how they knew how the amplifier—the crucial physical component of their demonstration—would behave. They read what was on the box. They trusted what they read, and they knew that in doing so, they were behaving like all their fellow engineers. They also knew that, if necessary, they could spot a rogue amplifier and take it back to the shop. They could afford to proceed in this way because they were the beneficiaries of a tradition of electrical engineering in which a reliable, shared, practical and theoretical understanding of amplifiers had emerged through trial and error. This tradition was the inescapable basis of their knowledge. It defined the nature of that knowledge and what counted as "full knowledge." The growth of that practical engineering tradition has been well illuminated by historians of technology.[84]

Is this interpretation of "full knowledge" really consistent with Taylor and Birmingham's argument? How can they invoke this pragmatic category without arguing in a circle? Their claim to "full knowledge" seems to apply to each amplifier in isolation, but on every occasion when engineers have built or used an amplifier, and on every occasion on which they have tinkered with its gain or experimented by adding or subtracting it from a circuit, the amplifier will have been functioning as part of a "system" of some kind. None of the other elements of the system (the rectifiers, thermionic valves, and connecting circuitry of resistors, capacitors, and induction coils) can be understood, element by element, in isolation from those systems any more than can the workings of an amplifier. Taylor made this point himself and was right to insist on it. Thus:

> In the last analysis, the behavior of a creature—or any physical object, for that matter—cannot be measured except through its effect on something else. In this sense, every measurement involves a system.[85]

It may appear that, by insisting on this point, Taylor was cutting the ground from beneath his own feet. But he was not. Rather, he was reminding us that knowledge of the parts of a system, and the role they are understood to play, can and must arise without ever taking those parts out of the various systems in which they exist. In this concrete and important sense, all knowledge is relative.

"Full knowledge" is the name that Taylor and Birmingham apply to the results of a protracted process of empirical probing that has attained stability and practical utility. Certain parts of that system of understanding, it may be supposed, have become taken for granted. These parts are taken for granted because they have become routine. Such routinization is achievable if, and only if, the knowledge in question is integrated into the stable practicalities of a form of social and professional life. "Full knowledge" is therefore, ultimately, a *sociological* category. It is often and correctly remarked that "technical systems" are ultimately "social systems." Knowledge of amplifiers has a *pragmatic, inductive,* and *collective* character, and the resulting practical stability suffices when there is a need to make distinctions between one part of a system and another—for example, between the common factor of the robot amplifier and the other components involved in the three tracking experiments.

Bartlett had similar epistemological distinctions in mind when, in his last book, *Thinking*, he reflected on the way that "experiment must grow out of earlier natural observation." From this beginning, Bartlett said,

> early experimental results will usually lead on to more specific and defined natural observations, and from these again will develop further steps of experiment with improved control; and so this kind of process may go on until it seems that the limit of control for the kind of subject-matter concerned has been reached.[86]

These considerations were not put forward with any reference to Taylor and Birmingham's argument, but Bartlett was in fact describing the only realistic candidate for what Taylor and Birmingham counted as "full knowledge." Bartlett was telling us that these words refer to an accepted social condition where it collectively *seems* that a limit has been reached, and where his use of the modest category of "seeming" reveals that we are, of course, in the realm of the relative, not the absolute.

In the wartime context, neither Bartlett nor Bradford Hill would have claimed to possess a "full knowledge" of fatigue in any sense of those words, whether unrealistically absolute or realistically relative. Both men were making urgent inferences on what they would have known to be an unavoidably narrow basis. Both would have been aware that their results were not directly comparable because of the different conditions under which the respective tests were carried out. Crucial variables were not under full control: fighter cockpits and bomber cockpits were different environments, and the degree and character of the training of the pilot-subjects in the two studies differed. In simple, practical terms: The two results needed further scientific work to clarify the apparent discrepancy. And, if there had not been a war to fight, events would probably have unfolded in just such a fashion.

To come back to the question of principle: If it could be said that the "full knowledge" needed to make routine inferences already existed in a field such as electrical engineering, then it could surely exist in the field of experimental psychology. Taylor and Birmingham had already admitted as much. The practical difficulty of resolving the Landing-Accident Anomaly could therefore be seen in its familiar aspect *as an insufficiency of accepted background knowledge.* The process of rectifying this deficit needed more peaceful conditions. Of course, these conditions gradually returned in the postwar years but, by then, the urgent and highly specific demands posed by the Landing-Accident Anomaly had dissipated. Broader programs of applied research came to dominate the scene, and it was to help meet these new demands that the Cambridge Cockpit was methodologically "dismantled."[87]

A Journey in Retrospect and Prospect

Bartlett spoke of the Cambridge Cockpit work as the beginning of a journey. It was a journey, he said, whose destination could not have been envisaged from the outset. He was, however, in no doubt that the Cockpit experiments, "more than any others, gave us the guiding principles," and the postwar work "assumed the character of exciting and engrossing adventure."[1] I have already described some of the central results of the postwar work reported in 1958 in *Perception and Communication*. I shall now argue that, even after that landmark publication, the guiding principles of Cambridge psychology were still those that came from the Cockpit experiments. This link is corroborated by the numerous references to the Cockpit experiments in A. T. Welford's 1968 monograph *Fundamentals of Skill*.[2] But I shall make the case primarily by reference to Broadbent's second, important, and comprehensive study *Decision and Stress*, published in 1971. Like his 1958 report, this book contains a review, not only of Broadbent's own work but also of the work of many of his colleagues in the Applied Psychology Unit. He also introduces some penetrating contributions from the United States. We shall see that the argument in *Decision and Stress* culminates with Broadbent outlining a neo-Jacksonian theory of levels of control in the central nervous system. I shall begin by showing that Bartlett's reference to the wartime and postwar research as an "adventure" was not merely a literary flourish.

Bartlett's Last Book

For Bartlett, the word "adventure" signaled an important psychological category, namely, the category of bold inductive thinking as it expressed itself in laboratory as well as real-life conditions. Bartlett always opposed the tendency amongst

psychologists to form "systems" and the corresponding tendency for their thinking to circulate, deductively, within the confines of their preferred theoretical system. He vowed never to let Cambridge psychology assume the form of a system.[3]
For him, genuine scientific thinking should not be system-bound precisely because it should always be probing the frontiers of inductive generalization. Properly understood, an inductive methodology is a way of making scientists sensitive to the unpredictable and hence to the adventurous. These methodological
themes provided Bartlett with the subject matter of his last book.

Thinking: An Experimental and Social Study was published in 1958.[4] In
style, it stood in sharp contrast to other psychological work being carried
out by Cambridge psychologists at that time. We have seen that Broadbent's
Perception and Communication, published in the same year, was packed with
experimental data and detailed analysis. *Thinking* was altogether looser in
structure and more personal, speculative, and self-referential. For a large part
of his own life, said Bartlett,

> I have been watching experimentalists at work, sometimes on questions which
> I have been able to suggest, and I have myself worked mainly experimentally
> in the field of psychology. In a way all this watching and working can be re-
> garded as a sort of prolonged experiment on experimental thinking.[5]

In *Remembering*, Bartlett's "experiment on experimental thinking" was informed by his methodological rejection of the proposition that one can only
be said to have "remembered" X when X really happened. In *Thinking*, he
adopted the corresponding stance toward the study of thinking and problem
solving. Thus:

> There is no psychological sense in saying that thinking takes place only if a
> "right" issue is reached. There is thinking whatever issue is reached so long
> as an attempt is made to carry further the evidence, or information, that has
> been made available.[6]

As long as the same inductive and cognitive mechanisms are in play, the question of whether the outcome is judged right or wrong by the investigator is irrelevant to the causal analysis. There is not one mechanism for right answers
and another for wrong answers. Obviously, Bartlett's stance does not preclude
a role for normative judgments in the professional interaction between psychologists themselves. Nor does it preclude those judgments subsequently
becoming the object of empirical study in their own right—for example, by a
later generation of psychologists, or by historians of science. In both *Remembering* and *Thinking*, Bartlett explicitly invited a developing, dialectical, and
reflexive process of precisely this kind.[7]

Bartlett's central thesis in his last book was that thinking was a form of skill. The implication was that thinking should be understood by analogy with what psychologists, working in the Cambridge laboratories, had discovered about skill—in particular, what they had discovered during WWII and in the immediate postwar years. The opening chapter of *Thinking* is therefore replete with references to the work of Bartlett's postwar Cambridge colleagues and students such as Richard Conrad and Alfred Leonard, and in later chapters there are additional citations to published and unpublished work by Richard Gregory, Christopher Poulton, and Margaret Vince. Thinking, said Bartlett, is not "a sort of gift."[8] Rather, the need is "to put thinking in its place as a natural development from earlier forms of bodily skilled behaviour."[9]

The comparison with skill enabled Bartlett to stress the opportunism and risk-taking character of the adventurous thinker—the very properties he admired in the good experimental psychologist. Following the analogy between skill and thought, he drew attention to the importance of timing and anticipation. The thoughtful experimenter discerns a certain "drift" in the scattered data, just as subjects in Poulton's tracking tasks discerned a pattern and learned how to anticipate. The good experimenter knows when to start a new line of experimentation or call a halt to an existing program of research. Thus:

> An original mind, never wholly contained in any one conventionally enclosed field of interest, now seizes on the possibility that there may be some unexpected overlap, takes the risk whether there is or is not, and gives the old subject matter a new look.[10]

It is difficult to resist the conclusion that, at least in part, Bartlett was talking about himself and his own risk-taking response to the wartime need for a deeper understanding of fatigue. Bartlett's thinking was never enclosed entirely within any one psychological field of specialization. This breadth of vision was why Bartlett was able to see an analogy, invisible to others, between the methods of Ebbinghaus and Kraepelin, and it is why he was able to put the insight to work to create the Cambridge Cockpit as a new form of experiment on the problems of skill, work, and fatigue.

It is easy to underestimate the subtlety of Bartlett's argument in *Thinking*. For example, he was clearly developing an argument based on analogy but denied any commitment to an "elaborate analogy" between thinking and motor skills.[11] Why the qualification? Was this mere equivocation, or the reflex habit of academic caution? I suggest that these words were wise and not merely cautious. They were expressions of good inductive practice. In both pursuing an analogy and warning against an "elaborate analogy," Bartlett was warning against what, in the field of data analysis, is called the fallacy of over-fitting. Given a number of data points, it can be better to fit them loosely to a simple equation that only passes through a few actual data points than to seek an elaborate equation that passes through all of them. Simple equations can give a better inductive projection than an over-fitted equation.[12] Shortly we shall see that Broadbent's version of this inductive principle was expressed through his preference for working with a number of simple but different mathematical models rather than committing himself to the elaboration of one preferred axiom system. Despite the many insights of this kind, the argument in *Thinking* lacked the sustained vigor of *Remembering*. The results seem thinner. This sense of unrealized potential was confirmed some five years later, when Broadbent pursued a similar analogy but took the argument to a deeper level.

Broadbent on Common Principles

In 1962, at a NATO conference on defense psychology in Washington, Broadbent delivered an address where he set out to identify the "Common Principles in Perception, Reaction and Intellectual Decision."[13] Both Bartlett and Broadbent were confident that there were "common principles" across different psychological functions, but while Bartlett had started with motor skills and worked (analogically) upward to more intellectual decisions, Broadbent started with a discussion of intellectual decisions and worked (analogically) down toward perceptual-motor responses. Fundamentally, however, the two approaches were the same. In both cases, the aim was scientific unification, and in both, the perspective was causal and the methods were reductionist.[14]

Broadbent's audience included representatives of the military and of various branches of the psychological profession drawn from the countries of the NATO Alliance. "I want to consider," Broadbent told them,

> the way in which approaches originally devised for the analysis of problem-solving, of intellectual decision, are beginning to be relevant to the study of simple functions such as sensory thresholds and reaction times.[15]

He indicated that the common principles he was about to identify had important implications at the social level. It appears, he said, pointedly addressing the military contingent in his audience, that

> the attitudes current in a particular military society may conceivably have an important effect upon the efficiency of these functions which some people have regarded as purely mechanical and unconnected with personal relationships.[16]

Turning now to the psychologists in the audience, he explained that he would describe some experiments coming from his own laboratory, although he was well aware that other laboratories, in both the United States and Europe, were working along similar lines. He expressed the hope that even psychologists from what he called the "more human side" of the discipline would "find something of interest in this part of the work of those of us who use brass instruments and follow the tradition of Wundt."[17] Of course, Cambridge psychologists neither lingered in the era of "brass instruments," such as the kymograph in Myers's textbook, nor did they owe conspicuous allegiance to the indefatigable pioneer Wilhelm Wundt. This phrasing was Broadbent's self-deprecating way of telling his audience that he was aware of the hostility sometimes directed at the experimental tradition by social psychologists. Social psychologists needed to know that adherents to the experimental tradition had some surprising insights to offer about social phenomena. After these preliminary hints, he set out the first steps of his argument.

Consider, he said, an officer on the battlefield who must make a decision to blow up a bridge. His own troops are retreating across the bridge, but the enemy is advancing toward it, and if the bridge is lost to the enemy, the battle is lost. But blow up the bridge too soon and comrades will be trapped; delay too long and the bridge will fall into enemy hands. Timing is of the essence, but the relevant probabilities are unknown—and there is no objective metric with which to calculate the desperate trade-off between losses and gains. What data is at hand? Are the retreating troops reduced to a trickle? Are there already telltale signs of enemy activity? Is there, as Bartlett might say, a discernable drift in the data? Although Broadbent does not make the comparison explicit, one can see at once that the structure of the situation is precisely that envisaged by Bartlett when he was thinking about the skills demanded of a scientific investigator. When should a line of experimentation be abandoned? What hints can be discerned in the data? Give up too soon and potentially fruitful results may be missed; give up too late and valuable resources will have been squandered and ground lost to more nimble scientific competitors.

Broadbent was aware of the extensive research on decision-making and risk-taking of this kind. He knew of the arguments surrounding the interpretation of laboratory experiments in this field, but he recorded his belief that, overall, research showed that humans have a natural tendency to choose risky courses of action. Perhaps, he said, senior officers should be more aware than they are of the tendency to opt for improbable but large gains rather than probable but small gains. "If commanders have this natural bias," he said, "towards actions which, however tempting, are unlikely to be successful, it may be rather serious."[18] After this characteristic understatement, he revealed the crucial generalization that he wanted to make:

> The process of decision between two alternative actions, each of which has a doubtful outcome, would be recognised by all of us as a familiar situation: but we would not perhaps see that it has anything in common with the simple sensory processes which occur when a man detects a faint light or a quiet sound.[19]

Reference to faint lights and quiet sounds will have alerted many in his audience to the direction of the argument: Broadbent was going to talk about men and women scanning radar screens or listening to the signals generated by sonar equipment. Each and every perceptual act, he was saying, must be understood by analogy to the decision to blow up the bridge. All perception, however immediate, direct, and compelling, actually involves a decision and a gamble.

Signal Detection Theory

What was the theoretical basis of this counterintuitive comparison? Broadbent outlined to his audience a model of perceptual processes that had recently been developed by the Electronics Defense Group in the University of Michigan. The central idea was described in a paper published in 1954 by Wilson Tanner and John Swets.[20] They made bold, simplifying assumptions to pave the way for a mathematical analysis. They assumed that, within the brain of someone watching a radar screen, there was a neural process whose intensity varied continuously from moment to moment. The variations were present both when there was a signal and when the screen was showing no signal. These variations oscillated around a mean value. The probability of occurrence of any given value was assumed to form a "normal" or "Gaussian" distribution. When a signal occurred, the neural process continued to oscillate with the same standard deviation but around a higher mean value.

Tanner and Swets then assumed that there was a degree of overlap in the distribution of the two sorts of oscillation: that with and that without a signal. The highest value of the variation without any signal was assumed to be higher than the lowest value of the variation with the signal. The situation is pictured in figure 9.1. The two distributions shown in figure 9.1 are to be thought of as representing, on the left, the background of merely random neural "noise" (N), and on the right, the distribution of "signal plus noise" (SN). The difference between the means of the two distributions is known as the signal-noise ratio. When expressed as a ratio of the standard deviation of

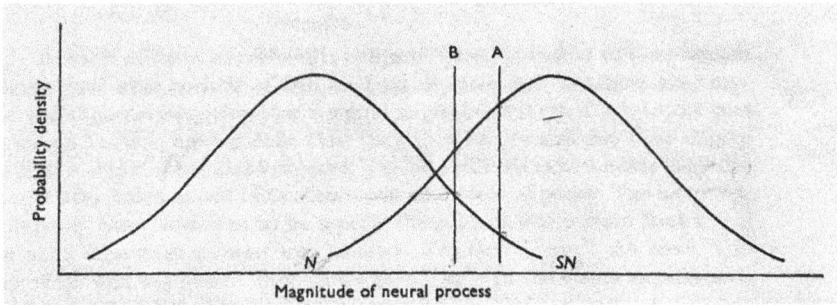

FIGURE 9.1. The theory of signal detection. A random background of neural noise is assumed, and the additional presence of a signal is assumed to shift the mean of the distribution. A signal is experienced whenever the neural process, whether noise (N) or signal plus noise (SN), exceeds a certain criterion level. On this model, it is possible that a shift in the criterion can achieve a large increase in detections at the cost of only a small number of false alarms. From Broadbent and Gregory 1963, 312.

the distribution it is conventionally designated by the symbol d'. Given any specified magnitude of the neural process in the brain, as represented along the horizontal x-axis, there will be a chance that it is caused by the background noise and a chance that it is caused by the presence of the signal. It could then be said that the problem for the brain is to decide whether or not it is confronted by a signal.

Suppose that an observer, looking at the radar screen, in effect "decides" that above some definite level of neural activity, they will declare that a signal is present, and below that level they will declare that no signal is present. Call that level the "criterion" implicitly used by the person watching the screen. This criterion is conventionally represented by the symbol β (beta). The use of the word "decision" to describe the placing of the criterion β is not wholly satisfactory in commonsense terms, and Tanner and Swets say that, psychologically, the criterion is simply a mental "set" or disposition. Depending on the set or disposition of the observer, the criterion will be more or less cautious. A cautious observer would demand a higher level of neural activity— that is, more evidence—before declaring the presence of a signal. A less cautious observer would take more risks and would be prompted into action by a lower level of neural activity—that is, by less evidence. Those disposed to take risks would declare more correct identifications but also more false alarms. The lines A and B in figure 9.1 represent two possible criterion settings, with β = A the more cautious setting than β = B.

Broadbent acknowledged that there is nothing new in saying that a cautious observer will strive to avoid false alarms but might pay the price of missing some real signals. But Tanner and Swets's model goes beyond this truism. Suppose that the criterion is placed so that about half the signals are being detected—as is the case with line A in figure 9.1. Now a small shift in the criterion, for example a move from the relatively cautious A to the riskier B, would make a big difference to the frequency of correct reports but a small difference to the false-alarm rate. This divergence is because the criterion would be traversing areas near the *mean* of the signal-present distribution but near the *edge* of the signal-absent noise distribution. This result can be read off figure 9.1. The change in the probability of detection produced by a change in the position of the criterion is proportional to the area under the curve that lies between the old and new criterion. The area between A and B under the left-hand "no-signal" curve is much smaller than the area between A and B under the right-hand "signal-present" curve. According to the Tanner and Swets model, then, the increase in the efficiency of detection due to the criterion shift can be large, while the increase in the number of false alarms is very small.

The implication, noted Broadbent, was that it might be bad policy for officers to discourage radar watchkeepers or sonar operators from reporting what turned out to be false alarms. Rather than expressing disapproval it might be better to tolerate a few more false alarms. Even a small shift toward more caution might produce a drastic reduction in the rate of successful detections. It would be good, said Broadbent, if military authorities were aware of this "unexpected effect of social attitudes upon the efficiency of the senses."[21]

Confidence and Fatigue

Broadbent also had some unpublished findings to report to the conference, and these results brought the theme of fatigue back into the discussion. Rather than employing the highly selected subjects typically used in experiments on signal detection in the laboratory, and rather than conducting the typical psychophysical experiments where the time of arrival of the signals was fairly well known, Broadbent wanted to see if the theory was applicable to more realistic vigilance situations. He therefore selected a task closer to the signal detection problems likely to be encountered in a real-world military context, and he used as his subjects ratings from the Royal Navy.[22] The watchkeeping task that he gave them lasted for forty-five minutes and the signals were few and far between and unpredictable. The signals were very faint sounds, and each lasted for just one second. Subjects were instructed to report any signal they thought they heard, even if they might be reporting a sound when none was present. In addition, his subjects were required to register their degree of confidence using a scale of one to four.

The results showed that the subjects had an accurate sense of the reliability of their own reports. Detections that were made with confidence were nearly always right, and there were very few confident but false claims to have detected a sound. Those reports about which the subjects were less confident were proportionately less accurate. This pattern held for every single individual subject. Even more striking was the relation that emerged when Broadbent studied the effect of criterion shifts. He did this by counting up only the most confident results and comparing the outcome in terms of correct detections and false alarms with the outcome of counting all of the results—that is, both the confidently reported results and the doubtfully reported results. Broadbent found that admitting the less confident results greatly increased the percentage of correct identifications while only slightly increasing the number of false alarms. In his experiment, he found that "one can get a rise of almost .2 in the detection rate at a cost of only .016 in the false report rate."[23] This high gain for low cost was exactly what the exponents of the signal detection model had predicted.

Broadbent also noted that when both the confident and less confident reports were counted up, there was no fatigue effect over the course of the watch. There was a change over time, but it was a change in confidence rather than detection rate. The change in confidence was both statistically significant and marked. "Almost everybody," said Broadbent "regards their own responses with more scepticism when fatigued."[24]

How do the experimental conclusions relating confidence and fatigue map onto the findings of the Cambridge Cockpit experiments? Bartlett spoke of fatigue as causing a "lowering of standards" in flying skill and said that fatigue was the enemy of good timing. A fatigued pilot may do the right thing but do it at the wrong time, with knock-on effects for subsequent maneuvers. At the behavioral level, it is easy to see that there are similarities between Broadbent's and Bartlett's experimental results. Fatigued subjects at the end of a signal detection experiment lose confidence in their observations, and fatigued pilots lose confidence in their flying ability: They keep checking their instruments and then hesitate to initiate the necessary maneuvers. Does this empirical similarity point to a deeper, underlying theoretical analogy? If so, how should we state Bartlett's conclusion in the language of signal detection theory? Broadbent's unconfident watchkeepers were operating with a cautious, that is high-β, setting of their decision criterion. It is therefore tempting to declare that Bartlett's "low standard" equates to "high β." This simple equation is right as far as it goes, but it cannot be the full story. It is necessary to remember that, at the other end of the scale, a low-β setting can also cause a low standard of flying. The situation now would be one in which risky and over-rapid decisions are taken. Now pilots are flying badly not because they have too little confidence in their own judgments but because they have too much confidence.

More will need to be said about the significance of measures, like the scale along which the criterion β ranges back and forth, and where inefficiency can be present at both the high and low ends. Meanwhile, I want to introduce a further complication that Broadbent had to confront in his analysis of vigilance phenomena.

The Jerison Anomaly

In *Decision and Stress*, Broadbent wanted to "follow the way in which knowledge had developed historically."[25] This method gave immediacy to the narrative but did so at a price. The historical approach lacks the logical simplicity of an after-the-fact reconstruction. Consider the way in which signal detection theory was introduced to the reader of *Decision and Stress*. In the Common Principles Lecture, Broadbent simply presented his audience with the argument to be found in one of Tanner and Swets's publications, but now he traced a different route—under the heading "Some Personal History."[26] This section formed the link between chapter 2 of *Decision and Stress*, which dealt with traditional measures of vigilance, and chapter 3, which dealt with performance measures derived from statistical decision theory. The change in the performance measure was justified as a necessary response to a puzzling experimental anomaly. The anomaly in question arose from a series of technical reports that American psychologist Harry Jerison and his colleagues had submitted in the mid-1950s to the Wright Air Development Center.[27]

Jerison's experiments used a variant of the Clock Test apparatus originally introduced during WWII by Mackworth to study submarine detection. This link between Jerison's WADC reports and the previous Cambridge experiments was probably due to postwar policies encouraging interaction between British and American psychologists whose work was of potential military interest.[28] Jerison repeated Mackworth's experiment twice: once with a single clock dial and once with three clock dials. In the single-clock experiment, there was a drop in the efficiency of detection over the course of the watch, In the three-clock experiment, by contrast, there was no decrement over the course of the watch. These results seemed to suggest that, counter to intuition, visual fatigue would afflict those given a single radar screen to monitor rather than those given three radar screens to monitor.

Broadbent had noted Jerison's results in *Perception and Communication*, but he only confronted the full force of the problem later, in *Decision and Stress*. In the earlier work, Broadbent had accepted the need to explore a number of different theoretical approaches to vigilance (for example, the expectancy approach, which stressed the role of stimulus probability; the arousal approach, which stressed the general state of activation of an organism; and his own filter analogy). All these approaches had their strengths, but none was wholly adequate. Like Bartlett, Broadbent was aware of the dangers of theoretical dogmatism and wanted psychologists to learn how to move back and forth between different theoretical languages. Broadbent's first line of attack on the Jerison anomaly was based on precisely such a point of contact between different theoretical languages.

We have already seen how Adams, in his attempt to build a "fatigue theory of our own," tried to harness the Hullian idea of reactive inhibition (I_R) to visual processes by speaking in terms of "visual responses" whose behavior and patterns of inhibition were assumed to be analogous to motor responses. Adams's idea was to count the number of "visual responses" by requiring his subjects to make a motor response each time they noticed a tracking error when observing another subject's performance on a Pursuit Rotor. Broadbent was not sympathetic to Adams's disregard for alternative explanations, but he was sympathetic to the attempt to find public indicators of inner, private events. Broadbent wanted to find better ways to study the inner fluctuations of the perceptual filter that was the vehicle for his own lines of theoretical analysis. His first response to the Jerison anomaly was to explore a technique for studying the so-called observing responses of subjects confronting a vigilance task. Broadbent adopted this method in his 1963 report "Some Recent Research from the Applied Psychology Research Unit, Cambridge."[29] He eventually discarded the attempt, but this unsatisfactory excursion had important consequences. It is described in the section on "Some Personal History." Referring to himself in the third person, Broadbent said:

> When there are many sources of information, it is obvious that filtering (to use Broadbent's term) or the occurrence of observing responses (to use Jerison's) may be distributed in a variety of ways over the different possible locations from which information may come. If, like Broadbent, one has the theory that decrement is due to a failure of filtering after a period of continuous work, one is very much hampered by inability to measure that process directly.[30]

Broadbent's apparatus consisted of three buttons positioned on the corners of a triangular frame. Each button was connected to a neon light, and the three corresponding lights were positioned side by side on a vertical panel in front of the subject. Each button press caused one of the lights to emit a flash—except, that is, when a "signal" was present. A signal was indicated by the absence of a corresponding flash. Over the course of a watch, a small number of "signals" of this kind were inserted by the experimenter, and it was the job of the subject to search them out and then report the detection. The number and distribution of the signals were identical to those used by Jerison in his comparison of the one-clock and three-clock tasks.[31]

The problem with Broadbent's new experiment was that the button pressing seemed to become automatic, detached from the subject's actually paying attention to the primary task of detecting the presence or absence of the signals. More button pressing meant more observing—so one would expect a corresponding increase in the efficiency of detection. But, as Broadbent noted,

> the purpose was frustrated by the fact that the rate of button pressing increased throughout the work period while simultaneously the time taken to detect each signal was also increasing. The subject would press a button, a signal would be displayed to him, and yet he would press the observing button several more times before he would report the presence of a signal. This process was happening more frequently towards the end of the work period and manifestly made the rate of button pressing useless as a direct measure of the efficiency of attention.[32]

At first glance, it looked as if the subjects simply adopted a mechanical routine of pressing the buttons in the sequence "left, center, right, left, center, right, etc." On closer inspection, the behavior of the subject did not fall into two, disconnected components of this kind—that is, a button-pressing task did not exist independently of the detection task. The timers and counters embedded in the new apparatus told a different story. These refinements were able to pick up errors due to the repeated pressing of the same button. Broadbent explained:

> When a signal occurs on the left-hand display, the sequence of button pressing may be 'centre, right, left, left, left,' followed by the report of a signal. Thus the sequence of observing responses does not continue automatically after the signal appears, but is interrupted and several more observations are made on the particular display which is showing the signal. In some sense therefore the signal must have produced an effect before it was reported, but the subject seemed to require additional information before he felt really justified in making a report.[33]

The repetition of the button-pressing response, and hence the repeated exposure to the signal, was now read as the subject's search for more information

about the signal—that is, for more evidence of its reality. The principal difference between Jerison's two tasks might therefore lie "in the level of confidence which they caused the observer to set himself before he could report
a signal."[34]

To an inductivist like Broadbent, these reflections exposed an area in
which more facts and more experiments were badly needed. If confidence is
the issue, what determines the level of confidence? If signals are reported, but
with low confidence, what happens to the number of false alarms? Jerison had
reported that there were numerous false claims to have detected a signal in
his three-clock runs but hardly any in the one-clock runs. An obvious experiment to perform would be to give one group of subjects a vigilance task and
instruct them only to report a signal when they were sure, but to ask another
group to report a signal even if they were doubtful. Alternatively, the experimenter could ask subjects to report all possible signals but then rate their
confidence after each report. Whichever method one uses, said Broadbent,
"one can plot the results in the form of a graph of correct detections against
false alarms."[35] It was, of course, precisely an experiment of the second kind,
responding to both clear signals and doubtful signals but recording the degree of confidence, that Broadbent reported to the audience of his Common
Principles Lecture. In that lecture, however, Broadbent made no mention of
the fact that he was trying to resolve a serious anomaly that confronted the
stance originally developed in *Perception and Communication*. That motive
only became clear in the quasi-historical narrative of *Decision and Stress*.

Signal Detection and Symmetry

What would the desired plot of the probability of detections against the probability of false alarms look like? From the armchair, said Broadbent, it was
difficult to predict the shape of the curve—but there were two broad possibilities, and each implied a distinct psychological theory. The two basic possibilities are (i) that the relation will take the form of a straight line indicating
that the number of false alarms is proportional to the number of detections,
or (ii) it will be more complicated than a straight line, thus indicating a more
complex causal link:

> If the graph takes the straight line form, then the assignment of confidence to
> any particular report about the presence of a signal is merely an additional pro-
> cess added on to the true perceptual mechanism, and has no intimate connec-
> tion with it. If on the other hand the graph is curved in form, then the confidence
> which a man feels in his report truly tells us something about its perception.[36]

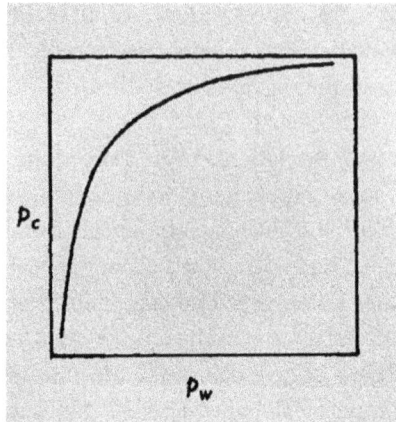

FIGURE 9.2. Detections plotted against false alarms. For a fixed signal-noise ratio, the probability of a correct detection (P_c) can be plotted against the probability of making a wrong claim (P_w) to have seen a signal. When this process is repeated for different values of the criterion, the resultant plot sweeps out a curve known as the Receiver Operator Characteristic, which takes the form of a downward concave curve. Broadbent says these empirical curves embody the central insight of signal detection theorists and show why large changes in the detection rate can coexist with small changes in the false-alarm rate. From Broadbent and Gregory 1963, 311.

Data from the experiment that Broadbent had described in his Common Principles Lecture turned out to support the curved option rather than the straight-line option. Its general form is given in figure 9.2. The implications of this figure run deep, and it is no wonder that Broadbent referred to "the explosive consequences of this very simple result."[37] The downward concave curve is fatal to the concept of a threshold. The concept of a threshold implies a qualitative difference between reporting a signal when it has actually been perceived and making such a declaration when the signal is below the threshold. At best, such a declaration could only be a random guess—and guesses are, or traditionally have seemed, qualitatively different from perceptions.

Broadbent's stance toward the theory of signal detection was complex. "I do not myself necessarily agree with all the mathematical details of their theory," said Broadbent in reference to Tanner and Swets, "but something of the sort must be true."[38] He was not saying that he had found an error in their calculations; rather, he was expressing his doubts about some of the assumptions behind their particular mathematical model. Unlike Tanner and Swets, Broadbent was not inclined to accept that the criterion placement in a decision process was determined by the long-term ratio of costs and benefits. To make such a rationalistic assumption at the outset of the analysis was, for Broadbent, bad inductive practice. It smacked of building a psychological "system." Broadbent was conscious of the wide range of contingencies that might have an impact on the complex process of signal detection: stimulus frequency, signal frequency, signal probability, signal strength, prior exposure to this or that signal rate, degree of arousal, and the impact of drugs on incentives and disincentives, as well as the long-term payoff matrix. He explored the insights that might be gleaned, opportunistically, from a range of different mathematical formalisms and axioms, provided only that they would yield the characteristic downward-concave curve relating the probability of successful detections to the probability of false alarms. This experimentally established curve, Broadbent argued, was the central, empirical finding of the work of signal detection theorists.

If he had chosen to adopt Bartlett's wording, Broadbent could have expressed himself by saying that there was no psychological sense in thinking that "signal detection" only took place when a signal was correctly identified, while "guessing" was something else entirely—a process of a different nature. Detections and false alarms could no longer be considered as different natural kinds. I have previously described the corresponding aspect of Bartlett's position as one of "methodological symmetry." For Broadbent, it was the methodological symmetry of the signal detection approach that was its real strength, but he admitted that this claim would strike many as "unorthodox, and in some ways curiously wrong-headed."[39]

Where did these developments leave the Jerison anomaly? Why did the three- clock task yield a better performance than the one-clock task? Insofar as the problem received a definitive resolution, it was that "the difference between Jerison's two tasks is to be explained by the difference in ß induced by the much higher signal rate in the multi-source task."[40] The process that actually induced the criterion shift was attributed, tentatively, to some form of arousal—the three-clock task indirectly generated a higher arousal, though, as we shall soon see, the appeal to the concept of arousal itself generated further anomalies and challenges.

Continuity and Change

In 1958, in *Perception and Communication*, the interpretive framework was provided by appeal to the mathematical theory of information. The basic model was of a communication channel.[41] In 1971, in *Decision and Stress*, the interpretive framework had shifted to statistical decision theory and the general-purpose computer. The picture of cognition was one in which imperfection and contingency were endemic and ineradicable: Every cognitive process was probabilistic. The only biological resource to mitigate this

predicament was for organisms to have evolved in ways that detect and exploit the probabilistic texture of the environment. Broadbent's aim in 1971 was
to explore this theme by generalizing the signal detection analysis introduced
by Tanner and Swets. Signal detection theory was one of the few developments that Broadbent was prepared to describe as "revolutionary." He usually
preferred less dramatic metaphors of scientific growth that stressed continuity rather than discontinuity.[42] Despite the suggested discontinuity of the
"revolution" metaphor, there were important continuities between information theory and signal detection theory—and hence between *Perception and
Communication* and *Decision and Stress*. Both theories embodied the insight
that the response to any given stimulus depended on the other stimuli that
might have occurred but, in the event, did not occur. This insight explains
why Broadbent said that, despite important theoretical shifts, his stance in
Decision and Stress could still be described as "psychology from the information processing point of view."[43]

One further element of continuity was that Broadbent retained the concept of a perceptual filter, but he now acknowledged that sometimes information from a blocked channel could influence subsequent behavior.[44] The
blocking was not absolute but statistical and context sensitive. In signal detection terms, filtering improved the signal-noise ratio and increased the value
of the parameter d'. Filtering still provided a basic strategy for selectively responding to events possessing a particular physical feature, "such as location
in a particular point in space."[45] A particularly interesting example of how
filtering retained its role in *Decision and Stress* is provided by Broadbent's
discussion of a series of experiments carried out by his research student Robert Hockey.[46] In these experiments, the subject was presented with a primary
tracking task and a secondary vigilance task of monitoring a set of signal
lamps. The aim was to study the effects of loud noise on the efficiency of the
two simultaneous performances. Broadbent reported that Hockey had shown
how the arousal due to the noise "increases the tendency to select information from sources which deliver more rather than fewer signals."[47]

Whether by accident or design, Hockey's apparatus took the form of a
skeletal cockpit (see fig. 9.3). Notice the long lever that the subject had to use
in the tracking task. Adjusting this lever was analogous to adjusting the joy-
stick lever controlling elevators and ailerons. Detecting the signal lights was
analogous to responding to the instrument on the display panel, while the
wide span of the signal lights corresponded to the fact that, although the important flight instruments are directly in front of the pilot, other instruments
that are no less vital are placed in less accessible positions on either side of the
main instruments. The deepest link with the Cambridge Cockpit, however,

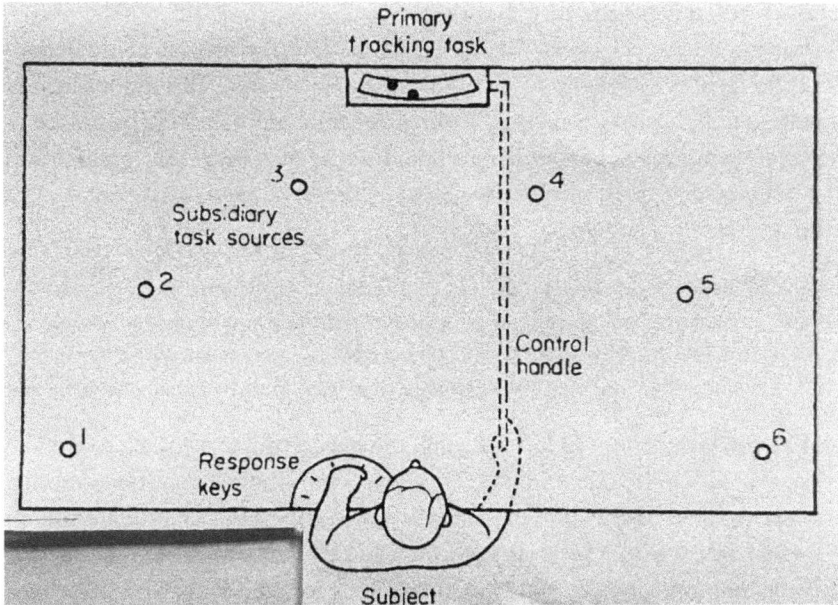

FIGURE 9.3. Hockey's apparatus: a skeletal cockpit? Hockey's apparatus, combining a central tracking task and a peripheral vigilance task, wittingly or unwittingly retained central features of the Cockpit experiments. His results reinforced the importance of what Bartlett had called the "splitting of the task," and they showed the continued relevance of perceptual filtering processes. From Hockey 1970a, 31.

was that Hockey was implicitly studying what Bartlett and Drew had called the "splitting of the task." This was their term for the important fact that fatigued pilots neglected peripheral instruments and peripheral tasks. Recall that Broadbent had isolated this result for consideration in *Perception and Communication*.[48] But what had been attributed to fatigue in the earlier discussions was now being explained as an expression of high arousal.

In *Decision and Stress*, Broadbent identified two more mechanisms that worked alongside filtering and helped relieve the load on the central processing capacities. These mechanisms were known, respectively, as "categorizing" and "pigeonholing." Categorizing was a strategy for reducing interference between stimuli by only selecting and processing stimuli that belonged to a certain class of events and that distinguished that class from other classes. To take a pertinent example, the pilots who took part in the Cockpit experiments were not only learning to fly, they were also learning to detect, and act on, the visual difference between friendly and hostile aircraft. They did this, of course, by learning to recognize the details of the various shapes that discriminated the members of the two classes. Categorization, said Broadbent, "is simply an extension of the idea that only information the person finds essential is processed."[49] But there was an important difference between filtering and categorization. The setting of the filtering process could be rapidly changed, for example by a verbal instruction, while altering a category system needs repeated practice "before a person can detect the presence of the member of a category with minimal interference from non-members of the category."[50] Once a category had been set up, explained Broadbent, a third strategy becomes possible:

> This is called *pigeon-holing*, and is like filtering in that it selects some events for further processing. Instead of selecting the events by a unique feature, however, this strategy operates by applying a bias to certain categories, so that they will be triggered off by less evidence than they would normally need.[51]

Unlike categorization but like filtering, the bias settings of the pigeonholing process could be changed rapidly. Thus, to give a simple illustration, filtering was exemplified by the human ability to pay attention to what the person on our right side was saying while ignoring what the person on our left was saying. Pigeonholing, by contrast, was exemplified by the ability, when hearing a medley of voices, to listen out for the mention of some target word, regardless of whoever spoke it. Filtering, categorization, and pigeonholing were the trio of postulated mechanisms now deployed to interpret the experimental evidence that had accumulated since 1958. Together they could begin to explain how the "systematic selection of information takes place at the entrance to the limited capacity system."[52]

Later, reflecting on the changes introduced into *Decision and Stress*, Broadbent emphasized the importance of not confusing stimuli with causes. "In attention just as in motor performance," he asserted, "many of the key processes predate the arrival of the stimulus."[53] Broadbent was underlining the significance of the internal state of an organism in preparing it to respond to the environment. At any given moment, these internal dispositions and causes were as important as the outer causes, the stimuli, in determining a response. This claim was Broadbent's way of saying that the complexity of a response was a function of the complexity of the responding organism, not the complexity of the stimulus. Bartlett's insights were thus built into the very fabric of *Decision and Stress*.

The changes introduced into *Decision and Stress* represented both a consolidation of the Cambridge perspective and a recognition of the need to reassess some of the processes assumed to be at work within the boxes of the 1958 flow diagram. Consider, for example, the relation between the length of a choice reaction time (RT) and the number, N, of the stimuli that might demand a reaction. Experimentally, it was agreed that the reaction time was approximately proportional to log N. Thus $(RT) = k \log N$, where k is a constant. If reaction time is plotted against log N, the resulting graph is a straight line of slope k. This finding was explained in different ways in the two books. In the older information-theoretic analysis, the increase in reaction time as N increased was explained by the differences in the length of code necessary to get perfect transmission of information. Broadbent called this explanation the "doctrine of optimum coding." In *Perception and Communication*, he accepted the doctrine of optimum coding as a simple fact, while in *Decision and Stress* he rejected it.[54] The new approach was based on the idea that the reaction time was the time taken to accumulate the evidence required to surpass a given criterion setting in the decision to react. To make the new approach viable, assumptions had to be made about the mechanics of this accumulation of evidence. Nevertheless, both approaches permitted the retrieval of the experimental result relating RT and N, though the derivation was based on different premises.[55]

If the two theories both yield an explanation of the choice reaction-time results, then why prefer one to the other? The reason was that problems had been mounting for the information-theory approach. Attempts to quantify the "channel capacity" of a human information-transmitting system faced deep problems, and these problems showed no sign of being resolved. For example, in the postwar period, reaction time itself was often used as an indicator of capacity. This procedure had been central to the pioneering paper "On the Rate of Gain of Information," published in 1952 by Cambridge psychologist W. E. Hick.[56] Hick was the first psychologist to offer a quantitative theory that related reaction time to the information transmitted by the response. But, as Broadbent argued, channel capacity is a measure of a maximum rate of work, while reaction time is merely a measure of the lag between input and output. Lag and work rate are quite different things, so treating the former as a measure of the latter is problematic.[57]

The problem can be illustrated by reference to the slope of the straight-line graph relating choice reaction times to the log of N. Under the right experimental circumstances, the slope can be reduced to zero—in other words, the reaction time is the same for different numbers of reactions from which to choose. This point was demonstrated in a number of ways in different laboratories. In Cambridge the point was established by Leonard, who performed reaction-time experiments with high stimulus-response compatibility—that is, the response was made very easy and natural.[58] Leonard found that, the higher the compatibility, the lower the slope of the straight-line graph. It became plausible that the slope could be reduced to zero. The problem for the user of information theory was that a slope of zero, that is, a graph with a horizontal line, would imply an infinite channel capacity, which is physically impossible.

In Broadbent's opinion, the problem lay in the tendency to see the mathematics as a substitute for an analysis of the actual psychological mechanisms in play. In reality mathematics was merely a language for talking about these mechanisms.[59] For the engineers who pioneered information theory, it was their tool for engaging with the effects of cable length, transmission loss, resistance, inductance, capacitance, and techniques for modulating pulses and waves to optimize the encoding of input. The lesson that Broadbent wanted his readers to draw was that psychologists should meet the problems of using information theory by becoming *more* like communication engineers, not less.[60]

The Problem of Stress

Fatigue is a form of stress, but in *Perception and Communication* the only stress that was discussed in detail was the stress of working in a noisy environment. In the years after the publication of *Perception and Communication*, the interest in noise became part of a program of Cambridge research encompassing many different forms of stress. These stresses included the effects of working in intense heat, the effects of breathing different atmospheres, the effect of irrelevant spoken speech as distinct from loud noise, the impact of various drugs (including alcohol), the effects of visual glare, the effects of high signal and event rates, and the effects of various potentially stressful incentives. One important line of stress research concerned experimentally enforced sleeplessness. The problem of lack of sleep had of course raised its head during the war, particularly in Bomber Command. Bartlett had warned the Flying Personnel Research Committee against assuming, too readily, that fatigue due to loss of sleep was the same as the fatigue of a skill, such as the skill of flying an aircraft.[61] Nevertheless, many psychologists, including Cambridge psychologists, found sleep loss a useful surrogate for fatigue in general.

The aim of the stress research was to find a general theory of stress. Broadbent had a vested interest in this search because his pioneering work on the impact of noise might have provided the basis for a general theory. The question was: could the assumed mechanics of noise stress be generalized to the other stresses? In 1958, Broadbent had assumed that a loud and continuous noise tended to be selected by the perceptual filter and hence took control of the response. Rather than responding to a task stimulus, the subject became increasingly inclined to respond to the interposed-noise stimulus. But while filter penetration and consequent distraction might have seemed plausible for the case of noise, they did not invite easy generalization. Sleeplessness, like noise, had a negative effect on performance (for example, in the Five-Light Test), but where was the strong, filter-piercing stimulation? The main symptom of sleeplessness was a general tendency to become unresponsive rather than a tendency to respond to non-task stimuli. A similar question arose in the case of the stress from high incentives. What were the intrusive stimuli, and why, in that case, was there a tendency to overrespond rather than under-respond? The difficulty of generalizing the original account of the impact of noise also raised another disconcerting possibility. Perhaps noise itself did not work in the way supposed and, all along, noise had been causing trouble for reasons other than its being an intrusive stimulus.

The drift in the data was toward the idea that noise influenced the over-all state of the responding organism—for example, its level of activation or arousal. Physiologists had already introduced the idea of an arousal mechanism through their discovery of the nonspecific reticular activating system.[62] Sensory stimulation, they had found, does not simply carry specific information to the cortex; it also has a nonspecific effect on the general reactivity of the organism. For some psychologists, part of the attraction of arousal theory was that it derived respectability from its roots in neurophysiology. Wouldn't that ensure that its concepts were truly scientific? Broadbent, however, warned against this attitude. By all means make an appeal to arousal, if the analysis squares with the facts of psychological experiment. We have already seen that Conrad's work indicated that arousal may be at work when increasing speed stress can have a beneficial effect on performance. But there were dangers in attempting to make premature physiological models of psychological mechanisms of this kind. The psychologist's arousal may not be the physiologist's arousal.

Jack Adams asserted that the concepts of arousal theory were more sharply defined than other current psychological concepts and less prone to obscurity and ambiguity.[63] Far from it, responded Broadbent: arousal was neither more nor less conceptually tricky to handle than other concepts. The idea of arousal was beset from the outset by some intrinsic ambiguities. Intuitively, sleepiness was a reduction in arousal, but sometimes physiological measures told the opposite story.[64] Again, should psychologists assume that a loud and prolonged noise was arousing because of the intensity of the distraction, or un-arousing because of its boring monotony? It was difficult to decide which stresses increase arousal and which decrease it. Nevertheless, the idea of arousal was a potentially useful starting point for the construction of a general approach to stress.

Arousal and Signal Detection

If arousal and signal detection were central themes in Cambridge research in the 1950s and 1960s, how were they related? Broadbent pointed out that both themes dealt with performances that had an optimum mode of operation. This broad analogy suggested that there might be a simple and direct relationship between them. Following the approach taken to the Jerison anomaly, we might suppose that the subject's level of arousal causally determined the criterion setting they would adopt in the performance of any task involving elements of signal detection. The higher the arousal, the lower (and riskier) was the value of the decision criterion β; conversely, the lower the arousal, then the higher (and more cautious) the β setting.

As long ago as 1908, experimenters had shown that if laboratory animals were given a task that required them learn to make a visual discrimination—perhaps to gain a reward or avoid a punishment—their learning improved if they were given a mild incentive, but would deteriorate if they were given too high a level of incentive. The empirical curve relating incentive (and hence arousal) to performance roughly took the form of an inverted U. The

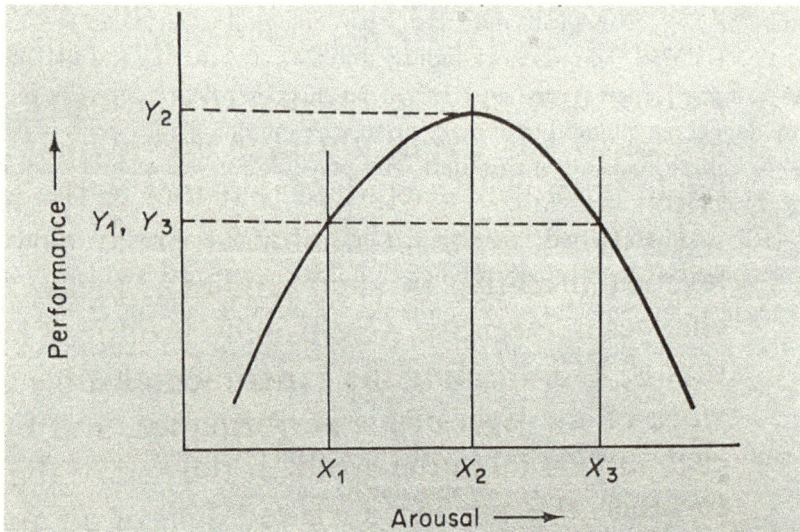

FIGURE 9.4. The Yerkes-Dodson law. In 1908, Yerkes and Dodson found that the relation between incentive and performance roughly took the form of an inverted U-shape. Both too-low and too-high levels of incentive led to non-optimum efficiency. Assuming that incentives generate arousal, the law can be re-expressed by saying that both high and low levels of arousal can be inefficient. But why, asked Broadbent, is high arousal inefficient? From Broadbent 1971, 407.

experimenters established that the optimum level of arousal for a difficult task was lower than for an easy task (see fig. 9.4). "This paradoxical relationship," said Poulton, "is known as the Yerkes-Dodson law."[65]

Broadbent did not follow Poulton in calling the Yerkes-Dodson law "paradoxical," but he certainly saw it as posing a problem. It is easy to see why very low levels of arousal yield inefficient levels of performance; but why, Broadbent asked, should high levels induce inefficiency? He accepted that if the level of arousal determined the various criterion settings that were involved in a complex task, and if the efficiency of the performance depended on the precise location of these criteria, then perhaps the problem of explaining the mechanics of overarousal would be solved:

> If criteria for all reactions are set rather cautiously, performance will be slow but accurate. If the criteria are set to allow response on relatively little evidence, performance will be fast but inaccurate. At the cautious end of the scale, the main effect may be the occurrence of a few very slow reactions, due to the decision to react having been preceded by an earlier decision that the evidence was inadequate. As the criteria move from a very cautious to a very risky position, they will pass through some optimal point where the elimination of slow reactions is balanced by the increase in error reactions. Any shift of the criteria in one direction or the other from this optimum point will produce a deterioration of performance, but the type of change will be different.[66]

This argument, said Broadbent, looked as if it might yield a simple and general explanation of why overarousal could cause deterioration. Attractive though the argument seemed, Broadbent ultimately rejected it—and the reasons for the rejection were Jacksonian.

The Return to Jacksonian Levels

The inverted U-shape of the Yerkes-Dodson law, relating performance to arousal, is based on the assumption that, at any given moment, a single dimension of arousal characterizes the state of the central nervous system as a whole. In *Decision and Stress*, Broadbent defended the basic idea of the Yerkes-Dodson law because he saw it as a useful approximation. But though it was a real overall effect, it was part of a more complex truth. Thus:

> The single dimension of arousal is not sufficient: nevertheless, it does embody a certain amount of truth since a mild level of stimulation may improve performance and a high level may cause deterioration.[67]

But, said Broadbent, "there must be some other factor coming in."[68] It was only in the final, speculative chapter of *Decision and Stress* that Broadbent revealed that this other factor was a Jacksonian factor. The reason why this further, Jacksonian step had to be taken was the difficulty, and perhaps impossibility, of accommodating the experimental data about the *interaction* of stresses within any model that only recognized a single dimension of arousal.

Broadbent discussed numerous examples of the interaction of stresses, but I shall simplify his argument by considering just two important stresses—namely, noise and sleeplessness. The facts concerning their interaction were studied experimentally by two of Broadbent's students and colleagues, D. W. J. Corcoran and R. T. Wilkinson.[69] The essential but problematic facts that they uncovered in their extensive and detailed experiments were as follows. First, when the effects of noise on the performance of tasks such as the Five-Light Test were studied, it emerged that noise had a negative effect, but that effect revealed itself only toward the end of the work period. Secondly, when the effects of sleeplessness were studied in the same way, that stress also had a negative effect on task performance that emerged late in the work period. Thirdly, when the combined effects of being sleep deprived and working in noise were studied, it was found that the two sources of stress counteracted one another. The increased arousal attributed to one of the stresses counteracted the decreased arousal attributed to the other. The two stresses had an opposite effect, but when applied singly, only revealed their effects late in the work period.

To appreciate the oddity of these results, consider first the case of sleep loss. Intuitively, the drowsiness caused by sleep loss might be identified as a case of low arousal. On this plausible basis, the fact that noise counteracts the effects of sleep loss points to the conclusion that noise increases arousal. But if loud and continuous noise increases arousal, noise stress, when applied singly, would be expected to have its primary and negative impact at the beginning rather than the end of the work session. As Broadbent argued:

> A condition which is overarousing ought surely to produce its greatest effect when first applied and give less impairment at a later time when habituation causes arousal to fall to the point when the inverted U is at a peak.[70]

How, then, could Corcoran and Wilkinson's experiments show that the negative effects occurred at the *end* of the work sessions? Broadbent explained these facts by supposing that there were two levels in the nervous system, levels he designated "Upper" and "Lower" (see fig. 9.5). He suggested that the Lower level was "concerned with the execution of well-established decision processes" and dealt with patterns of activity where there was an optimum. The Upper level, by contrast, was concerned with "monitoring and altering the parameters of the Lower in order to maintain constant performance."[71]

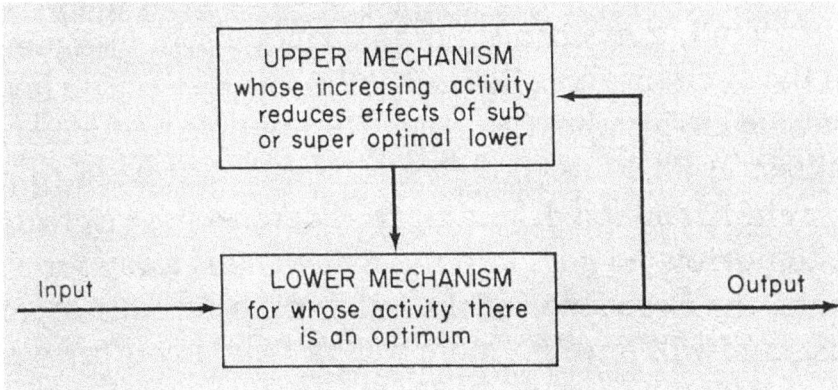

FIGURE 9.5. Broadbent's two-level theory. The puzzling interaction of noise stress and sleep-loss stress led Broadbent to argue that there was a need to postulate an updated form of the Jacksonian theory of levels in the nervous system. How, otherwise, could the finding that performance decrement occurred late in the work session be explained? From Broadbent 1971, 446.

The problematic experimental results on noise and sleeplessness were then analyzed as follows. In these terms, said Broadbent,

> it is the Lower which is affected by sleeplessness and noise. Those conditions shift the decision criteria between a very cautious and unreactive extreme and a very risky and hyper-active extreme . . . So long as the Upper is in an efficient state, however, the consequences of Lower inefficiency do not become evident. If we suppose that the Upper is efficient at the start of a work-period and becomes less so after prolonged monotonous work, then it would follow that noise and sleeplessness would show their effects only at the end of the work-period. The effects of those two conditions would however be opposite once they did appear.[72]

Is there is a logical problem here? Broadbent invoked two different loci of fatigue. First, there was the fatigue generated by sleeplessness, whose effects were assumed to operate on the Lower level. Secondly, some form of fatigue was clearly involved in the assumption that the Upper level starts off its work of compensation in an efficient manner and then becomes less efficient by the end of the work period. Hadn't Broadbent just offered an explanation of fatigue by saying that it was caused by fatigue? Aren't we being taken back to the circularities that prompted Muscio to expunge "fatigue" from the lexicon of science?

This objection is less cogent than it may seem. What is wrong with the idea that there is more than one kind of fatigue? Why not say that each Jacksonian level has its own characteristic kind of fatigue? It is a commonplace amongst physiologists to say that there is more than one kind of sleep, so why not explore the thought that there is more than one way to be fatigued? In his

1942 report no. 488 to the Flying Personnel Research Committee, Bartlett had differentiated muscular fatigue, mental fatigue, and skill fatigue.[73] Broadbent was merely following in Bartlett's footsteps in treating sleep-loss fatigue as analytically and causally different from skill fatigue. Skill fatigue is an affliction of the Upper level; sleeplessness, like the stress produced by noise, is an affliction of the Lower level. Broadbent's analysis was not vitiated by logical circularity; it was just an inductive inference using the resources of the Jacksonian tradition.

The appeal to an Upper and Lower level not only solved the sleep-noise interaction anomaly, it also allowed Broadbent to retrieve the classic Jacksonian analysis of alcohol intoxication. The fact that alcohol is a physiological depressant but sometimes seems to act as a stimulant was another problem confronting the user of a simple one-level analysis of arousal. "As a stress, alcohol is something of a paradox," said two puzzled researchers at the Applied Psychology Unit.[74] Now Broadbent was able to remind his colleagues that, in neo-Jacksonian terms, perhaps alcohol depressed the Upper level so that the push and pull of arousal at the Lower level could not be adequately regulated. Such a mechanism, noted Broadbent, would also explain why experiments had shown that alcohol can act in a way that, despite it being a depressant, increases the effect of certain incentives—incentives such as giving subjects knowledge of the results of their performance tests.[75]

Jackson's "inhibition" of lower levels by higher levels had now become the cybernetic control of the criterion setting of the Lower level by the Upper level. Jackson's "release" had become the uncontrolled state of over, or under, arousal of the Lower level. "Release" was the uncoupling of the feedback loop between the Upper and Lower levels or, in engineering language, when the Lower servo ran "open-loop." In his discussion of the two-level model, Broadbent quickly sketched a number of other areas of research where problematic results might be clarified by employing some form of the two-level model. He cited areas of biochemical research, such as the relation between the adrenergic and cholinergic systems, as well as research on sleep and sensory deprivation. He noted that in certain kinds of sensory deprivation experiments (those involving steady noise and translucent goggles), there had been reports of subjects exhibiting "a disruption of behaviour which takes the form of excited but uncontrolled activity rather than unarousal." Could this disruption, he asked, be described as "Upper inefficiency with Lower activity unimpaired?"[76] Although Broadbent did not remark on the fact, this description immediately calls to mind the strange reports of the cable-breaking violence sometimes visited on Craik's Cockpit by the pilot-subjects, and the disruptive refusal of some of them to submit to the experiments.

Up to this point, Broadbent's discussion, though highly empirical, was both post hoc and schematic. The neo-Jacksonian form of his conclusions was clear, but he had not yet offered any engineering details concerning the feedback loops indicated in his diagram, nor had he described the actual character of the servomechanisms assumed to be at work. Later, in another important publication, Broadbent supplied some of the required detail, and this material will be introduced in the next and final chapter. I shall end that chapter by taking a look at how a cybernetic Jacksonian pilot would land an aircraft.

Levels, Hierarchies, and the Locus of Control

I shall now describe Broadbent's final attempt to articulate and extend what he called "the Craik-Bartlett model"—the neo-Jacksonian ideas that had been used to interpret the Cambridge Cockpit experiments. I shall base my account on a lecture that Broadbent gave in London in January 1977 to the Experimental Psychology Society.[1] Figure 10.1 provides a glimpse of the Broadbent of those years. The title of the lecture was "Levels, Hierarchies, and the Locus of Control." His aim was to assess Bartlett's achievements, and the argument of the lecture was both subtle and compressed. Broadbent took the opportunity to present some novel experimental results that illustrate the continued relevance of the neo-Jacksonian perspective. These results must be examined in some detail. They show how the study of skill could be opened out into the study of cognition in general. I shall also use the present chapter to describe what Broadbent had to say about the relation between postwar Cambridge psychology and the "cognitive revolution" that had taken place in American psychology. This theme was not discussed in the lecture itself, but it runs parallel to and augments the argument of the lecture. My first task will be to use Broadbent's lecture to address a problem that, from the outset, may have disturbed some readers of the present book. It was a problem that Broadbent knew would be in the minds of his own audience at the Experimental Psychology Society.

The Problem of Ubiquity

Broadbent said that, as a Cambridge undergraduate, he had the good fortune to have been taught to approach psychology through the principles of the

FIGURE 10.1. Donald Broadbent (1926–1993). Broadbent volunteered for the Royal Air Force during WWII and trained as a pilot. After the war, he studied experimental psychology in Cambridge under Bartlett and became a researcher in the Applied Psychology Research Unit. In 1958, he became director of the unit, a position previously held by Craik, Bartlett, and Mackworth. © Dr. Nik Chmiel.

Craik-Bartlett model. The ideas in the model, he said, were the product of out-standing minds; they were special to Cambridge and were far ahead of their time. They had been "flung out as conjectures ahead of the general body of psychology."[2] Though taken for granted in Cambridge in the 1940s and 1950s, Broadbent indicated that the Craik-Bartlett model had now fallen into neglect. Given the context of Broadbent's remarks—a celebratory lecture—it is right to feel a degree of skepticism. Was he exaggerating out of respect for his teacher? There are two central questions to be addressed: First, was there really a the-ory based on the idea of levels in the nervous system that was special to Cam-bridge? Secondly, and assuming that there was such a theory, had that theory really fallen into neglect? Regarding the first question, it would be obvious to the psychologists in Broadbent's audience that assumptions about the hierar-chical structure of mind and behavior are commonplace in psychology.[3] If so, can Cambridge psychologists and the Craik-Bartlett model really be so special?

Historical considerations seem to reinforce skepticism about Broadbent's confident claims about Cambridge exceptionalism. In his *Inhibition: History and Meaning in the Sciences of Mind and Brain*, the historian Roger Smith gave a detailed account of the concept of inhibition and its relation to ideas about the nervous system.[4] The physiological study of inhibition, said Smith, was established by the 1860s, but it was based on a preexisting discourse of regulation and order. The idea that the "higher" must regulate the "lower" had a long history. "A hierarchy between mind and body, or a hierarchy within the body itself," said Smith, "was a standard feature in nineteenth-century psychological and moralistic writing."[5] John Hughlings Jackson was not alone in formulating a theory of levels, and Smith established this fact with vivid quotations from numerous sources. No one who reads Smith's book can be left in doubt that notions of level and hierarchy in the nervous system were, in some form or another, ubiquitous. If such ideas were ubiquitous, how could Cambridge thinking have been "ahead of the general body of psychology"?

One further aspect of Smith's argument deserves attention. He identified a second, less salient concept of inhibition. Here the assumption was not hierarchical but depended on the distribution of finite resources. Psychologists have often assumed that a person has limited amounts of available "mental energy."[6] The more "energy" to be directed along one channel, the less goes to alternative channels. Mental processes that are denied the requisite energy are said to be "inhibited." There were thus two different social models of inhibition to be traced by the historian of psychology. The finite-resource approach, Smith noted, drew on "the competition of the economic marketplace."[7] Both hierarchies and markets have provided the psychologist with social metaphors for the working of the mind. The interest of this part of Smith's analysis for my argument is that the concept of attention and the necessity of filtering can be seen as examples of this second economic model.

Broadbent's Answer

With regard to levels and hierarchies, however, Broadbent's answer to the problem of ubiquity was to argue that, in Cambridge, the neo-Jacksonian tradition had been given a precise specification. These specifications distinguished it from the mass of similar ideas to which, admittedly, it bore a family resemblance. Broadbent's position was that, once these differences are properly appreciated, the apparent ubiquity evaporates and what is special about the Cambridge version of the model becomes apparent. To bring the necessary clarity into this potentially cloudy issue, Broadbent gave a succinct formulation of the Craik-Bartlett model. There were five Cambridge assumptions:

(i)   Human processing of information takes place on many levels.
(ii) Some of these levels modify or control the operation of others.
(iii) Systems subject to control deal with parts of the activity of the person.
(iv) Upper levels operate on longer timescales than lower levels.
(v)   Lower levels operate even when the higher centers are otherwise occupied.

Broadbent then patiently worked through a series of recent examples where psychologists had made a general appeal to various postulated "levels" or "hierarchies." His aim was to show his well-informed but potentially skeptical audience that psychologists still had something to learn from Bartlett and Craik's particular formulation. He examined (i) accounts of levels of processing in memory, (ii) the identification of hierarchies in perceptual analysis, (iii) hierarchic phrase-structure in language, and (iv) the hierarchical structures involved in the programs that enabled computer problem-solving. His claim was that the current theories in these fields captured some of the insights of the Craik-Bartlett model, but none captured all of them. Those that came closest dealt with computer problem-solving. Even so, he reported that the most plausible of these candidates invoked either "stages" of information processing or "transfer of control" from one part of a program to another. Neither of these mechanisms, he asserted, exactly corresponds to those specified in the Craik-Bartlett model. In particular:

> They do not include the idea of simultaneous operation over different timescales; and above all they do not include the idea of one processor altering the operation of another.[8]

Broadbent pointed out that a model of "stages" depends on the assumption that information about an event proceeds through a number of different processors. The output from one processor provides the input to the next stage. It might be said that, as a consequence, the information becomes more abstract as it is treated in different ways at different stages. Thus, one could speak of "levels" of increasing abstraction. Broadbent also acknowledged that the idea of stages allows that information about one event may be processed while information about another event is kept in store. These ideas do indeed capture part of what Craik and Bartlett wanted to convey, but only part.

Broadbent made similar observations about the models of the brain that depend on the transfer of control. Here, "a single processor is supposed to carry out one operation, store the result, and then carry out a different set of operations in response to instructions from a different region of the program."[9] The Craik-Bartlett model, however, demanded more. One part of the program would have to *control* another part of the program, not merely hand information over to it or command it to stop or start. The long-term memory would have to include rules for rewriting another part of the long-term memory. Surveying the recent developments in artificial intelligence—that is, those that came closest to the Craik-Bartlett model—Broadbent concluded that

> the notion of hierarchy falls away, and we find rather a network of possible transitions between states, in which levels are very difficult to distinguish. It becomes hard to say that one part of the program controls another, merely that at any moment one part of the program is in control of the processor, and that this control will be transferred through the possible transitions allowed within a very complex network.[10]

We are left, he said, with the feeling "that one part of the traditional scheme has disappeared; the notion of control of one level by another."[11] By the "traditional" scheme he meant, of course, the five neo-Jacksonian principles that he had identified at the beginning of his lecture. Broadbent added:

> I am naturally not so naïve as to suppose that we need multiprocessor computers to simulate the control of one process by another. The key point is that theories based on transfer of control should include the possibility that one section of program can alter other sections of program.[12]

Such a simulation would not call into question the claim that there was something specific and perhaps unique to the Craik-Bartlett view of the nervous system. Having cleared away the confusions that might arise from the problem of ubiquity, Broadbent then made his argument for the contemporary relevance of neo-Jacksonian ideas and the Craik-Bartlett model.

The World as a Cockpit

Shakespeare said: "All the world's a stage, and all the men and women merely players."[13] An updated Cambridge version of this insight might read: "All the world's a cockpit and all the men and women merely pilots." Broadbent didn't cite Shakespeare but argued that the Cambridge vision of the human actor was of someone engaged with the problems of display and control. After the war, Cambridge psychologists did not need to work with a real cockpit to give practical expression to this standpoint, but the legacy of the Cockpit still defined the terms in which they thought. In his lecture, Broadbent captured the point nicely:

The contemporary equivalent of the Spitfire pilot does not control course, height and airspeed; he rather controls buildings, equipment, manpower and consumable supplies. Like the pilot, he has to perform all these tasks simultaneously. It is not satisfactory, for example, to complete a hospital building when you have no staff to fill it, or to have numbers of trained teachers when there are no jobs for them. Actions concerned only with one part of the system need therefore to be coordinated with actions elsewhere. Again like the pilot, the administrator takes some actions whose consequences take a long time to emerge, so that any decision taken at one instant needs to be properly related to events in the distant past and future.[14]

While Craik took his metaphor for the brain from the organization of Fighter Command, Broadbent shifted the metaphor to that of peacetime governance—though, one should note, to a particular, recognizable, ideological conception of proper government and efficient civic administration. To those on the political right, Broadbent's words, with their assumption of a rationally planned society, will no doubt seem repugnant and dated. His words are indeed redolent of the successful postwar settlement in Britain when, under an impressive Labour administration, the economic doctrines that held sway were those of Keynes rather than Hayek.[15]

The comparison between the skills of the Spitfire pilot and the skills of the administrator was meant seriously and provided the basis for an inductive generalization. It implied, said Broadbent, that the tasks of social organization and coordination might well be expected to show behavior similar to that observed by Bartlett and Drew in the Cambridge Cockpit. Broadbent readily admitted that there were differences between the two cases. The pilot's skill is expressed through movements of hands and feet; that of the planner through words and symbols; but, he argued, it would be wrong to treat the two as qualitatively different. It is mere snobbery to treat intellectual processes expressed through the manipulation of symbols as irreducible:

> The use of words and other symbols in decision making has encouraged the idea that such decisions are taken merely by familiar logical processes, or, if they are not, then they should be.[16]

"Familiar logical processes" are the sort of inferences that are codified by logicians, or the inferences that are involved in following the rules of arithmetical or algebraic calculation. Broadbent was saying that these intellectual processes depend on a deeper level of informal and intuitive response. The intellectualist picture does not take us to the heart of human cognition. He then described to his audience a small experiment to illustrate this point. As befitted the modern, postwar, and post-Spitfire world, the experiment concerned town planning. The subjects were asked to imagine that they had the task of managing a transport system in a city. Obviously, the problem had to be given a simplified form, though Broadbent insisted that his experiment was still more complicated than most current laboratory investigations.

From Cockpits to Town Planning

Subjects were told that they could manipulate two features of a city's transport system:

(i)   the interval of time (t) between the buses entering the city, and
(ii) the fee (f) for using municipal car-parking facilities.

The purpose of manipulating these two parameters was to control the load (L) on the buses and the number of empty spaces (S) in the car parks. The load was specified in terms of the number of passengers per one hundred buses entering the city. Broadbent assumed that the urban reality to be administered was governed by two simultaneous equations linking the four quantities t, f, L, and S. Thus:

$$L = 220t + 80f, \text{ and}$$
$$S = 4.5f - 2t.$$

The subjects in the experiment were not informed of these relations, nor did they operate with symbols and equations. They had to explore the dynamics of the system by trial and error until they got a feel for its structure and could control it in a practical but intuitive manner. The experiment began by informing the subjects of the current, stable state of affairs. They were told the current values of the load (L) and the interval (t) between buses as well as the current parking fee (f) and the number of empty spaces (S) in the car parks. The subjects were then informed that it was necessary to alter the bus load and the number of empty car-park spaces, and they were given the new target values of L and S that were to be achieved. Their task was to alter times and fees to hit these targets. Subjects would put forward a proposal for a new time interval and parking fee, and the experimenter (using the equations) would immediately tell the subject the consequences of their proposal. In the light of this feedback, the subject would then make another proposal, and so on, until the targets had been achieved. At this point, the subject was set new targets and began another experimental run. Both before and after the experimental runs, the subjects were asked what effect they thought an alteration in the time interval between buses would have on the load and the car-parking spaces, and also what effect an increase in car-parking fees would have.

The first, overall result of the experiment was that subjects were successful in learning how to administer the system. When set new targets for bus load and vacant parking spaces, they generally achieved the desired state of affairs after about five or six adjustments of the time interval between the buses and the size of the parking fee. The second overall result was to expose a disparity between the subjects' relatively successful behavior in manipulating the system and the reliability of the answers they gave to the beginning- and end-of-session questions. They often gave wrong answers and, sometimes, answers that indicated the opposite of what they actually did. On one level, the subjects knew what they were doing; on another level, they did not know. The lesson that Broadbent drew was that an investigator who really wanted to understand the skill should address the nonverbal level rather than waste time listening to justifications and post-hoc rationalizations.[17]

In order to probe into the operative mechanisms that were untouched by the surface rationalizations, Broadbent then examined the two components of the administrative task separately. He simplified the experiment by concentrating on the subjects' responses to one input at a time. First, subjects were tested on getting the bus load right and were allowed to neglect the effect of their decisions on the parking problem. They would be given a new target load (L) and asked to adjust the time interval (t) between buses, regardless of the effect on parking spaces; they would be informed of the effect but were not required to do anything about it. They would then be tested on the parking-space problem alone and allowed to ignore the impact of their proposed parking-fee changes on bus load.

Broadbent's strategy was to test the subjects in his transport experiment in the same way that a control engineer tests a servo system. Consider the bus load problem. Each of the subjects, Broadbent explained, "can be regarded as responding to a sharp step in one of the target values." It is worth considering, he said, "the various hypothetical mechanisms which might be postulated to explain response to step inputs of this kind."[18] He then moved on to examine the subjects' response to the parking-fee problem, and here he confronted his subjects with what engineers call a "ramp input," that is, a steadily increasing input. The use of step inputs and ramp inputs (along with sine-wave inputs) is stock in trade for engineers because they reveal the response characteristics of servomechanisms. Knowing the response to step, ramp, and sine-wave inputs provides a good guide to how the servos will behave in more complex practical situations.

A wide range of different classes of servomechanism have been subject to rigorous mathematical and empirical analysis by control engineers. Their characteristics are well known, and it is these engineering achievements that

represent the taken-for-granted background of Broadbent's conclusions. It
must be admitted, however, that in calling upon this background Broadbent
was demanding rather a lot of his audience as he moved rapidly through the
analysis of his experimental results.[19] I shall try to slow this process down and
point the reader to the assumptions that lie behind Broadbent's descriptions
and conclusions. I shall start with the step-function input results and then
look at the ramp-function case.[20]

The Step-Input Results

Broadbent's findings in the bus load experiment can be represented by figure 10.2. The figure shows one of the subject's responses to the task of finding the
bus interval that was needed to shift the bus load from a starting point of 3,800
to a desired target of 5,800. The graph, said Broadbent, was fully representative.

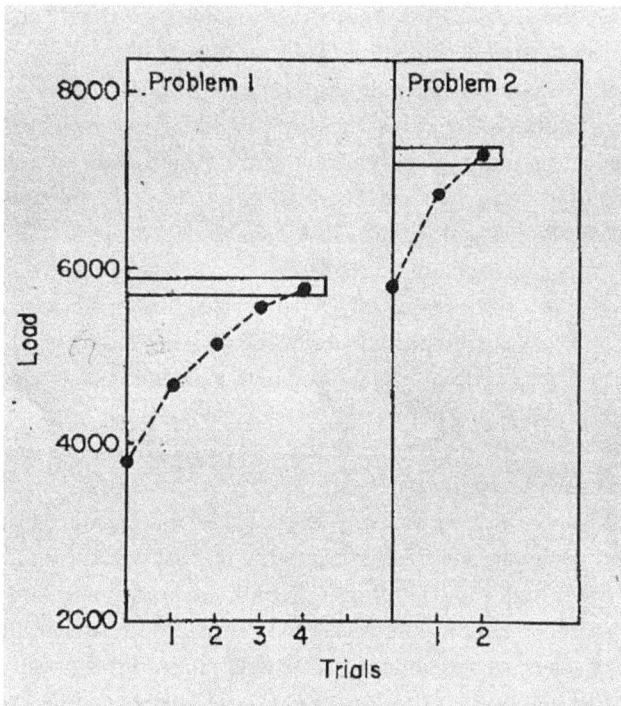

FIGURE 10.2. Response of a subject to a step input. If a subject is confronted by a sudden request to
achieve a specified goal, then this demand can be described as a step-input function. By studying how
subjects achieve that goal, it is possible to see revealing analogies and dis-analogies with certain classes of
servomechanism. Notice in the present case that the subject's approach to the goal shows no overshoot or
oscillation. From Broadbent 1977a, 195.

It is clear from the graph that the subject was not simply responding to the size of the target value—for example, producing a large response to a large target and a small response to a small target. In the language of the servo-engineer, the responses were not that of a deterministic "open-chain" mechanism—that is, a mechanism devoid of feedback loops. Open-chain responses are the kind that psychologists might attribute to stimulus-response conditioning. Had the response been open-chain in character, then it would have been correct on the first attempt, or as is more likely, it would be wrong and would stay wrong. This, then, was a mechanism that Broadbent immediately ruled out. Some other mechanism must be at work. The most obvious candidate for consideration was a mechanism in which the response is proportional not to the size of the target but to the size of the discrepancy between the target situation and the current situation. The response would be proportional to the error (ε) in the present output and thus equal to $K\varepsilon$. Broadbent referred to these devices as "simple feedback" control systems, but they are often called "proportional controllers," where K, the "gain," is the constant of proportionality.

A proportional controller would certainly be able to home in on the required bus load. If the first proposed bus interval overshot the target load, then in the next proposal the interval would be reduced; if the next attempt undershot the target, then the time interval would be increased again; and so on. All the corrections would be proportional to the error, and so the responses would get closer and closer to the target. But this consideration is not enough to justify the conclusion that the experimental subject is functioning as a proportional controller. The question is whether the target is achieved in the right way—that is, the way that the subjects actually arrived at the target as shown by the results graph. Broadbent argued that a proportional controller would not behave as the subject behaved. His reason was that his subjects actually achieved their target, whereas a "simple feedback system always ends up with a constant small error between its input and its output."[21]

Broadbent was referring to a standard result in control theory that deals with a phenomenon known as "offset" or "steady-state error." The unavoidable presence of steady-state error in the response of any proportional controller that is subject to a step-function input can be demonstrated by going back to the mathematical description of the system. The governing second-order differential equation of the servomechanism can be solved, either directly or by the use of a Laplace transformation applied to the overall ("closed-loop") transfer function of the system. The transfer function specifies the ratio between output and input. There is a theorem, known as Laplace's Final-Value Theorem, that then yields the magnitude of the steady-state error—the small error to which Broadbent was alluding.[22]

That Broadbent was indeed referring to the steady-state error is supported by the next paragraph in his paper. He imagined an objection. An objector might say that the subjects *were* behaving as proportional controllers and *were* displaying the inevitable steady-state error, but the experimental graphs were too crude to reveal its presence. Broadbent had a counterargument. Once again, it was the mathematical relationships that characterize the workings of proportional controllers that provided Broadbent with the basis of his answer. He did not explicitly state the principle, but the crucial point on which he relied was that the size of the steady-state error is inversely proportional to the gain of the servo—that is, the value of K. A very small steady-state error—the sort that might escape detection in the experiment—demands a very high gain.[23] But if that were true, the performance graph would, initially, take the form of a massive overshoot and then an oscillation of gradually diminishing undershoots and overshoots. This is not what the graph shows. As Broadbent put it: "The gradually decreasing value of error shown . . . is inconsistent even with a simple feedback system having an enormous gain."[24]

Integral Controllers

Because simple servos cannot explain the subjects' behavior, the next step in Broadbent's argument was to consider whether a more complicated feedback mechanism would correctly model the subject's behavior. Suppose the output of the system is proportional not to the error itself but to the integral of that error. Suppose their responses were proportional not to $K\varepsilon$ but to $K\int \varepsilon dt$. Servomechanisms that behave in this way are known as integral controllers (see fig. 10.3).

Integral controllers, like the experimental subjects, do not display a steady-state error when confronted with a step function. They are capable of

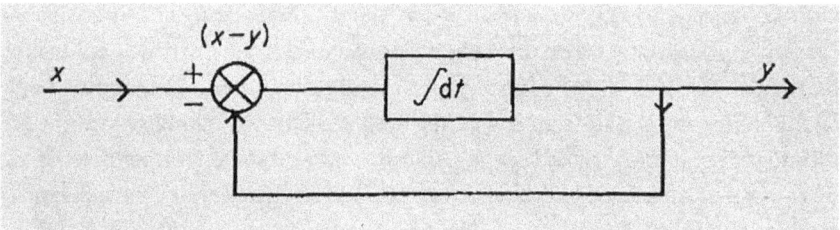

FIGURE 10.3. An integral controller. The output of an integral controller is proportional not to the error but to the integral of the error. When presented with a step input, the subject's behavior provides a good analogy with the behavior of an integral controller given a step input. But there are other tests to be passed. From Broadbent 1977a, 196.

approaching and achieving the goal set them in the same way as that shown in the graph of the experimental results in figure 10.2. While a proportional controller can be ruled out, an integral controller remains a possible candidate for a model of brain function as revealed so far by the experimental data. Nevertheless, Broadbent warned against jumping to conclusions. There are other tests to be made and other criteria to be met. It was at this point that Broadbent introduced a test having the form of a ramp input.

The Ramp-Input Results

The ramp input was created by setting some of the subjects a sequence of demands to provide ever increasing numbers of car-park spaces. First one target was set, and the subject was asked to fix a fee. As soon as a response was made and a fee proposed, the subject was told the number of spaces actually created and immediately given a fresh target to achieve. The targets rose steadily by the same amount after each decision, but to avoid making the task too easy, the numerical values given to the subject were chosen so that they disguised the simple character of the progression. This time Broadbent found a corresponding and steady rise in the values achieved by each subject (see fig. 10.4).

Going back to first principles, we know that an integral controller can perform this feat—that is, it can keep the output rising steadily in alignment with the input. Was the integral controller therefore the model of the nervous system that Broadbent was seeking? Once again he argued that it was not. Despite its previous success in modeling the human response to step functions, the integral controller did not in fact provide an accurate model of the human response to the ramp input. This sort of servo can keep the output rising steadily in pace with the input, but there was still a crucial difference between the response of the servo and the response of the subjects. In following a steadily increasing demand, Broadbent explained, an integral controller behaves in a very characteristic way: it "has to lag behind the demand by a constant amount."[25] Broadbent observed that "not one person lagged behind the target by a constant amount. Normally they overshot it and then oscillated about it within close bounds." He concluded that this result was therefore "inconsistent with the idea that people are acting as a simple integral controller."[26] Nor, he added, is the situation improved simply by adding more stages of integration.

Two plausible hypotheses have now been ruled out as adequate accounts of the behavior of the subjects in the experiment: humans operate neither as proportional nor as integral controllers. At this point, Broadbent paused to take stock. Should the conclusion be that, when humans perform a task, the simplest control-theory analogy varies from task to task? If that were the case, then the scientific value of such analogies for the psychologist would surely be called into question. The desirable aim of scientific generality would be compromised. And there were two further difficulties that had been passed over in the reasoning so far. Both of these concerned practice effects. After the first step-input test, a second test was given which now involved a jump from a load of 5,800 to finding the interval between buses for a target load of 7,300. Broadbent found that the response time for the second step task was considerably less than for the first. Five of the six subjects showed this effect, and the result was statistically significant. There was a practice effect that is not explicable on either of the servo models considered so far.

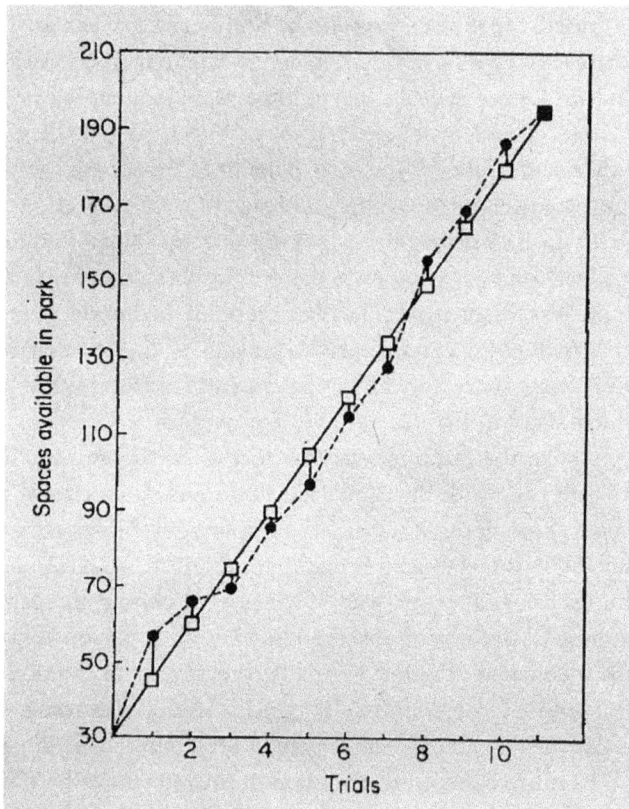

FIGURE 10.4. Response of a subject to a ramp input. A sequence of regularly increasing goals counts as a ramp-input function. Integral controllers show a lag when confronted with a ramp input. Notice that the subject's response to the steady increase in the targets they are to meet does not involve a constant lag behind the newly set goals. The subjects are not integral controllers. From Broadbent 1977a, 197.

The second difficulty arose from the ramp problem. After generating the results of the type shown in the ramp-function graph, each subject was given a further but lower target. Broadbent reported that "five out of six cases reached the correct target value on the first decision, and the one deviant person took two decisions."[27] This finding suggested that the level of practice acquired by the subject in the course of the ramp task had rendered feedback more or less irrelevant. The subject was no longer dependent on feedback and could achieve success simply by a firing out a reflex response. The closed-loop system had now become an open-chain system—the very thing that had been rejected at the outset of the investigation.

How could this change be explained? And how are these practice effects to be accommodated? The answer, said Broadbent, was to try an entirely new kind of cybernetic control system—one that would allow psychologists to draw on, and further articulate, the Cambridge heritage. The simplest servo that could embody the Craik-Bartlett model, argued Broadbent, was a two-level adaptive controller of the kind shown in figure 10.5.

A Jacksonian Servomechanism

The servo device shown in figure 10.5 is "adaptive" because, like the human subjects in Broadbent's experiment, it can change its own transfer function

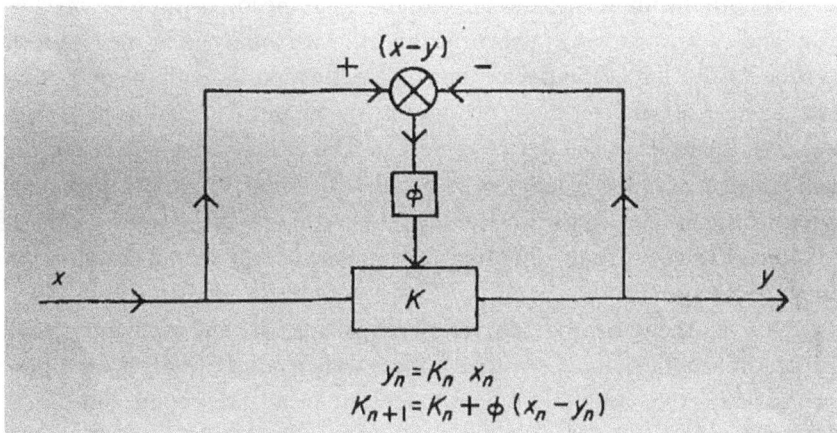

FIGURE 10.5. A two-level adaptive controller. The best match to the subject's behavior in Broadbent's experiment was provided by a two-level adaptive controller. Such a controller might be called a Jacksonian controller because it possesses two levels and meets the specifications that Broadbent set out when he defined what was special about the Craik-Bartlett model. From Broadbent 1977a, 198.

to cope with different problems. The adaptive controller can operate in both open-chain and closed-loop modes. When people start to acquire a skill, such as flying an aircraft, they are initially slow and deliberate in their responses. They depend on attending to the feedback from their responses. As they gain skill, they speed up and increasingly find that they can make some of the required responses automatically while anticipating what is to come next. They are now responding in open-chain mode.

To understand the mechanics of this process, consider the depiction given in figure 10.5. The lower level is identified by the input arrow labeled X and the output arrow labeled Y, where Y is some function of X. We can imagine a sequence of inputs X_n and a sequence of corresponding outputs Y_n where $n = 1, 2, 3$, etc. The input and output are connected by the lower-level mechanism inside the box marked K. This lower-level mechanism works on the open-chain principle and given a stimulus X_n causes the output response Y_n, according to the law $Y_n = K(X_n)$. The transfer function is that of a simple amplifier and merely determines the "gain" of the device—that is, the proportionality between the magnitude of the stimulus and the magnitude of the response. Suppose now the gain is not well-adjusted to the conditions of a test and the response is too large or too small. This circumstance calls for a change in the gain K. Here, the box marked by the symbol ϕ in figure 10.5 enters the story. The box represents a closed-loop device that operates on the gain of the open-chain mechanism at the lower level.

The upper level works as follows. It can be seen that vertical, arrowed lines feed off both the input (X_n) and the output (Y_n) of the lower response mechanism and come together in the comparator device indicated by a circle with a cross. It is the job of this device to compare the input X_n with the output Y_n. The divergence between these two things, the quantity (X_n minus Y_n), then serves as the input to the device represented by a box marked ϕ. When the gain is right, and the sequence of quantities (X_n minus Y_n) eventually equals zero, the upper level stops functioning, but the lower level does not stop. It continues to give the right, or optimum, response, assuming that the situation does not change.

In summarizing the difference between an adaptive and an integral controller, Broadbent pointed out that if there were a constant target for a performance, then the output of the lower level of the adaptive controller would change after every decision by an amount proportional to the error. But if the target were steadily increasing, the change in output is not simply proportional to the error but has a further component due to the change in the input: "Unlike the simple integral controller an adaptive system of this kind changes its transfer function when it is given a different kind of input."[28]

The central point that Broadbent wanted to make in his lecture was that the overall features of the adaptive controller fit the specifications of the Craik-Bartlett model. Rather than one process simply furnishing the input to the next, it actually modifies it, so the process is not just a series of stages:

> Once this kind of control mechanism has been exposed to a situation repeatedly, however, it will react more rapidly and efficiently. Indeed, because its improved performance is due to correct adjustment at the lower level, once learning has taken place the upper level is no longer necessary, and the abolition of feedback would not matter. It would also be possible for the upper level to devote itself to other purposes. The upper level is needed only in the stage of coping or adjustment.[29]

This feature of the adaptive controller thus sheds light on the previously noted improvements due to practice shown by Broadbent's subjects. It is also clear that the upper-level mechanism, the ϕ box, operates on a longer timescale than the lower-level K box. As Broadbent put it:

> the lower level reacts merely to the momentary input, whereas the value of gain set by the upper level is a function of the total error throughout the experiment.[30]

The upper level, Broadbent explained, is "concerned with modifiability and response to novel situations, rather than transmission of information according to one unchanging code."[31] The lower level of an adaptive controller, however, can function in the absence of the upper level, "so that the observations of Hughlings Jackson on epilepsy or alcohol or of Bartlett on fatigue become reasonable."[32]

A Compromised Revolution

The next step in my argument is to see what an appeal to two-level adaptive controllers might mean for understanding the Landing-Accident Anomaly. Before taking that step, however, I want to place these cybernetic developments in the context of the so-called cognitive revolution in psychology. As this term is usually used in the United States, it refers to the shift away from behaviorist theories such as that developed by Clark Hull. In their place, psychologists became more willing to speculate about internal cognitive processes that mediated between input stimuli and output responses. In the last chapter, we saw how Hull's approach to reactive inhibition and conditioned inhibition fell into the sort of confusion or "crisis" that can provoke a scientific revolution—a developmental pattern made familiar through the work of

Thomas Kuhn.[33] Things proceeded somewhat differently in Britain. The wartime work of Craik, as well as the prewar work of Bartlett on memory, can be seen as setting the scene for the corresponding British developments, or what might be called the British cognitive revolution. The revolution was a gentler affair on the British side of the Atlantic because psychologists had been less committed to militant forms of behaviorism. In Cambridge, behaviorism counted as precisely the kind of "system" that Bartlett always wanted to avoid.

In 1986 Broadbent was invited to write the foreword to a reissue of an influential American book, *Plans and the Structure of Behavior*, published in 1960 by George Miller, Eugene Galanter, and Karl Pribram.[34] Along with Ulrich Neisser's *Cognitive Psychology*, published in 1967, *Plans* was, justifiably, seen by many American psychologists as a founding document of their cognitive revolution.[35] One of the most widely quoted arguments from *Plans* concerned the cybernetic analysis of a simple motor skill—hammering a nail—in which the action is broken down into a sequence of operational components that fall under the headings Test, Operate, Test, Exit. This schema, abbreviated to TOTE, was described variously as a "plan" or a "computer program." The central insight was that "the operational components of TOTE units may themselves be TOTE units."[36] TOTE units were declared to be the new cybernetic unit of psychological analysis that would replace the old stimulus-response, S-R, unit.

These new TOTE units could be cast into the form of a nested hierarchy of operations. Each level in the hierarchy can be specified in terms of goals and sub-goals. The question then arises: Are these levels *Jacksonian* levels? To address this question, we may follow Miller, Galanter, and Pribram. On their analysis, "hammering" consists of "lifting" and "striking," and the program or plan was said to run off in a sequence of the following kind: "Test nail. (Head sticks up.) Test hammer. (Hammer is down.) Lift hammer. Test hammer. (Hammer is up.) Strike nail. Test hammer. (Hammer is down.) Test nail. (Head sticks up.) Test hammer" and so on.[37] This sequence is followed until the goal is reached. The head of the nail no longer stands proud of the surface, and Exit is triggered. The difference in approach with Cambridge psychology is evident. While TOTE units might indeed be described as standing in some form of "hierarchical" relation, they do not form a pattern of levels that interact in the way imagined in the Craik-Bartlett model. Rather, they represented a case of what Broadbent identified as "transfer of control."

When writing the invited foreword to the reissue of *Plans*, Broadbent took the opportunity to set out his own estimate of the American cognitive revolution. His tone was generous but could not disguise the fundamental criticisms that lay behind his words. To begin with, he warned against the danger

of introducing a false polarization into the recent history of psychology—
for example, "good'" cognitivists versus "bad" behaviorists. The insights of
men such as Hull, he said, should not be underestimated.[38] Furthermore, in
Broadbent's opinion, in America the revolution had lost its way and there was
a need to get back to fundamentals. Cognitive psychology had become too
formalistic and too focused on linguistic theory:

> After "Plans," psychology became cognitive not only in the sense of allowing
> explanations from processes inside the person, but also in confining itself to
> verbal and intellectual processes rather than the manual skilled ones.[39]

How could this accusation be true given the prominent role accorded to the
TOTE analysis of the manual skill of hammering? Broadbent noted that while
the TOTE schema was often mentioned by the admirers of *Plans*, it was rarely
the basis of subsequent work. It was a symbol of the cognitive revolution but
not a paradigm of experimental practice. Broadbent also praised chapter 6
of *Plans*, which was devoted to motor skills, and where landing an aircraft
provided the central example, but he made the same point: its potential was
not exploited and its promise not fulfilled. Instead of motor skills, it was the
study of language (and study from a very particular point of view) that had
become "almost the dominant field of human experimental psychology."[40] In
Broadbent's opinion, the revolution had been hijacked by Chomsky's ratio-
nalism and formalism.[41]

There were compelling grounds, said Broadbent, for moving on. Both
the work on parsing grammatical structures, and the work on the computer
simulation of purposive action, such as hammering, had run into the same
problem. In the codification of these procedures, it always proved necessary to incorporate elements of empirical knowledge into the formal rule-
governed sequences. In the case of hammering the nail, Broadbent argued
that the TOTE units relevant for the analysis of real-life action would never
typically fit together in a simple nested hierarchy.[42] In everyday terms, they
would never do so because any practical program would need to cope with
contingencies of the kind that always intervene and disrupt any abstract plan
of action. Broadbent deliberately used a very simple and commonplace coun-
terexample. Suppose the goal is hammering, but, let us say, the wood splits,
then the program must contain an extra, special-purpose rule of action—for
example, abandon the hammering and go back to the shop to buy more wood.

Broadbent was not adopting a generalized anti-reductionist stance; on the
contrary, he was in favor of reductionism and thought it much maligned.[43]
Nor was he saying that commonplace interruptions, like the wood splitting,
show that human behavior transcends anything achievable by a computer

program. He was making a point about the kind of program, or reductive analysis, that is called for. For the experimental psychologist, considerations involving contingencies and interruptions meant that realistic models of the nervous system must, at the very least, involve *both* open-chain *and* closed-loop processes. These processes had to be coordinated with one another in the machinery of thought and action. Human actors always proceed in a way that calls for analysis in terms of both the new units of action and the old units: the closed loop and the open chain. Consider a real human confronted with a real problem. Broadbent's point was:

> When we are faced with a fresh problem we adopt a heuristic without any feedback, without being certain that an action is actually taking us in the direction we want.[44]

Even granted the reality of all the goals, sub-goals, and sub-sub-goals of the TOTE analysis, the only way to avoid an infinite regress is to accept that the choice of the overall goal is an open-chain process. The triggering of the choice mechanism can be grounded in nothing more than a statistical decision process. Ultimately, said Broadbent, some things have to be done because they feel right, not because there is any plan that can be reported. This line of argument was not a rejection of hierarchical models. The real question was: What sorts of hierarchy are needed, and what nonhierarchical features are also present and at work?

As a pointer to where the answer lay, Broadbent made the same move that Bartlett had made in *Thinking*. He invoked the postwar Cambridge work on skill and tracking. Poulton had already shown that skilled motor action involved *both* closed-loop *and* open-chain components that were closely woven together. Put simply: A target had to be selected and, as it were, grabbed out of the air in an open-chain manner, before it was possible to track it, or close in on it, in some, appropriately engineered, closed-loop manner. In Britain, said Broadbent,

> Craik had gone on from *The Nature of Explanation* to launch a programme of research continued by Gibbs, Leonard and Poulton. As they showed, the achievement of a skilled movement involved a smooth interplay of closed-loop mechanisms with ballistic open-chain ones, with long-range anticipatory actions intelligible only in the light of the overall plan.[45]

Broadbent could have added the names of Craik's collaborators, Margaret Vince and Sheila Macpherson, to this list, but the point was made. And it was precisely this smooth interplay between closed-loop and open-chain processes that Broadbent had sought to capture in the cybernetic mechanism of the two-level adaptive controller.

Experimental Paradigms

One result of the formalistic bias in the American work, argued Broadbent, was that, following the publication of *Plans*, theory and experiment had drifted apart. There had been valuable changes in some of the basic assumptions about the working of the mind, but there had not been corresponding changes in experimental practice. It was true that the human mind was now widely assumed to be active, and some of Bartlett's early conclusions about the creative role of memory schema were now acknowledged in the American literature. Nevertheless, in much of the American work, the experimental subject was still treated as passive. Experimenters, said Broadbent, still stayed close to perception and allowed little initiative in the choice of responses. This limitation, he argued, applied both to *Plans* and to the results reported in Neisser's *Cognitive Psychology*. The experimental techniques and the conclusions drawn were often remarkably similar to those used in classic reaction-time experiments. The stimuli might now take the form of, say, two sentences, rather than two signal lights, but the subjects still sat waiting for a stimulus to which they had to respond by pressing a button. The new experiments were only informative about certain kinds of internal events, and from the outset they imposed serious restrictions on the mode of response. "Consequently," wrote Broadbent,

> while "Plans" is full of people moving from sub-goal to sub-goal until their purpose is achieved, the cognitive laboratory of the 1960s was equally full of people sitting passively until an experimenter fired a stimulus at them.[46]

Practical considerations sustained variants of what Broadbent called "Pavlov's original paradigm" of laboratory procedure.[47] This paradigm was grounded in the assumption that experiments would become too complicated if subjects were given the freedom to choose their own responses—recall the harnesses used by Pavlov to constrain his experimental subjects. Without these built-in simplifications, it would be impossible to keep track of what was happening—or so the argument went. This limitation, Broadbent argued, had now been removed by the availability of the electronic computer. Thus:

> There is no longer an insuperable problem of letting the person rather than the experimenter take the initiative, and yet being able to specify the consequences that will follow from any of a large number of chosen actions.[48]

Perhaps the major problem in the study of cognitive processes, the flexibility of human responses, could now be confronted. People adopt different strategies in different circumstances. The search for "the" way people parse sentences is vain, and so, insisted Broadbent, is the search for "the" way they do anything—whether hammering a nail or landing an aircraft. Twenty-five years of research, he said, had shown that different people do the same task in different ways, and the same person may perform the same task in different ways on different occasions. This simple reality is why the old, passive "Pavlovian" paradigms were a limiting factor—or, he might have added, the old, passive Kraepelin-style experiments.

Broadbent will have known that his mention of twenty-five years of research on this theme was an underestimate. The denial of a single optimum way to perform any skilled action was an old Cambridge theme. It would have been familiar since 1932 to any reader of *Remembering* and to anyone who had attended Bartlett's postwar lectures. It was the essence of Bartlett and Myers's prewar objection to the importation from America of the vulgar and provocative pretensions of "scientific management."[49] And it was the message of Bartlett's comments to the Flying Personnel Research Committee about landing an aircraft. The true timescale for the appreciation of different strategies of response was, therefore, not twenty-five years; it was fifty-years. The implication was that there had been two cognitive revolutions: an American cognitive revolution circa 1960 and a British cognitive revolution circa 1940.[50] Broadbent was well aware of the impressive strand of experimental and theoretical work on manual control generated in certain American laboratories. He did not deny the existence of these contributions; rather, he deplored their marginality in their native land. All too often, in Broadbent's view, academic psychologists saw the study of manual control as the esoteric preoccupation of mathematical specialists, rather than something lying, as it should, at the heart of the discipline.[51]

Broadbent's discussion of paradigms of experimentation carried the same methodological and historical message. What sort of experiment would break with the old precognitive paradigm? What sort of experiment gave the subject, rather than the experimenter, the initiative? What experiments let subjects choose their own responses? There was an obvious answer to all these questions. Broadbent was calling for experiments of precisely the kind that had been devised by Bartlett, enabled by Craik, and conducted by Drew for the Flying Personnel Research Committee. The injunction to hand over initiative to the experimental subject originally took the form of Bartlett's decision to leave the manifestations of fatigue an open question—to be settled by experiment rather than by a priori definition. Craik did not have the advantages of digital computing at his disposal for measuring a wide range of responses, but using Kelvin's integrating discs, he had nevertheless constructed an analogue device that served the purpose. Broadbent would have known that his specifications for a new experimental paradigm had been met, long ago, by Bartlett, Craik, and Drew, when they had planned and built the Cambridge Cockpit.[52]

I shall now return to the 1977 lecture to the Experimental Psychology Society and explore the relationship between the original Cockpit results and the conclusions that Broadbent offered his audience on that occasion.

Jacksonian Pilots and Jacksonian Accidents

Broadbent concluded his lecture on "Levels, Hierarchies, and the Locus of Control" by bringing the discussion back to the theme of pilot error and aviation accidents. He had generalized the analysis of the skills needed to fly a Spitfire and applied it to the skills needed in civic administration. He was now ready to close the narrative loop. Precisely how he did so is worth notice. The vital element of continuity in his lecture was the neo-Jacksonian, or Craik-Bartlett, model of a hierarchy of levels of control. Broadbent did not make any attempt to give a point-by-point mapping of his discussion of integral and adaptive controllers onto the process of skillful flying and landing, but the contours of that connection are easy to discern in his concluding remarks.

Suppose a pilot has the task of maintaining straight and level flight in turbulent conditions. Turbulence is a statistical phenomenon, and gusts of wind that threaten to disturb the flight will arrive randomly and vary randomly in size and direction. For simplicity, I shall concentrate on those that simply push the aircraft up or down. The gusts can be thought of as sudden step deviations from the target altitude, so they introduce an error signal that provides a step input. If the target altitude is to be maintained, the pilot must make a rapid response to these error signals but neither overreact nor underreact. In describing the two-level adaptive controller, Broadbent said: "If there is a constant target for performance then the output of the lower level changes after every decision by an amount proportional to the error."[53] The pilot whose reactions are modeled by the two-level adaptive controller will therefore be able, under the right conditions, to home in on an average degree of reactivity that is adjusted to the circumstances.

Now consider the shift from straight and level flight to a steady descent. Previously, I have explained that when the pilot begins the final descent to the runway, the task is that of flying down an imaginary straight line to a chosen point of touchdown. Under turbulent conditions, this task will present a similar problem to that encountered in straight and level flight, only now the problem will be complicated by the need to deal with a ramp input. Broadbent discussed ramp inputs where the target increased in size, while in the present case of landing an aircraft, the ramp to be traversed is downward. In the case of taking off, however, the situation would be exactly that modeled in Broadbent's experiment. But in both moving up a ramp and moving down a ramp, a two-level adaptive controller will ensure that, as Broadbent put it, "the change of output is not simply proportional to the error but has a further component due to the change of input."[54] A two-level adaptive controller can cope with the problem. What is the role of practice in this analysis? Broadbent found in his experiment that practice effects allowed subjects to perform in ways that were largely independent of feedback. Likewise, Jacksonian pilots, whose brains embody the capabilities and levels of an adaptive controller, will be able to perform certain familiar tasks with very little dependence on feedback. Their performance characteristics will be attuned to the properties of the environment. In the right circumstances, this attunement will make the responses rapid and accurate; it also makes intelligible how (putting weather and mechanical damage aside) the highly trained pilots of Bomber Command had so few landing accidents.

There are clear advantages of having a central nervous system that embodies the engineering features of the two-level controller. But, as all engineers know, advantages and disadvantages are interconnected. A trade-off is unavoidable. Broadbent addressed this important theme in his lecture when he showed how the model of the two-level adaptive controller might play a guiding role in future research. He argued that a pilot whose brain is described by the Craik-Bartlett model, with its Jacksonian levels, may have certain weaknesses as well as strengths. These weaknesses will be exposed when the pilot is faced with a range of contingencies that may present themselves during the process of taking off and landing. In short, a Jacksonian pilot may be prone to a certain class of characteristic errors and accidents. The errors in question are not the result of random processes that might occur at any time and under any circumstances. Rather, they have a characteristic and recognizable structure. Broadbent explained that they arise when (on the lower level) a person is performing routine and habitual tasks, while (at the upper level) they are preoccupied with and engaged by some unrelated matter. Broadbent cited a body of research on such mistakes that had been carried out by psychologists and that included an analysis of aircraft disasters that, plausibly, could be traced back to such processes.[55] Broadbent gave the example of an accident that occurred during the takeoff of a BAC 1-11 airliner. The aircraft suffered a failure of one of its two engines, but in response, the crew inadvertently switched off the good engine. They would have been attempting to follow routine but were probably distracted by the demands of communication between those present in the cockpit or by the activation of the stall-warning signals. These errors precipitated a disaster. Although Broadbent did not use this terminology, such episodes might be called "Jacksonian accidents." "They represent," he said, "open chain behaviour running off in the absence of higher level control."[56] The study of such episodes, he argued, showed the continued relevance and importance of the Craik-Bartlett model.[57]

In the final summary of the properties of these neo-Jacksonian mechanisms, Broadbent reiterated the points he made at the beginning of the lecture. The dominant feature of the Craik-Bartlett model was that one process alters another process, rather than merely supplying it with input. The appeal to the model of the two-level adaptive controller permitted the retrieval of the three important facts that are difficult to accommodate when merely choosing between simple open-chain or simple closed-loop mechanisms. First, the adaptive controller can be used to explain why lower-level processes can operate independently of the higher level. Second, it explains how the upper level mediates responses to novelty. Thirdly, it shows how the upper level operates on a longer timescale. It was the recognition of these psychological features and their coexistence that made Cambridge thinking special. Work in many fields, said Broadbent, had confirmed Poulton's 1975 conclusions on tracking. Poulton, he reminded his audience, had shown "that practised performance can survive loss of feedback, but that people find it difficult to know only the error discrepancy between their own actions and the desired state of the world."[58]

Why, at that crucial, final moment of the lecture, did Broadbent mention these two results? The answer is that the two results were analogous to the two components of the Landing-Accident Anomaly. They encapsulated the point at issue. Both of the wartime studies contributing to the anomaly dealt with tracking problems, but while the Cambridge Cockpit posed a problem in *compensatory* tracking, the landing-accident study dealt with *pursuit* tracking. In the study for Bomber Command, the landing strip provided a more or less visible goal in addition to the information—the error signals—displayed on the instrument panel. In the Cockpit study, the only information available to the subjects were the readings on their instruments and their conformity to or deviation from the printed schedule of maneuvers strapped to the pilot's knee.[59] The compensatory display in the closed Cockpit only provided an error function that was the resultant of the input and the pilot's response movements. Compensation tracking is harder than pursuit tracking, and the two wartime studies reflected and amplified the difference between them. Poulton's two results were the Landing-Accident Anomaly in miniature. In citing them, Broadbent was implicitly invoking the Landing-Accident Anomaly, and he was using it to epitomize the problems that had confronted postwar Cambridge psychologists.

Summary and Conclusions

I now want to survey the main steps in my argument and draw some conclusions. I shall also take a brief look at some of the approaches to the problem of fatigue to be found in today's psychological and psychophysiological literature. The need to understand fatigue, particularly pilot fatigue, is still urgent. Today, that need derives primarily from ruthless commercial pressures, but war still takes its toll.

My account of the Cambridge Cockpit work carried out by Bartlett, Craik, Drew, and Russell Davis covered the construction of the apparatus, the conduct of the experiments, their results, and their contested evaluation. The description included the wartime role of the Flying Personnel Research Committee, which provided the immediate institutional context of this work. At the individual level, I have also identified the creative thought processes and the historical contingencies that gave rise to the remarkable form taken by the experimental apparatus—at once sophisticated and makeshift. The magnitude of the step taken by fatigue researchers, from Kraepelin's ergograph to Craik's Cockpit, is strikingly evident.

Bartlett was perhaps an unlikely person to have played a leading role in this enterprise, but I have traced his conception of the Cockpit experiments back to features that can be identified in his early work in social psychology and memory. Bartlett discerned an important analogy between Ebbinghaus's experiments on memory and Kraepelin's experiments on fatigue—and exposed their common shortcoming. Both Ebbinghaus and Kraepelin assumed that giving an experimental subject an apparently simple and repetitive task would reveal the elementary psychological processes out of which the subject's more complex responses were constructed. Both failed to understand that the complexity of a response is a function of the complexity of the responding organism rather than the complexity of the stimulus. Originally, it was Bartlett's anthropological orientation that gave him insight into the limitations of Ebbinghaus's work on memory. Once achieved, this insight gave Bartlett his model for the innovative Cambridge experiments on fatigue.[1]

I then took a step back and looked at the controversial status of some of the intellectual resources that Bartlett employed when interpreting the Cockpit results—in particular, his use of ideas about the structure of the nervous system derived from the work of the neurologist John Hughlings Jackson. The early expressions of Jacksonian psychology in Cambridge were associated with the work of Head and Rivers and their alleged discovery of the protopathic and epicritic systems of cutaneous sensibility. Some scholars have dismissed Rivers's and Head's Jacksonianism as ideological pseudoscience and have set it in contrast to the allegedly more scientific work of the London surgeon Wilfred Trotter, who reported a failure to replicate the Cambridge results. If Jacksonian theory was really ideology disguised as science, this fact might prompt doubt about Bartlett's later, Jacksonian account of the Cambridge Cockpit findings. I gave my reasons for dissenting from a dismissive assessment of Rivers and Head. If there were ideological overtones in the dispute with Trotter, it was true of both parties.

Coming back to the Cockpit experiments, the focal point of my discussion was an episode that also took place in the course of WWII. I am referring to the conflicting indications generated by the laboratory-bound Cockpit results and a field study of landing accidents based on statistical data from Bomber Command. The Cambridge experiments suggested that fatigue would be a major threat to the conduct of aerial warfare, but the Bomber Command study suggested that fatigue effects were operationally negligible. I have referred to this finding as the Landing-Accident Anomaly and argued that it played a pivotal role in shaping postwar research in Cambridge. The anomaly acted as an implicit challenge and led to the (methodological) "dismantling" of the Cockpit. By "dismantling" here I mean the component-by-component study of the structure of the skilled behavior originally demanded by the Cockpit tests. The result was a deeper understanding of perceptual filtering, information processing, the effects of the probability of stimuli, the role of anticipation in motor skills, the importance of the distinction between paced and unpaced tasks, and the role of "arousal"—the general level of activation of the nervous system.

To show how the resources of Broadbent's 1958 *Perception and Communication* might have been mobilized in an explicit effort to resolve the Landing-Accident Anomaly, I imagined what a direct response to the Bomber Command study might have looked like. I called that (hypothetical) response the

Filter Theory of Fatigue. It only constitutes the beginnings of an answer to the anomaly, but it provides a way to articulate Bartlett's subtle disquiet about the recalcitrant but limited data from the landing-accident study. The resolution of the anomaly depended on finding that the two results were compatible. What then is the resolution? Stated in its simplest terms, the reason for the divergence was that the paced performance demanded in the Cockpit was being compared with a relatively unpaced performance of a highly practiced landing procedure—and it is paced tasks that show dramatic declines in efficiency. Speaking of paced tasks, Welford observed that the "classical example of such a task used in the study of fatigue is the Cambridge Cockpit."[2] But the paced character of the Cockpit task cannot be the only factor at work because there was an element of pacing in both of the conflicting results. Landing an aircraft always has a paced dimension to it, though the impact of the pacing is moderated by training, practice, routine, predictable conditions, and the stability and flying qualities of the aircraft.[3] The other salient factor is that the Cockpit posed a tracking task of an essentially compensatory nature, while the landing-accident study involved the richer information available in pursuit tracking.[4]

A fully recognizable neo-Jacksonian response to the Landing-Accident Anomaly had to await the elaboration of ideas first broached in Broadbent's 1971 *Decision and Stress*. Here fatigue took its place as just one manifestation of stress amongst many others. The Arousal Theory of Stress replaced the Filter Theory of Fatigue. Broadbent described a range of different lines of research that might be amenable to analysis in terms of a neo-Jacksonian, two-level model, but I concentrated on just one of these lines: the problem of the interaction of noise and sleep loss. In his 1977 lecture "Levels, Hierarchies, and the Locus of Control," Broadbent then showed how to re-express the Jacksonian conception of "levels" in cybernetic terms by using a model of the nervous system based on multi-level, adaptive servo systems.

If the recognition of the problem of pilot fatigue was transnational, the responses to it were not: they all bore the stamp of local circumstances, predicaments, and institutions with their particular histories. Evidence derived from tracing the very different wartime contributions of psychologists in Britain, Germany, and America pointed to the conclusion that constructing something like the Cambridge Cockpit was by no means an obvious course of action. I argued that there was no German Cockpit, although the work of applied psychologists in the Technische Hochschule in Berlin-Charlottenburg, the Kaiser-Wilhelm-Institut in Dortmund, and the Medizinisches Institut in Garmisch-Partenkirchen might have moved in that direction if circumstances had permitted. Particular attention was paid to the Dortmund Institute for

Arbeitsphysiologie because of its apparent similarity to the Applied Psychology Unit in Cambridge—a similarity noted in a British intelligence report. Indeed, Lehmann and Graf in Dortmund were at work on an early version of the Arousal Theory of Stress. I examined both their work on pilot fatigue and the administrative structures that ultimately hindered that work. Two main findings emerged: (i) the absence in Germany of a salient Jacksonian perspective for the interpretation of the experiments, and (ii) the protracted and futile struggle of the Dortmund institute to win the patronage of the higher echelons of military aviation. The contrast with the experience and situation of the Cambridge group was pronounced.

The American case was different from that of either Britain or Germany. Historians have convincingly documented the massive and consequential wartime mobilization of American psychologists whose work was then subordinated to the demands of military selection procedures. Fatigue was a matter of interest to those engaged in this program, but mainly because of its role as an interfering factor in the conduct of testing and training. American wartime psychology, however, was not wholly devoted to problems of selection and training. I also examine the work of aviation psychologists, such as Jack Adams, who engaged with Clark Hull's learning theory, as well as the team around Paul Fitts, in the Aviation Psychology Program, and psychologists, such as Franklin Taylor, associated with the US Navy.

Hull's quasi-axiomatic approach suggested that there were rigorous ways to bring coherence to the diverse factors involved in fatigue. It was this diversity that had threatened to bring the German followers of Kraepelin to a standstill. Ultimately, Hull's effort was itself a failure, though (Broadbent insisted) an honorable one. In contrast to Hull, Fitts adopted an approach to fatigue that was based on practical ergonomic improvements to the design of cockpits and their controls. The ergonomic approach was given a radical twist by Taylor and Birmingham and their colleagues, who showed (at least in mathematical outline) how to design a cockpit that made minimal demands on the pilot. They came close to "solving" the problem of pilot fatigue by a design philosophy that entirely erased the role of the pilot. None of these different approaches led American psychologists toward building anything comparable to the Cambridge Cockpit. The Cambridge Cockpit appears to have been historically unique.

I confess to having been worried by this conclusion, but perhaps I should not have been disturbed. The highly individual character and circumstances of the various national efforts to understand fatigue suggests that, rather than being improbable, such a result might have been foreseen from the outset. Perhaps the failure to find German Cockpits or American Cockpits was entirely predictable, but the predictability was shrouded by the frequently voiced slogan—and half-truth—that science is universal. Psychologists, like historians, know better. As Broadbent remarked in the foreword of his *Defence of Empirical Psychology*, "psychological research grows out of particular people in particular contexts."[5] The point had been anticipated by Bartlett in *Remembering*. Elements of culture, he noted, assume their final form as a result of passing through many interactions between the members of a group. He mentioned rumors, stories, and decorative designs in this connection, but he also included technological achievements and military artefacts. "Whether we deal with an institution, a mode of conduct, a story, or an art form," he said, "the conventionalised product varies from group to group, so that it may come to be the very characteristic we use when we wish most sharply to differentiate one social group from another."[6] I have instantiated Bartlett's observation by showing its applicability to the Cambridge Cockpit.

Bartlett described the Cambridge Cockpit work as the beginnings of an adventurous scientific journey. He was quietly taking for granted the success, rather than the failure, of the enterprise. We have seen enough of the vicissitudes of the Cockpit story to know that this presupposition should not be taken for granted. At some point, the question cannot be avoided. Was the Cambridge Cockpit project a success or a failure? What exactly did Bartlett achieve by the idiosyncratic Cockpit experiments? A case can be made for saying that nothing was achieved. The claim here is not that the experiments were faulty or the measurements unreliable: the point at issue is the utility of the work. The upshot of the work was not, let us say, a reorganization of squadron duty rosters, as was the case with Mackworth's work on fatigue in radar watchkeeping. Nor did it generate useful suggestions about when and in what order certain necessary tasks should be performed, as with Grindley's simple, practical, but well-observed suggestions about debriefing Coastal Command aircrews. On the contrary: When the time parameters for the onset of fatigue suggested by the Cockpit experiments were put to the test in the landing-accident study, they were found to be undiscriminating. They were therefore immaterial to important military decisions that had to be made. The only wartime result of the Cambridge work was the distribution to aircrews of an Air Ministry pamphlet: *Notes on the Prevention of Flying Fatigue in Flying Personnel.*[7]

There was nothing wrong with the pamphlet, but it offered little more than commonsense reflections on the subject of fatigue. The advice to be alert to the symptoms of fatigue was worth repeating, but it only told thoughtful persons what they already knew. Consider all the time and energy expended on the Cambridge experiments—hours and hours of tests involving much-needed pilots—and for what? A pamphlet! Measured in terms of immediate, practical utility, the Cockpit experiments must be deemed a failure. Despite Bartlett's creative thinking, Craik's ingenious engineering, and the novelty of the approach, the problems posed by fatigue were too complicated to sort out in a way that generated immediate practical results. Perhaps a sense of this failure contributed to the frustrations that brought an exhausted and depressed Craik to say, in 1942, that in his view:

> most people insufficiently appreciate the impermanence and relativity of things. . . . So I view life as an affair of unutterable complexity in which activity and inactivity, dogma and science, are all alike finally unsatisfied.[8]

Now suppose that the framing of the question about success and failure is altered, and that, along with the initial Cockpit experiments, the postwar development of Cambridge research is brought into the assessment. If we take seriously Bartlett's metaphor of a scientific journey, then the question of success or failure assumes a somewhat different aspect. Bartlett's buoyant description becomes more intelligible. Craik never lived to see the consequences of the wartime work of the Cambridge Psychological Laboratory, or the postwar output of his own Medical Research Council's Applied Psychology Unit. Death denied Craik the experience of reading *Perception and Communication* or *Decision and Stress*. The strange complexity of fatigue was still evident in these two works, but they showed that this complexity was being confronted with vigor and a measure of success. Kraepelin's stultifying list of unmeasurable variables was a thing of the past. Cambridge psychologists were at the center of an active field of research and there was a sense of progress. Not everyone agreed with one another, but the disagreements were spurred by the forward movement of discussion and argument.[9]

Using medium to long-term measures, rather than short-term wartime measures, the Cockpit experiments could be called a success. They were a success because the journey that started with those experiments reached a point where many of the practitioners themselves felt that they were succeeding.[10] This definition of success may sound circular or self-serving, but no other credible or competent measure is available. It must be remembered, of course, that the feeling of success within a scientific group is not an arbitrary or subjective thing. It is based on an absence of shared frustration, on the emerging routines of laboratory activity, a collective sense of what experiments and tests to do next, and the experience of participating in and sustaining a puzzle-solving tradition.[11]

Though the feeling of success has an objective social basis, nevertheless the success that is signaled will only be a relative success. Broadbent was frank in admitting the presence of enduring gaps in both the data and the argument.[12] A fully adequate synthesis of the cybernetic and signal-detections dimensions of fatigue had yet to be achieved. Craik's earlier anxiety had not been irrational, but no reader of *Decision and Stress* could doubt that, now, scientists both knew more facts about fatigue than they did before and were grappling with the problem with greater sophistication. The question of the success or failure of the Cockpit work must be posed against this background. An adequate answer to the question must take both the short term and the longer term into account. Granting the equal validity of the wartime and the postwar perspectives, the conclusion must be that, within the timescale of this study, the Cambridge Cockpit research on fatigue was *both* a failure *and* a success. Expressed fancifully, this conclusion might be counted as yet another of the paradoxes of fatigue—though in reality, it is no paradox at all.

Bartlett's methodology was inductive and analogical. He was opportunistic, tolerant of complexity, and suspicious of abstraction and simplification. In many respects these qualities were transmitted to subsequent generations of Cambridge psychologists, though with the passage of time, there was an increased awareness of the importance of mathematical models and statistical analysis. I introduced the word "symmetry" to describe another striking feature of Bartlett's methodological stance. In Bartlett's case the word refers to his decision to treat both accurate and inaccurate memories as having been generated by the same causal mechanisms. Both kinds of memory were indicative of a mode of biological adaptation. Broadbent's scientific curiosity also had this important structural feature. Broadbent boldly expressed the same methodological symmetry when he said:

> It is still not widely realized in our culture that a man can see something which did not happen, and that he does so precisely through the workings of the system which in other cases makes him perceive accurately.[13]

This symmetry was implicit in the measuring practices originally encouraged by the impressive development of Signal Detection Theory in the United States, though it was Broadbent who give that methodological message both the precision and the prominence it deserved.

Bartlett's metaphor of a scientific journey prompts the question: Where are we now? Stepping outside the time frame of my case study: How do things stand today in the field of fatigue research? Does the neo-Jacksonian tradition still play a role? When Broadbent, as an ex-RAF pilot, went to Cambridge to study under Bartlett, those neo-Jacksonian ideas, he said, were taken for granted. He later came to feel that they had fallen into a degree of unjustified neglect and that it was time to reaffirm their value. How successful, or unsuccessful, was Broadbent in this effort? Has there been an enduring revival of the Craik-Bartlett model, or has it once again almost disappeared from view?

I know of only one attempt to replicate or test Broadbent's transport-system experiment. As part of a larger study, Chmiel, Totterdell, and Folkard (1995) ran a small experiment using precisely the same problem-solving scenario.[14] They introduced two modifications: (i) they studied the effect of sleep deprivation on the pattern of responses, and (ii) they replaced the step and ramp character of the goals by variations that fluctuated in the form of a sine wave, and that were presented visually. The second change was merely a matter of technical convenience, but introducing sleep loss established a useful link between Broadbent's 1971 discussion of the two-level approach, in *Decision and Stress*, and his 1977 transport experiment. The results of their experiment led Chmiel and his colleagues to doubt the adequacy of Broadbent's 1971 suggestion that sleep loss had an impact on the Lower level, while fatigue, generated by time on task, had its impact on the Upper level. They were led to this conclusion by examining both the efficiency of the problem solving and the work rate of the subjects. According to Broadbent, efficiency would be the responsibility of the Upper level while work rate was the responsibility of the Lower level. By contrast, Chmiel, Totterdell, and Folkard found that sleep loss had a detrimental effect on *both* performance measures and hence implicated both levels.

The number of their subjects was small ($N = 6$), and, accordingly, the experimenters did not want to place too great a weight on their empirical results. Nevertheless, they argued that those results were sufficient to expose a methodological weakness in what they called "the levels approach"—that is, the neo-Jacksonian approach. The problem, they argued, was that conclusions about the role of any particular level were under-determined by any of the performance measures that might be available to an experimenter. This fact made it unwise to follow Broadbent in "attempting to relate particular performance indices to particular levels."[15] Chmiel, Totterdell, and Folkard were posing the question: How can an experimenter tell which aspects of a subject's performance are to be explained by which level? This challenge, they said, was "inherent in the levels of control notion and illustrated . . . that the explanatory constructs act together to control behaviour, and hence outcomes."[16] The levels approach, they concluded, might provide an "interpretive framework," but experimentation needed a stronger paradigm. They cautiously recommended the approach provided by the theory of connectionist networks.[17]

Chmiel, Totterdell, and Folkard are surely right that further experiments, like theirs, are likely to generate empirical anomalies with which to challenge the neo-Jacksonian approach. But aspects of their methodological criticism are open to question. They were presenting an argument whose logic was the same as that already encountered, in chapter 8, in Taylor and Birmingham's paper on the confounding of system variables within a "man-machine system."[18] The two parts of the former example, the man and the machine, correspond to the two levels of Broadbent's neo-Jacksonian system. The logic of the problem of untangling their respective contributions is the same. But we have seen that Taylor and Birmingham themselves indicated a way out of the conundrum. They identified the circumstances under which these considerations amount to a threat, and those in which the threat can be, and routinely is, circumvented. It all depends on the depth of the background knowledge. What Taylor and Birmingham called "full knowledge" can solve the problem. Underdetermination, in Broadbent's proposal, was therefore not, in itself, an immediately compelling reason for halting the line of the research. On the contrary: it could be a reason to pursue the research with renewed vigor. This was the implicit lesson drawn in comparable circumstances by Bartlett in the case of the Cambridge Cockpit.

Robert Hockey's 2013 book *The Psychology of Fatigue: Work, Effort and Control* provided an authoritative twenty-first-century update of the state of fatigue research.[19] As befits a former student of Broadbent's, Hockey grounds his work in the two-level model of *Decision and Stress*. Thus far, the neo-Jacksonian tradition still appears to be a significant resource, although Hockey makes no mention of Jackson's name, nor does he discuss Broadbent's transport-system experiment. The explanation for these omissions may lie in the fact that Hockey's aim was to offer "a new perspective," and that perspective was based on a social metaphor different from the one that Jackson favored.[20] Hockey's insight was that fatigue can be seen as performing a useful biological function: fatigue isn't always a bad thing. The experience of fatigue prompts the reappraisal of goals, the selection of new strategies, and new ways to manage and distribute effort. In itself, this stance is thoroughly Bartlettian, but recall that the historian of psychology Roger Smith identified two different concepts of inhibition: one drawing on the model of hierarchies, the other on the model of market processes. A market is a social mechanism for dealing with matters of supply and demand and the distribution and flow of scarce resources. Hockey's book is full of expertly marshaled experimental data that he interprets in theoretical terms based on market metaphors. Thus, the reader encounters effort budgets, opportunity costs, models of resource

management, and cost-benefit analysis.[21] For Hockey, the psychological journey had a neo-Jacksonian starting point, but the new perspective perhaps
indicates a change of direction.

A further challenge to the Craik-Bartlett style, and to Broadbent's desire
to retain that style, comes from another body of researchers who proceed
with relatively little concern for psychological theory but with a great deal of
concern for sophisticated techniques of measurement. I am referring to the
use of electroencephalography (EEG) to study the brains of pilots as they un-
dertake various flying tasks and maneuvers. The subjects usually go through
the motions of "flying" a flight simulator, though sometimes recordings are
made during actual flights. The aim of the experimenters is to correlate states
of fatigue and workload with characteristic patterns of the subject's brain activity as revealed in the EEG trace. Surveys of this active field of work, cover-
ing some two hundred publications, are to be found in Borghini et al. 2014
and van Weelden et al. 2022.[22]

Work of this kind may well yield useful warning technologies that inform
pilots that their brains are operating inefficiently—for example, if they are
falling asleep for short periods without realizing it. We have seen that during
WWII, Kornmüller, in Berlin, had tried to develop such techniques, though
in Britain, the Flying Personnel Research Committee did not think the project worth pursuing.[23] Today, the Kornmüller project looks more plausible because artificial neural networks and other sophisticated forms of data analysis
can be used to interpret the complex statistical structure of the brain waves.
There are, of course, challenges to be overcome. The neural networks used to
detect the characteristic features of the brain's response to various demands
have to be "trained" on an individual subject and are less reliable when used
to detect corresponding patterns in the EEG records of another individual,
or the same individual at a later time. Despite these problems, perhaps the
elusive "fatigue test" is about to be unveiled.

Electroencephalography may also throw light on the underlying brain
mechanisms involved in the response to stresses such as fatigue and overload.
The precondition for success in this enterprise is that there are interesting
theories that generate predictions of the kind for which EEG data is relevant.
Today, EEG machines and flight simulators are widely available, at least in
technologically advanced economies. One might almost say that everybody
now has their own Cambridge Cockpit. Bartlett long ago put his finger on
the methodological pitfalls that can attend research under such conditions.
There is the danger of doing experiments simply because a piece of attractive
apparatus is conveniently at hand. That mistake, he said, had plagued fatigue
research in the early years. There are, in fact, dangers in conducting any line

of psychological research that uses a very narrow range of data. This point was the basis of Bartlett's criticisms of Ebbinghaus, Kraepelin, and Bradford Hill. Whether those currently pursuing EEG research in aviation physiology will avoid these traps, only the future can tell. But there are ominous signs.

In *Decision and Stress*, Broadbent had criticized an earlier generation of postwar EEG researchers. He objected to their premature physiological theorizing when they used the percentage of alpha waves in an EEG recording as a measure of arousal.[24] That criticism was made fifty years ago. Does it still apply? Today, the proposed operational definitions that link psychological states to neurological processes are more sophisticated. They now combine alpha-wave data with data from the beta and theta range in the EEG power spectrum. (Alpha [α] waves comprise the 8–13 Hz range, while beta [β] waves are from the 13–30 Hz band, and theta [θ] waves are from the 4–8 Hz band.[25]) In their above-mentioned 2022 paper, "Aviation and Neurophysiology: A Systematic Review," van Weelden et al. express confidence that, by using these new, combined measures, both fatigue and the effect of workload can be defined neurologically.

The term "workload" refers to the amount of work *demanded* of the pilot in different phases of a flight. Thus, workload is high on approach to landing and low when cruising at constant speed and altitude. Workload may be measured in various ways. A simple way would be to count the number of control movements required per minute, which will be greater in landing than cruising. Workload, as a demand, must be distinguished from the amount of work actually *delivered*. Typically, when more work is demanded of the pilot, more work is delivered as the pilot engages with the task at hand. In principle, however, the work actually performed may be either more or less than the required workload. One pilot may make more control movements than another and so, in this sense, perform more work. EEG recordings convey information about aspects of the pilots' internal response, so some of these recordings, or some combination, may indicate and correlate with the amount of work delivered. But any link between the EEG trace and the external demands on the pilot, or between the EEG trace and the pilot's behavioral response, must be established and validated independently. The links are entirely contingent.

Van Weelden et al. report that modern methods of identifying and analyzing the EEG waves yield two results that are important for understanding and predicting pilot behavior. They present and label these results as follows:

$$\text{EEG engagement index} = \frac{\beta}{\alpha + \theta}$$

$$\text{EEG mental-fatigue Index} = \frac{\alpha + \theta}{\beta}$$

The measure of "engagement" is, necessarily, a measure of the work delivered, not the work demanded. "These equations," say van Weelden et al., "could serve as neurometrics for workload and fatigue in research and in ecological applications in aviation."[26] "Ecological" here means in real-world field conditions.

Mathematically, the fatigue index and the engagement index, as given above, stand in a close logical relation to one another. One index is the reciprocal of the other. The two indices are so defined that the greater the fatigue index, the smaller the index that correlates with the amount of work that the pilot actually delivers—whether it be measured by the number of control movements, or in some other plausible manner. Although expressed in novel, symbolic terms, the equations are a restatement of an old and widely taken-for-granted conclusion—namely, that fatigue is a decreased capacity to perform work. Van Weelden et al. formulate and endorse this conclusion explicitly when they say, in their table of main findings: "The term 'mental fatigue' refers to a workload and task-related state that decreases performance and vigilance."[27] The quoted sentence does not simply represent a finding about the preferred terminology of the community of researchers whose work is under review. The reviewers endorse it because it is an implication of the two equations given above—and the reviewers endorse these equations.

In chapter 1, when I described the conception and construction of the Cockpit, we saw Bartlett challenge the assumption that fatigue can be defined as a reduced capacity to deliver work. The question he put to Craik in 1939, in that hazardous but fruitful car journey, was direct: Does fatigue necessarily imply that less work is being done, or can fatigue be accompanied by *more* work? Do these two things really stand in a reciprocal relation, or not? The Cambridge Cockpit experiments were designed to answer that question. And the answer was that fatigue can sometimes lead to more, not less, work being done. This fact was read off the recordings of the pilots' movement of the controls, and it was these movements that constituted the work in question. The two equations cited above, the engagement index and the fatigue index, are inconsistent with that empirical finding.

If the Cockpit result was right, then either the engagement index is wrong, the fatigue index is wrong, or both are wrong. Can this negative conclusion be dismissed on the grounds that both indices will have been independently validated? The fatigue index is validated by its correlation with factors such as time on task; the engagement index by correlation with measured output of work. But attempts to bypass the problem of the Cockpit result in this way merely postpone the confrontation. The implication of the Cockpit result is that the two indices must have been erected on an inadequate inductive basis. Their prior validation must be called into question.

Writing in 1958, and referring to work on fatigue conducted around 1900, Bartlett (by now emeritus professor, but still a consultant to the Applied Psychology Unit) observed that the notion of fatigue, both bodily and mental,

> was so strongly linked with that of decreasing amount of work that people ever since have never for long got away from the view that when there is fatigue there is bound to be work decrement, and very often they have at least implied, if not asserted, that where there is work decrement, there must also be some kind of fatigue.[28]

With his uncanny prescience, Bartlett could have been talking about the definitions and interrelations of the fatigue index and the engagement index, given above, where the tacit implication is made explicit. It is unclear why the possibility that fatigue can produce more, rather than less, work appears to have been overlooked once again. For Bartlett, his appreciation of that possibility and his embrace of the Jacksonian perspective went hand in hand. I have traced the story of that embrace in the preceding chapters, but for now, it looks as if Bartlett's warning and Broadbent's own fears about disciplinary amnesia and premature physiology are about to be realized.[29]

I have mentioned the methodological "dismantling" of the Cambridge Cockpit in the postwar years, but what was the fate of the Cockpit as a material object? Was it literally dismantled, or is it still in the Psychological Laboratory in Cambridge? Or is it in the Science Museum in Kensington? Have the remains of the old Spitfire perhaps found a resting place in the Imperial War Museum at the Duxford aerodrome near Cambridge? John Rolfe, an expert on aircraft simulators who worked in the Institute of Aviation Medicine at Farnborough, recalls that, around 1986, he met Richard Gregory at a symposium organized by the Royal Aeronautical Society. Rolfe explained to Gregory that he was working on a paper analyzing the use of flight simulators during World War II:

> Knowing that he was at Cambridge during some of the time when the Cockpit was in use I asked for any recollections he might have of it. His reply was totally unexpected. He told me that what remained of the Cambridge Cockpit was at Bristol University. Richard arranged for me to see the Cockpit, which was stored in the basement of a house owned by the University.[30]

The Bristol location might be explained by the fact that both George Drew and Derek Russell Davis worked at the University of Bristol after the war.

One or the other of them may have been responsible for the move. Rolfe duly visited Bristol and found that the Cockpit had suffered from the attentions of other researchers, who had stripped it of components, presumably for use in their own apparatus. This encounter is described in an article that Rolfe wrote on Craik and the Cambridge Cockpit, published in *The Psychologist* in 1996. Rolfe included in his article a grainy black-and-white photograph of what he had found in the Bristol basement. There was little left. The photograph could have portrayed the detritus left by an aviation accident rather than an ingenious aid to scientific research. It was a sad sight. In December 1996, Richard Gregory told me in conversation that the Bristol cellar had suffered a flood, and the remains of the Cockpit had finally been disposed of. I was too late to pay my own respects.

The fate of the Cambridge Cockpit symbolized that its life as a piece of scientific apparatus had truly reached its end, but the spirit of the enterprise lived on—at least for a while. Like Bartlett, Broadbent was in no doubt about the continuity between the wartime work and the postwar work. We have seen Broadbent drawing an analogy between controlling a Spitfire and controlling our economy, but the continuity went deeper. For some individuals, it informed their overall view of science and its role in society. It contributed to a certain mentality. My mind, said Broadbent,

> is not organised in a way which makes air raids and political systems irrelevant intrusions on the pure niceties of science. On the contrary, I would hold that, if warfare is the continuation of diplomacy by other means, then equally there is a sense in which science is the continuation of warfare by different techniques. . . . I would certainly, however, classify my activities in the laboratory as part of a framework which includes the era of armed resistance to Fascism.[31]

In the introduction, I began this book with what I termed a "war story." The story dealt, on some half-remembered level, with the significance of the Cambridge Cockpit and the role it was alleged to have played in the armed resistance to Fascism. In the subsequent chapters I have traced that story back to its historical basis. I have tried to tell the story in a way that can withstand historical and sociological scrutiny. The radically truncated version circulating in the Cambridge laboratory, when I first heard mention of the Cockpit, was overdramatized, oversimplified and above all, and perhaps mercifully, radically false. Psychologists could do many things, but they could not tell who would be shot down. None of these unreliable qualities should surprise a reader of Bartlett's *Remembering*. Even though the first version of the story that I recounted was little more than laboratory gossip, Broadbent's remarks remind us that a more adequate narration of the Cockpit experiments would still have to locate them within a framework of military and ideological conflict. But though rumor has now been replaced by a more responsibly documented account, for better or worse, the result is still a war story.

Acknowledgments

First, I want to express my gratitude to my wife, Celia. Throughout the entire project, I have depended on her unstinting help and support and, above all, her sound critical judgment. *The Cambridge Cockpit and the Paradoxes of Fatigue* has been written since I retired from the University of Edinburgh, and since the Science Studies Unit (my old department) was absorbed into the Institute for the Study of Science, Technology, and Innovation (ISSTI). I must thank Professor Robin Williams, the director of ISSTI, for his generous and collegiate provision of office space and facilities. I have much appreciated contact with this lively and friendly group.

Because of the comparative dimension of the analysis that I give in *The Cambridge Cockpit*, it would have been impossible to write the book without the support of the Max-Planck-Institut für Wissenschaftsgeschichte (MPIWG) in Berlin. In particular I must thank Dr. Dagmar Schaefer, the head of Abteilung III, where, during much of the research, I had the privilege of being a visiting scholar. My project on fatigue was a contribution to one of the department's research programs which addressed the theme of "The Art of Judgment." I benefited beyond measure from the chance to meet, talk, and interact both with fellow members of Abteilung III and with the members, junior and senior, of the other research groups within the institute. The collective fund of knowledge and experience, and the willingness to share and explore ideas across a diversity of perspectives and projects, make the MPIWG a truly remarkable institution.

No one who works in, or in association with, the MPIWG can fail to be impressed by the institute's library or the help they receive from the head of the library, Dr. Esther Chen, and its expert and friendly staff. Specifically, I must thank Frau Urte Brauckmann for advice on a copyright problem and

Ms. Uebele of the Image Library of the Max-Planck-Gesellschaft for locating
the photograph that became my figure 7.2. I must also extend a special thanks
to Herr Matthias Schwerdt. During the pandemic, with its restrictions on
work and travel, I had to cancel an intended visit to Berlin, but, thanks to his
good offices, I was able to receive copies of reports and papers that, otherwise,
would have been inaccessible. Indeed, without his help, progress on the book
would have come to a halt.

Concerning the archival material that I consulted in more normal (i.e.,
non-COVID) times, I want to express my thanks to the directors and staff of:

(i)   the Archives of the Max-Planck-Gesellschaft, Berlin;
(ii)   the Archives of the Technical University of Berlin;
(iii)   the Archives of the Deutsches Museum, Munich;
(iv)   the Bundesarchiv, Koblenz; and
(v)   the National Archive, Kew (formerly the Public Record Office).

I am most grateful to Dr. Annette Vogt for her expert advice regarding the
archive of the Max-Planck-Gesellschaft and for her insights into the situation
of German wartime science, to Dr. Cornelius Borck for his help and guidance in locating material in the Bundesarchiv, and to Dr. Florian Schmaltz for
valuable discussions regarding the Kaiser-Wilhelm-Institut für Arbeitsphysiologie. I am also conscious of how much, during the writing of this book,
Celia and I have benefited from extensive discussions with Dr. Ursula Klein,
Dr. Wolgang Lefèvre, Dr. Dieter Hoffmann, Dr. Annette Vogt, the late
Dr. Thomas Kuczynski, and Dr. Peter Beurton, on the history and politics of
postwar Europe.

Whilst writing this book, I gave talks on the Cambridge Cockpit in the
University of Constance, the University of Vienna, the München Zentrum für
Wissenschafts- und Technikgeschichte, the Goethe Universität, Frankfurt, and
the Max Planck Institute in Berlin. I benefited greatly from the stimulus of these
invitations and from the discussion and questions that followed the talks.

On completing the first draft of the book, I asked a number of close friends
if they would be kind enough to read and criticize the entire draft. Here I
must express my deep gratitude to Barry Barnes, Jonathan Harwood, Ursula
Klein, and Christopher Martyn. Though generous in their encouragement,
they identified numerous places in the exposition that called for clarification
and improvement (and, at one point, mercifully rescued me from committing a statistical fallacy). I have also benefited greatly from discussions with
Donald Mackenzie and Andrew Barker. After extensively modifying the draft
in response to comments from these sources, I then had the benefit of reports
from two anonymous publisher's readers of the University of Chicago Press.

Their well-aimed and constructive criticisms quickly convinced me of the need for extensive further revision and some radical reorganization. I want to express my appreciation of the work carried out by all these readers—both known and unknown. The demands made upon them were considerable, but *The Cambridge Cockpit* has been greatly improved as a consequence of their commitment, expertise, and insight.

Throughout the various phases of the process of publication I have also been conscious of needing and benefiting from the guidance of Karen Merikangas Darling and Fabiola Enríquez Flores of the University of Chicago Press. I want to express my gratitude for their great patience and professionalism. I also want to thank Jessica Wilson, who performed heroic deeds at the copyediting stage. The reader can be assured that all the remaining inconsistencies are my responsibility alone.

Abbreviations

This bibliography includes unpublished technical and research reports—for example, the reports of the British Flying Personnel Research Committee, as well as comparable US and German reports. Archival details of meetings, minutes, and correspondence are given in full in the relevant endnotes.

AMPG   Archiv der Max-Planck-Gesellschaft, Berlin-Dahlem
DMM   Deutsches Museum München, Munich
FPRC   Flying Personnel Research Committee, National Archives, Kew
TNA   The National Archives of the UK, Kew
WADC   Wright Air Development Center, Dayton, Ohio

Notes

Introduction

1. Boas 1916, 393. On war stories, see p. 536.

2. The quotation from Bartlett that I have used as an epigraph to the book is from Bartlett 1958b, 512. Two informative obituaries of Bartlett are Broadbent 1970 and Welford 1970. There is an extensive literature on Bartlett, but most of it deals with his memory experiments and sheds little light on the Cockpit experiments. Some commentators adopt a negative stance toward the applied aspect of Bartlett's work. See, e.g., Costall 1991; Costall 1992; Douglas 1986; and Douglas 2000. My aim is to show that Bartlett's social orientation was integral to his seemingly more technical and applied work.

3. Bartlett 1946; Zangwill 1980.

4. Adrian, Craik, and Sturdy 1938.

5. For a representative sample of his work, see Craik 1938a, 1938b, 1940b, 1943a, 1943b; and Craik and Vernon 1941. For a bibliography of Craik's publications, see Sherwood 1966, 179–81.

6. Some of Craik's work from 1943 and 1944 was carried out for the Armoured Fighting Vehicle Sub-Committee. It was published after the war as Craik 1963.

7. Craik 1947–1948. A "servomechanism" in the sense used here can be formally specified as a control system that is "fully automatic, error actuated and which employs power gain." See Pitman 1966. An example that illustrates this definition would be a radar-controlled anti-aircraft gun. A source of power is used to move the heavy gun in ways that depend on the error signal, which indicates the disparity between the current direction of the gun barrel and the direction of the target, as detected by the radar system. The power is used to reduce the magnitude of the error. Such automatic-fire control systems would also need to be able to make corrections that allow for both the movement of the target and the time of flight of the shell. The history of these systems is analyzed in Bennett 1963, chapter 6; and Mindell 2002, chapters 2 and 3.

8. Craik 1943c. The quotation is from pp. 120–21. The Kelvin example is from p. 51. For a sympathetic and insightful critical appraisal of Craik's 1943 book, see Gregory 1983.

9. Craik 1945.

10. Some of the fragmentary material that was meant to be included in this second book was gathered together and published in Sherwood 1966.

11. The premonition story is in the introduction to Sherwood 1966, x. For a pertinent

account of the role played by Craik's reputation in defining the postwar discipline of British psychology, see Collins 2012 and Collins 2013.

12. Bartlett 1936, 40–41. Bartlett wrote informative obituaries of both Rivers and Myers. See Bartlett 1923b, 1948a.

13. Bartlett 1923a, 61. See also Bartlett 1920a, 1920b, 1924.

14. Bartlett 1932.

15. The aircraft-detections systems in question were sound locators. On this work, and Bartlett's discreet references to it in *Remembering*, see Bloor 2000. During the Great War, Bartlett had also worked on the selection of sonar operators engaged in submarine detection; see Bartlett and Smith 1920.

16. Russell Davis 1948.

17. Dearnaley and Warr 1979.

18. In 2003, the documents in the Public Records Office and the Historic Documents Commission were amalgamated to form the National Archives of the UK. Material dealing with the work of the Flying Personnel Research Committee are categorized as TNA: AIR 57. Documents AIR 57/1 to AIR 57/9 were closed until 1972; AIR 57/10 to AIR 57/11 until 1978; and AIR 57/12 to AIR 57/16 until 1979.

19. Hence the title of Franz Borkenau's famous book *The Spanish Cockpit* (1937), an eyewitness report from the Spanish civil war. This vicious confrontation with the forces of Fascism set the stage for WWII.

20. Bracken 1956.

21. Craik 1943c, 4. In *The Nature of Explanation*, fatigue was not identified as a central concern, but it was mentioned in connection with the theme of paradox on p. 13. See also pp. 4, 21, 22, 24, 82, 94, and 121.

22. On the evolving historical relationship between British physiologists and psychologists, see Young 1970; Daston 1978; and Danziger 1982. On the more fraught relations between these disciplines in Germany, see Ben-David and Collins 1966; Ash 1980; and on the relation between psychology and philosophy in Germany, see Kusch 1995, 1999.

23. Myers 1911a, 174. See also Reid 1928.

24. In using the term "revolution" I am obviously hinting at the analysis of scientific change given by T. S. Kuhn (1962) in his *Structure of Scientific Revolutions*. I shall make use of other Kuhnian terms of art in what follows, such as "anomaly" and "normal science." I am aware that Kuhn's analysis has shortcomings, but I continue to believe that, as a readily comprehensible overall picture of science, it is the best available. I also believe that it must be understood as a *sociological* analysis and that the central idea of the work is best seen as that of normal science. I also think that for present purposes it is best to avoid the massive overlay of highly partisan "philosophical" commentary to which the work has been subjected. The correct way to carry forward Kuhn's sociological work was clearly set out some forty years ago by Barry Barnes in his *T. S. Kuhn and Social Science* (1982). This book has been largely ignored by philosophers.

25. Rabinbach 1990; Kehrt 2010.

Chapter One

1. The organization of wartime research in the United Kingdom is displayed in the *Nature* editorial of February 20, 1943, 206. The links between the Medical Research Council, the Privy Council, and the Cabinet are clear, as are the links between the Royal Society and the Cabinet.

2. Harsch 2004. There is a photograph of the meeting between Livingston and Strughold on p. 26.

3. I am following Gibson and Harrison 1984, chapter 4; and Livingston 1962, 157–72. Livingston went on to play a significant role in the Vision Committee of the FPRC; see Hartcup 2000, 34.

4. Roughton 1949.

5. Welford 1970, 113.

6. Mellanby was central to the discovery of vitamin D and its role in the prevention of rickets. See Platt 1956. On Mellanby's political contacts, see Dale 1955, 214.

7. Most of the reports submitted to the FPRC indicate the name of the person responsible for the report as well as a title and date—although the presentation was not entirely uniform. The few reports that I will mention that do not indicate the name of the author, such as reports 144, 321, and 384, deal with administrative matters or the German war effort. They may be assumed to have been authored and submitted by Whittingham and are entered under his name in the bibliography.

8. The topics to be researched included (i) hypoxia, (ii) increased gravity and blackouts, (iii) aircrew selection, (iv) reducing cockpit noise, (v) improving visual standards, (vi) causes and consequences of fatigue, and (vii) accident investigation. Minute Book of the FPRC, TNA: AIR 57/40. Details of the dates of meetings of the FPRC, and the minutes taken at those meetings, are cited in full in these endnotes. Details of the reports generated by the FPRC are included in the bibliography.

9. Whittingham 1939a.

10. Bartlett 1946. Bartlett described Craik as religiously opposed to the scientific-materialist world view. Margaret Vince, who worked with Craik during the war, insisted that Craik was totally committed to a materialist outlook. She says that Bartlett knew this but wanted to avoid distress to Craik's grieving mother. See Vince 1996, 68.

11. On Craik's absent-minded professor stance, see Vince 1996, 68. On the otherworldly and unhappy Craik, see, e.g., the brief note with the title "Claustrophobia," in Sherwood 1966, 171.

12. Myers had already challenged the assumption that fatigue would necessarily generate a work decrement: "The higher nervous levels appear normally to exercise an inhibitory influence over the lower, which may disappear in fatigue and under the influence of alcohol. Such a loss of higher control may manifest itself temporarily in an increase in the amount of muscular work performed" (1924b, 322). Bartlett raised the question with regard to skill rather than strength, but the link is a significant one: It is indicative of the Cambridge character of the idea.

13. Minutes of the Flying Personnel Research Committee, meeting 6 (October 4, 1939), TNA: AIR 57/40.

14. Much of the work commissioned by the FPRC was carried out at the RAF Laboratory. There is a lively account of its foundation and work in Gibson and Harrison 1984.

15. Minutes of the Flying Personnel Research Committee, meeting 6 (October 4, 1939); meeting 8 (January 29, 1940); meeting 10 (May 15, 1940); meeting 16 (April 3, 1941); and meeting 18 (June 4, 1941), TNA: AIR 57/40.

16. The mechanism was invented by James Thomson and then refined and developed by William Thomson (Lord Kelvin). See Thomson and Tait 1896, appendix B. In the United States, Kelvin's integrator was later modified by Vannevar Bush and used to solve problems in electrical engineering. See Mindell 2002, 157–61.

17. See Hartree 1935. Craik will almost certainly have read Hartree's article in *Nature*.

18. Craik 1940a.

19. Craik 1940a, 1–2.

20. Diagrams were handed round at the FPRC meeting, but I have not found them in TNA.

21. Craik 1940a, 2.

22. Craik 1940a, 2.

23. Minute 68g, FPRC meeting 7 (December 20, 1939), TNA: AIR 57/40.

24. On the Link trainer and its relation to Craik's Cockpit, see Rolfe 1995, 1996.

25. Drew 1940.

26. Drew's prewar work includes Drew 1937a, 1937b, 1938, 1939a, 1939b.

27. Drew 1940, 1.

28. Drew does not say whether any of the same pilots who took part in the preliminary, aborted experiment went on to take part in the subsequent, revised experiment.

29. Drew 1940, 141. In saying that the Cambridge Cockpit was "unstable," Drew meant that its responses simulated the responses of an unstable aircraft. He did not mean that the apparatus itself was prone to tilt or shift about.

30. Vincenti 1990; Abzug and Larrabee 1997.

31. Craik also simplified the display-control dynamics by removing the secondary control effects. For example, in a real aircraft, a banking maneuver will not only tilt the aircraft but lead to a drop in the nose. The drop is the secondary effect.

32. Whittingham 1940.

33. FPRC minutes, meeting 9 (March 29, 1940), TNA: AIR 57/40.

34. FPRC minutes, meeting 10 (May 15, 1940), TNA: AIR 57/40.

35. FPRC minutes, meeting 11 (June 28, 1940), TNA: AIR 57/41.

36. Drew 1940.

37. FPRC minutes, meeting 15 (February 6, 1941), TNA: AIR 57/41.

38. Drew 1940, 16.

39. Drew 1940, 16.

40. Drew 1940, 15.

41. Drew suggested that the pilots who thought they were feeling cramp interpreted this sensation as a sign that they were getting tired, and it spurred them into making a greater effort. He did not offer a systematic explanation for the fact that the performance of both fresh and fatigued pilots improved, but a conjectural account could be offered in terms of the arousal effects of the stimulation. More will be said about arousal later.

42. Drew 1940, 11.

43. Drew 1940, 11.

44. Drew 1940, 11.

45. Drew 1940, 21.

46. The unreliability of reports after an aerial battle was well-established during WWII. See Terraine 1985, 185–88.

47. Drew 1940, 14.

48. Drew 1940, 14.

49. Drew 1940, 22.

50. Broadbent 1977a, 183.

Chapter Two

1. The minutes of FPRC meeting 4 (June 7, 1939) record that the director of medical services "impressed on members that all official information, statistics and confidential reports

concerning the Royal Air Force, which was communicated to the committee, must be regarded as secret" (29). At meeting 5 (July 26, 1939), in response to a request for information, the minutes record a resolution "that a reply be sent stating that the proceedings of the Flying Personnel Research Committee are secret and cannot be divulged in private correspondence between members of the committee and outside individuals" (54). TNA AIR 57/40.

2. Bartlett 1943b.

3. Bartlett 1943b, 249.

4. On Kraepelin, see Shepherd 1995; Weber, Burgmair, and Engstrom 2006. On Jackson and his background, see Young 1970, chapter 6; and Danziger 1982. In 1893, Rivers visited Kraepelin in Heidelberg and co-authored an experimental study, *Ueber Ermüdung und Erholung*, with Kraepelin. Bartlett made no mention of this link in his Ferrier Lecture. See Rivers and Kraepelin 1896.

5. Bartlett 1943b 248.

6. Bartlett 1943b, 248.

7. Bartlett 1943b, 249.

8. Bartlett 1943b, 249.

9. Bartlett 1943b, 254.

10. Bartlett 1943b, 252.

11. Bartlett 1943b, 256.

12. Bartlett 1943b, 256–57.

13. Bartlett 1943b, 250.

14. In 1915, the Ministry of Munitions set up the Health of Munitions Workers Committee. This committee was dissolved in 1918 and, at the behest of the new Department of Scientific and Industrial Research and the Medical Research Committee, was replaced by the Industrial Fatigue Research Board. The chairman was C. S. Sherrington, and the board included Myers, then director of the Psychological Laboratory in Cambridge.

15. Vernon 1916. There is also a summary of work on industrial fatigue in Myers 1924a.

16. Bartlett 1958b, 510.

17. Bartlett 1958b, 510.

18. Muscio 1922, 45.

19. It is true that no criterion is "independent" of the opinions of the scientific community, but the consensus needed to identify a test of fatigue does not need such a criterion. The problem lies in constructing the required consensus. The general form of this counterargument is to be found in Kuhn's important 1961 paper.

20. Hancock, Desmond, and Mathews 2012, 64.

21. Bartlett 1927.

22. FPRC minutes, meeting 5 (July 26, 1939), p. 17, TNA: AIR 57/40.

23. An analogy of a different kind between Kraepelin and Ebbinghaus is noted in Jaeger 1985, 95. Jaeger's concern, however, was the pedagogic and political significance of the projects in which Kraepelin and Ebbinghaus were engaged. Bartlett's work, therefore, did not enter into Jaeger's analysis. Kraepelin himself drew attention to some similarities with Ebbinghaus's results on nonsense syllables and his own findings on mental fatigue. He noted that both investigations had shown the superiority of distributed practice over massed practice. The purpose of Kraepelin's comparison, of course, was orthogonal to Bartlett's. Kraepelin was seeking corroboration while Bartlett was drawing attention to a shared weakness. See Rivers and Kraepelin 1896, 674.

24. The expedition was organized by Alfred Haddon (1855–1940) and defined the reputation of Cambridge in the field of anthropology. See Herle and Rouse 1998.

25. For a useful source dealing with the background of this debate, see Barnard 2000, 52–54.

26. The evolution versus diffusion debate was beset by many complications and compromises. On Rivers's role, see Slobodin 1997, 148–60.

27. Ebbinghaus 1885 (translated as Ebbinghaus 1964).

28. On the enduring value attributed to Ebbinghaus's methodology, see, e.g., Zusne 1984, 114; Boneay 1998, 64.

29. Bartlett was not saying that Ebbinghaus denied the reality of the higher functions of the mind. Ebbinghaus's commitment to their reality is evident, e.g., in Ebbinghaus 1908, 155–91. Bartlett's objection was that Ebbinghaus's experimental methods did not give these functions scope to operate.

30. As Ryle put it in *The Concept of Mind*, "recall unsuccessfully" and "recall incorrectly" are illegitimate phrases. See Ryle 1949, 278.

31. Oldfield and Zangwill 1943, 113. Later, Oldfield was to hold the chair of psychology at Oxford and Zangwill the chair at Cambridge.

32. Bartlett 1932, 16.

33. Bartlett 1932, 63.

34. Bartlett 1932, 65.

35. Ebbinghaus was aware of the kinds of fact adduced by Bartlett, but he saw them as problems to be overcome. His quantitative treatment of time intervals was radically different to Bartlett's.

36. Bartlett did not offer any account of what the story meant to the Kathlamet peoples from whom it was taken. He also adapted the story slightly and removed some gruesome detail. For the original, see Boas 1901, 182–84.

37. Bartlett's experiment is usefully discussed, from a wide range of different points of view, in Saito 2000. The evidence in this book shows that Bartlett's results can be replicated.

38. Bartlett 1932, 126.

39. Bartlett does not use the label "conscientious objector" but expresses the point by saying, "These reproductions were all effected in the early days of the War, and this type of reason for shrinking from a fight was very effective in the kind of social group to which nearly all of my subjects belonged" (Bartlett 1932, 129).

40. Bartlett 1932, 173.

41. Bartlett's caution is expressed in the lengthy, critical discussion of themes such as that of the "collective unconscious" and the "group mind" in the final chapters of *Remembering*.

42. Writing in 1938, the American psychologist R. S. Woodworth said that the accusation of artificiality directed at nonsense-syllable experiments "is less serious than it seems." He treated Bartlett's work as merely the study of one type of memory, namely, memory of form rather than content (see Woodworth 1938, 69). Examples of the distortion reading of Bartlett's results are Anderson 1980, 54; Atkinson, Atkinson, Smith, and Bem 1999, 324; Leaking 1987, 279; Miller 1951, 221. For a historical analysis of the psychological study of memory that shows a fuller awareness of the significance of Bartlett's work, see Danziger 2008.

43. Kraepelin 1902. A systematic criticism of Kraepelin's work on both empirical and statistical grounds was put forward in Thorndike 1912 and Thorndike 1923. A more sympathetic view of Kraepelin's complex account of the work curve was taken in chapter 22 of Bills 1934.

44. Bartlett 1951.

45. Bartlett 1951, 210.

46. Head 1926. For an overview of Henry Head's work, see Jacyna 2000, chapter 5. See also Myers's 1941 obituary.

47. Bartlett 1958a, 146.

48. Bartlett 1951, 211.

49. Bartlett 1958a, 146.

Chapter Three

1. Ward 1920, 148.

2. Rivers and Head 1908, 324.

3. Head, Rivers, and Sherren 1905, 102.

4. Rivers and Head 1908, 325.

5. Head, Rivers, and Sherren 1905, 104.

6. Head, Rivers, and Sherren 1905, 106.

7. Trotter and Davies 1909, 1913. On Trotter, see Elliott 1941 and Holdstock 2004.

8. The quotations in this paragraph are all from Trotter and Davies 1913, 104.

9. Trotter and Davies 1913, 124.

10. Trotter 1923.

11. The negotiable character of replication has been the subject of systematic sociological investigation. See the empirical studies of H. M. Collins 1992. The methodological and logical basis of this negotiability was first identified by the physicist Duhem. See Duhem 1962.

12. Swets 1961. An alternative to the concept of threshold is discussed in chapter 9.

13. Sharpey-Schafer 1929.

14. Myers 1911a, 13.

15. On Adrian's work and character, see Hodgkin 1979.

16. Adrian 1928, 6; Parsons 1927, chapter 10. Parsons was also a member of the Flying Personnel Research Committee.

17. Adrian 1935, 47.

18. Adrian 1949a, 14.

19. During the Great War, Head and Rivers were also members of the Medical Research Council Air Medical Investigation Committee. See Head 1918; Rivers and Rippon 1920.

20. Myers was also deeply involved with the study of shell shock and appears to have been the inventor of this controversial term; see Myers 1940. For valuable discussions of this subject, see Young 1995 and Shephard 2000.

21. Rivers 1924.

22. Rivers 1924, 27.

23. Head and Riddoch 1917, 218, 229.

24. Rivers 1924, 29.

25. Phillips 1974.

26. Walshe 1942. I have not been able to establish whether Walshe's reiteration of his anti-Rivers polemic was prompted by Bartlett's 1941 Ferrier Lecture or whether the timing was a coincidence.

27. Walshe 1921, 428.

28. Walshe 1921, 428.

29. Walshe 1922.

30. Walshe 1922, 219.

31. Walshe 1922, 220.

32. The complex story of the relation between Spencer and Darwin, and their relation to Hughlings Jackson, is described in chapters 5 and 6 of Robert Young's (1970) important study *Mind, Brain and Adaptation in the Nineteenth Century*.

33. Walshe 1922, 222.

34. Jackson 1884, 660. The lecture was reprinted at greater length in Taylor 1932, 45–75, with the quoted passage on 58.

35. Miller 1978. An earlier, shorter version of the argument was presented in a BBC broadcast and published as Miller 1972. The broadcast had the revealing title "The Dog Beneath the Skin."

36. Miller 1978, 320.

37. Myers 1920b, 339. Myers's disagreement with Rivers went beyond this point. Rivers wanted to unify psychology and biology, while Myers wanted psychology to be an independent science.

38. Jacyna 2000, 169–70.

39. Trotter 1916.

40. The more thoughtful aspects of Trotter's argument had originally appeared in the *Sociological Review*; see Trotter 1908, 1909.

41. Trotter rightly says that the herd instinct is at work even amongst scientists and cites Darwinism as an example; see Trotter 1916, 39.

42. Trotter suggests that different national forms of gregariousness are not inherited biologically but are transmitted by "organized State suggestion" (1916, 198). This is as far as he goes in offering an explanation.

43. Trotter 1916, 139–40.

44. Freud discussed Trotter's book in *Massenpsychologie und Ich-Analyse* of 1921. He argued that Trotter had underestimated the importance of group leaders in creating social order in the primitive horde. See the translation in Freud 1940, chapters 9 and 10.

45. Bartlett 1926, 585.

46. "Roughly, we say that there is a gradual 'adding on' of the more and more special, a continual adding on of new organisations" (Jackson 1884, 662).

47. Walshe 1961, 63. Walshe dismissed Spencer as a "now-forgotten philosopher."

48. The anti-Spencerian point insisted on by Bartlett is actually present in some of Head's own writings on aphasia, but Head equivocated. This equivocation allowed Walshe to turn what might have been an escape route for Head into evidence for the further crime of logical inconsistency. Walshe makes great play with this aspect of Head's writing by quoting some blatantly contradictory statements from Head. See Walshe 1942, 64–65.

49. Bartlett's thinking "drew little from an evolutionary source," said Broadbent in his obituary (Broadbent 1970, 8). In the light of the controversy over Rivers's and Head's work on afferent sensibility, it is reasonable to suspect that, rather than a mere lack of interest, Bartlett had learned the wisdom of keeping away from such themes.

50. Sherwood 1966.

51. Sherwood 1966, 38.

52. Zangwill 1980, 10.

53. Koffka's impressive *Principles of Gestalt Psychology* was published in 1936. Koffka subscribed to what I have referred to as the principle of symmetry that distinguishes the scientific from the philosophical approaches to cognition. As Koffka put it: "This distinction between two kinds of perception, normal and illusory, disappears as a psychological distinction as soon as one becomes thoroughly aware of the fallacy which it implies, much as it may remain as an epistemological distinction. For each thing we have to ask the same question, 'Why does it look as it does?' whether it looks 'right' or 'wrong'" (Koffka 1936, 79). On this methodological point, Craik and Koffka were in agreement.

54. Craik and Zangwill 1939, 148.

55. Sherwood 1966, 176.

56. On the engineering approach that informed much of Cambridge psychology, see Gregory 1963. But not all Cambridge psychologists were happy with this trend, e.g., Zangwill. I shall return to Zangwill's doubts in chapter 10.

57. On the transition from a morally charged nature to a morally neutral nature, see Douglas 1966, 91–92.

58. In the light of these past controversies, it is perhaps surprising that Spencer's name still found its way into Zangwill's 1950 *An Introduction to Modern Psychology*. When characterizing the background to modern psychology, Zangwill said: "Herbert Spencer's conception of psychology as that aspect of biology concerned with the continuous adjustment of the organism to its external relations is nowadays generally accepted." However, this Spencerian aperçu is consistent with a rejection of Spencer's account of evolution—which had been the point at issue in the controversies. See Zangwill 1950, 11–12.

Chapter Four

1. Bartlett 1941b.

2. Bartlett 1941b, 3.

3. Bartlett 1941b, 3.

4. Bartlett 1942c.

5. Bartlett 1942c, 3.

6. Bartlett 1942c, 3.

7. Bartlett 1942c, 5.

8. Bartlett 1941a, 5.

9. Bartlett 1941a, 4.

10. Bartlett 1941a, 4.

11. Bartlett 1942c, 8.

12. Bartlett 1941a.

13. Bartlett 1941a, 2.

14. Bartlett 1941a, 2.

15. Bartlett 1941a, 3.

16. Bartlett 1923a, 248–49. Bartlett directed his readers to Farmer 1921 and Myers 1920a.

17. Bartlett 1941a, 8.

18. Bartlett 1942b.

19. On Chambers and Farmer, see Burnham 2009, chapter 3.

20. Bartlett 1942b, 7.

21. Bartlett 1942b, 7.

22. FPRC minutes, meeting 27 (September 3, 1942), minute 246, p. 3, TNA: AIR 57/42.

23. FPRC minutes, meeting 19 (July 6, 1941), minute 170 (a), p. 2, TNA: AIR 57/42.

24. At that time, Bradford Hill was reader in epidemiology at the London School of Hygiene and Tropical Medicine. For his textbook, see Bradford Hill 1937. On his life and career, see Doll 1994. For a lighthearted account of his flying days, see Bradford Hill 1983.

25. Bradford Hill and Williams 1943. The references to "special circumstances" and "ordinary straightforward landings" are on p. 2 of the report.

26. In the opinion of the instructors, length of flight was not the only factor causing fatigue: formation flying, bad weather, and strong opposition by defending forces were also deemed important (Williams 1942).

27. FPRC minutes, meeting 13 (October 30, 1940), minute 121(f), p.6. TNA: AIR 57/41.

28. Bartlett 1942a.

29. Grindley 1940.

Chapter Five

1. Russell Davis uses Bartlett's work as resource and reference point in, e.g., Russell Davis 1951 and Russell Davis and Cullen 1958.

2. The metaphor of the liaison officer comes from Russell Davis 1952. For biographical information see Russell Davis 1957; Russell 1993; and Hooper 1993.

3. Russell Davis 1946b, 7.

4. Russell Davis 1948, 2.

5. Russell Davis 1948, 33.

6. Russell Davis 1948, 32.

7. Russell Davis 1948, 25.

8. Russell Davis 1948, 3.

9. Russell Davis 1948, 2.

10. Russell Davis 1948, 29.

11. Russell Davis 1948, 33.

12. Russell Davis 1948, 30.

13. Bartlett's focus on normality was a central part of his methodology in *Remembering* and was central to his suspicions about Ebbinghaus's methods. The word "normal" occurs four times within the first two pages of the preface to *Remembering*.

14. FPRC minutes, meeting 33 (November 16, 1945), minute 2, TNA: AIR 57/42.

15. Bartlett 1942c, 6.

16. FPRC minutes, meeting 29 (February 23, 1943), minute 267 (b), p. 3, TNA: AIR 57/42.

17. Air Ministry 1947.

18. FPRC minutes, meeting 29 (February 23, 1943), minute 267 (b), p. 3, TNA: AIR 57/42.

19. Reid 1942. Reprinted in Dearnaley and Warr 1979, 43–62.

20. FPRC minutes, meeting 29 (February 23, 1943), minute 267 (b), p. 3, TNA: AIR 57/42.

21. On Bartlett's policy of keeping Freud out of the syllabus in the Cambridge Psychology Laboratory, see Forrester 2004 and Forrester and Cameron 2017.

22. Bartlett 1958b, 512.

23. Bartlett and Mackworth 1950.

24. Mackworth 1950, 12.

25. The Official History of the war at sea indicates the magnitude of the problem. Losses due to submarine attack were: for 1941, 432 ships, 2,171,754 tons; and for 1942, 1160 ships, 6,266,215 tons. See Roskill 1961, 479.

26. Mackworth 1944.

27. Mackworth 1944, 2.

28. Cambridge was the home of the No. 1 Initial Training Wing, and the laboratory had an agreement with the RAF that it would furnish experimental subjects. See Whittingham 1939b.

29. Mackworth 1944, 7.

30. A full account of these further experiments was published after the war. See Mackworth 1950.

31. Mackworth 1948–1949.

32. Hill 1936. On the centenary of Pavlov's birth in 1949, E. D. Adrian also paid a glowing tribute; see Adrian 1949b.

33. Bartlett 1932, 210–11.

34. Oldfield 1937.

35. Oldfield 1937, 37.

36. Oldfield 1937, 37.

37. Craik 1943c,114.

38. Pavlov 1927, 395.

39. Broadbent 1980, 51.

40. Broadbent 1953a.

41. Broadbent 1953a, 335.

42. Broadbent 1953a, 331.

43. Normally, the greater the energy of a stimulus, the greater the effect, but there is a "paradoxical phase" that Pavlov defined as follows: "In a special phase during the transition from the waking state in the direction toward sleep this normal relation of the effects disappears. Either all the effects become equal (the equality phase), or the relation becomes reversed so that the effects from weak stimuli are greater than those from the strong, or the latter may even remain without any effect (*paradoxical* phase)" (Pavlov 1928, 356). The emphasis is Pavlov's.

44. Pavlov 1927, 20.

45. Pavlov 1927, 143.

Chapter Six

1. Russell Davis 1946c.

2. Broadbent 1980.

3. Broadbent 1980, 48.

4. Broadbent 1980, 49.

5. "It seems fair to say that the periods of genuine scientific growth in psychology have synchronized with periods in which men have had to solve intensely practical problems, and that in this field especially the necessity for applying knowledge has been the main spur to its advance" (Bartlett 1945, 45).

6. Broadbent 1980, 49.

7. Woodworth 1938.

8. Hilgard and Marquis 1940.

9. It is important to distinguish between the austerity of the postwar Labour administration and the severe economic austerity imposed upon the country by subsequent Conservative administrations. The former was a necessity due to the war; the latter was ideological choice. See Addison 1994.

10. Broadbent 1980, 41.

11. Broadbent 1977a, 181.

12. Broadbent 1980, 49.

13. Broadbent 1980, 181.

14. Broadbent 1980, 49–50.

15. Bartlett 1945, 46.

16. Bartlett 1947, 1948b, 1948c, 1949, 1950.

17. Bartlett 1948b, 33.

18. Bartlett 1948b, 32.

19. The same argument is advanced in Bartlett 1945, 45.

20. Bartlett 1948b, 31.

21. On Stevens (1906–1973), see Woodward 1990. On the Harvard Psycho-Acoustic Laboratory, see Capshew 1999. For an overall survey of the Harvard work, see Rosenzweig and Stone 1948 and Miller 1951.

22. Russell Davis 1948, 17–19.

23. Broadbent 1958, 92.

24. Broadbent 1957.

25. On Leonard (1922–1972), see Howarth 1972.

26. Broadbent says that the "sixth 10 min. was just significantly worse than the first, t = 2.32 while the 0.05 level is 2.31" (Broadbent 1953c, 296).

27. Broadbent 1953c, 298.

28. Broadbent 1958, 95.

29. Broadbent 1958, 88.

30. Broadbent 1958, 91.

31. Broadbent 1958, 88.

32. Thirty years after the episode just described, Poulton (1976) used a masking theory of this kind as grounds for declaring that Broadbent's Twenty-Dial and Five-Light Test results were experimental artefacts. Broadbent met Poulton's objection head-on and gave a detailed, point-by-point reanalysis and defense of the early experiments. See Broadbent 1976 and Broadbent 1977b.

33. Bills 1931, 1934.

34. Bills 1937, 437.

35. Bartlett 1942c.

36. Broadbent 1958, 133.

37. Broadbent 1958, 41–42.

38. Sherwood 1966, 44–45.

39. Broadbent cited evidence from Conrad and Hille 1955.

40. Broadbent 1958, 136. Broadbent's references here are to Bartlett's Ferrier Lecture (Bartlett 1943b) and Drew's FPRC report 227 (Drew 1940).

41. Broadbent cites other sources of experimental evidence for the perceptual locus of fatigue effects: Siddall and Anderson 1955; Whittenburg, Ross, and Andrews 1956; and unpublished work by E. Saldanha, carried out at the Applied Psychology Unit.

42. On Poulton, see the obituary Anon. 2001, 23.

43. Poulton 1952a, 222.

44. Vince 1944, 1946a, 1946b.

45. Poulton 1957a. Further related papers include Poulton 1950, 1952b, 1957b, 1957c.

46. Broadbent 1977a, 184.

47. Poulton follows Bartlett when he decries attempts "to split up a skilled performance into elements, each of which is believed to depend purely on the elements immediately preceding it." This sequential understanding, he insists, cannot be the complete picture: "Present performance is always dominated by the subject's idea of the immediate future by what he expects to happen, and by what he is trying to accomplish" (Poulton 1950, 110).

48. Poulton and Gregory 1952, 65.

49. Poulton 1957b, 194.

50. Craik 1947–1948. Poulton found that sudden discrete corrections were less frequent than Craik reported (Poulton 1957b, 194).

51. Poulton made the distinction in his early papers, but the point is most extensively discussed, along with criticisms of the American experimental evidence, in Poulton 1974, chapter 9.

52. Conrad 1951, 1954. Under Broadbent, both Poulton and Conrad became deputy directors of the Applied Psychology Unit.

53. Conrad 1955, 175.

54. Broadbent 1958 (288) draws attention to this "unexpected" result.

55. Bartlett 1948b, 36.

56. Bartlett will also have been aware of a study by R. C. Browne that was commissioned by the FPRC. Browne looked into the performance of teleprinter switchboard operators working at RAF communication centers. He compared their day and night response times. The longest response times were in the early hours of the morning, but responses were quicker when the number of incoming calls was higher: "There was a tendency for performance to improve, the more there was to do" (Browne 1943–1944, 125).

57. Broadbent 1958, 297–301.

58. Broadbent 1958, 300.

59. Broadbent 1953b.

60. Broadbent 1980, 49.

Chapter Seven

1. Atzler 1927, 196–328 and 488–651. On Durig, see Burtscher et al. 2012.

2. Simonson 1935. Simonson also contributed the chapter "Arbeitsphysiologie" to the Bethe et al. 1930 *Handbuch der normalen und pathologischen Physiologie*. Simonson later worked in the USA and argued that the flicker fusion frequency provided the much-sought test of fatigue; see Blackburn 1975.

3. Benary, Kronfeld, Stern, and Selz 1919. See Kehrt 2010, 61–165; and Kluwe 1997, 209–11.

4. See, e.g., Koschel 1913.

5. Gundlach 1997, 2004, 2005.

6. Gade 1928.

7. As representative of an extensive and important literature on the psychotechnic movement, see Dorsch 1963, 80–93; Maier 1970; Chestnut 1972; Rabinbach 1990, chapters 7 and 10; Killen 2007. For the "Tod der Psychotechnik," see Dorsch 1963, 88.

8. See Geuter 1984, 296; 1985a; 1985b. For two brief, authoritative overviews of the transformation of German psychology under Nazism, see Ash 1995, 342–46, "Transformation of Psychology Under Nazism"; and Rabinbach 2018, chapter 5, "Psychotechnics and Politics in Weimar Germany."

9. On the shift from psychotechnics to characterology, see Herrmann 1966; Jaeger 1985.

10. Jaeger and Staeuble 1981; Bauer and Ullrich 1985.

11. On Moede, see Spur 2008; Geuter 1984; Haak 1996. Moede's copiously illustrated *Lehrbuch der Psychotechnik* gives a good indication of the apparatus-based approach to aptitude testing in large undertakings such as the German railway system. See Moede 1930, 415–25.

12. Schmidt 1938.

13. Everling 1926, 1934a, 1934b. Another product of Everling's department was Petersen 1939. For institutional background, see Koelle 1979.

14. Eschenbach 1938, 1941.

15. See Kehrt 2010, 371, for a discussion of the Everling and Eschenbach apparatus. Kehrt says that dummy cockpits were used for training in blind flying by the German Air Force at the end of the 1920s. There is no suggestion that such cockpits were used for research into fatigue.

16. Anthony 1941.

17. Anthony had joined the NSDAP in 1933 and had been a medical consultant at the Reichsluftsministerium in Berlin since 1940. See Klee 2003, 17. After the war, Anthony resumed his medical professorship at the University of Rostock.

18. Anthony and Schaltenbrand 1936; Anthony and Atmer 1936.

19. Anthony 1941, 1327.

20. All translations are by the author unless otherwise indicated.

21. Atzler 1927, 244.

22. Anthony 1941, 1328.

23. Anthony 1941, 1328.

24. Anthony 1941,1329.

25. Anthony 1941, 1327.

26. Anthony 1941, 1327.

27. Eckert 2013, 213.

28. Unger 1994. For a vivid account of the widespread use of Pervitin, see Ohler 2016.

29. Graf 1950, 1080. The reference to the German Air Force regulation is on p. 1099. I shall say more about Graf in the next section.

30. On the development of electroencephalography in Germany, see Hagner 2001; Borck 2001, 2008.

31. Kornmüller was interviewed about his work by a US Army intelligence officer on July 20, 1945. The interviewer was Major Leo Alexander, who later achieved prominence as a medical advisor at the Nuremberg Trials. In his Combined Intelligence Objectives Sub-Committee (CIOS) report, Alexander discussed Kornmüller's 1944 book *Einführung in die Klinische Elektrenkephalographie* and expressed skepticism about Kornmüller's claims regarding his fatigue-detection device. See Alexander 1945.

32. Kennedy 1953, 49 and 153.

33. Some statistics on the distribution of the German research effort in aviation psychology and physiology during WWII is provided by Strughold (1950). In table I.4 (chapter IA, 8) of *German Aviation Medicine*, Strughold's statistical breakdown shows that between 1933 and 1945, under the heading of high-altitude effects, there were 45 studies concerned with flight, 132 concerned with air mixtures, 378 concerned with the decompression chamber, and 92 dealing with oxygen respiration, i.e., a total of some 647. There were 28 studies of fatigue, and even this total refers to the combined categories of fatigue and recreation.

34. Some of the German research on the effects of high altitudes and oxygen deficit was pursued using prisoners in Dachau as experimental subjects. On this appalling subject, see Roth 2000 and 2002.

35. Russell Davis 1946c, 2. On the background of the Dortmund institute, see Plesser and Thamer 2012.

36. Lehmann and Graf 1942.

37. Both Lehmann and Graf had published on the effects of Pervitin. See Graf 1939; Lehmann, Straub, and Szakáll 1939.

38. On Graf's career, see Lehmann 1963.

39. Lehmann was a member of the NSFK (the National Socialist Flying Corps) and from 1937 a member of the NSDAP (National Socialist German Workers' Party). See the archives of the Max-Planck-Gesellschaft (AMPG) in Berlin-Dahlem, AMPG Abt. II, Rep. 1A.

40. Lehmann and Graf 1942, 183.

41. Lehmann 1950, 1104.

42. In the UK, this phase of the war is known as the Battle of Britain. The military situation was dominated by four facts: (i) France was under German occupation, (ii) Germany and Russia were acting as allies, (iii) the United States was neutral, and (iv) the English Channel was twenty-five miles wide. See Wood and Dempster 1969; Addison and Crang 2000.

43. Lehmann and Graf mention A. V. Hill by name but do not give any references. They will have had in mind such papers as Hill and Upton 1923.

44. Anthony 1940; Rein 1941. On Rein (1898–1953), see Klee 2003, 486–87.

45. Lehmann and Graf 1942, 197.

46. Graf 1933, 195.

47. Graf 1943, 450.

48. Lehmann 1950, 1107.

49. Graf 1943, 456.

50. Graf 1933, 172.

51. Graf 1950, 1081.

52. Lehmann and Graf 1942, 195.

53. Graf 1950, 1081.

54. Graf 1950, 1080.

55. Graf 1950, 1098.

56. Strughold 1950. On Strughold, see Klee 2003, 610; and Harsch 2004.

57. On the work of the Dortmund institute for the German Army, see Schmaltz 2012.

58. Erhard Milch to Max von Cranach, May 13, 1935, AMPG I Abt. Rep 1A. Akte 1367/5. On Milch, see Weiß 1998, 320.

59. Max von Cranach to Edgar Atzler, May 14, 1935, AMPG I Abt. Rep 1A. Akte 1367/5. On Hippke, see Klee 2005, 258.

60. Edgar Atzler to Max von Cranach, May 27, 1935, AMPG. I Abt. Rep. 1A. Akte 1406/3.

61. Edgar Atzler to Max von Cranach, September 30, 1935, AMPG. I Abt. Rep. 1A. Akte 1367/6.

62. One study by Irene Raehlmann, and another by Alexander Neumann, have already established that the Dortmund institute was the loser in a competition for preferment and funding. I am adding detail relevant to the Lehmann and Graf experiments. See Raehlmann 2005, section 5.1, especially 98–99; and Neumann 2012, especially 185–86.

63. Strughold's name first appeared on the title page of *Arbeitsphysiologie* in 1937, vol. 9.

64. Klee 2005, 184.

65. Strughold's categorization shows the marked contrast between the number of studies on subjects, such as oxygen deficit, that were receiving scientific attention and those (such as fatigue) that weren't. The ratio was roughly 20 to 1.

66. The six research centers were (i) Institut für Flugmedizin bei der Luftwaffe, Rechlin (headed by Dr. T. Benzinger); (ii) Institut für Flugmedizin bei der Deutschen Versuchsanstalt für Luftfahrt, Berlin-Adlershof (headed by Dr. S. Ruff); (iii) Institut für Luftfahrtmedizin am der Universität Hamburg (headed by Dr. H. Lottig, later Dr. H. Schwarz); (iv) Institut für Luftfahrtmedizin, Munich (headed by Prof. G. A. Welz); (v) Luftfahrtmedizinisches Forschungsinstitut, Berlin, with branches in, e.g., Göttingen (headed by Dr. H. Strughold); (vi) Sanitätsversuchs- und Lehranstalt, Jüterbog (headed by Prof. H. von Diringshofen, later Prof. W. Knothe).

67. Russell Davis 1946c, 13.

68. A number of historians have argued that, between 1933 and 1945, Germany is best seen as

having a "polycratic" rather than "totalitarian" structure. Their aim was not to diminish the evils of fascist dictatorship but to challenge the claim that such systems can operate with a unique and ruthless degree of efficiency. See Hüttenberger 1976; Harwood 1997; Macrakis 2000; Neumann 2005. The present case study is consistent with this broader thesis.

69. Goldie (1943) had placed the emphasis on fear rather than fatigue as the source of performance decrement during operational flights.

70. Goldie and Gilson 1945.

71. All of the quotations in this paragraph are from Goldie and Gilson 1945, 2. The specifications and history of the Bachem Ba 349, Natter, are given in Kosin 1988, 192–93.

72. Smith 1992, 171.

73. Kraepelin 1902, 489.

74. Graf 1933, 172. Graf cites a paper by the engineer Hallgebauer (no first name or initial given). See Hallgebauer 1927. Hallgebauer mentions the similarity between driving a motor vehicle and piloting an aircraft.

75. Oswald Schmiedeberg (1883, 35–41) argued that the stimulating effect attributed to alcohol is really the reduction of an inhibitory process, but Jackson's name is not mentioned. On Schmiedeberg, see Meyer 1922. Foerster (1936) was an admirer of Jackson. Though a distinguished scientist, he did not form a school of followers. On Foerster, see Zülch 1973.

76. The chapter on alcohol in Atzler's *Handbuch* was devoted to a criticism of the release model of intoxication and a defense of an alternative approach derived from Kraepelin's work. It was written by Georg Joachimoglu (1887–1979), who was the acting head of the Institute of Pharmacology in Berlin (Eichholtz 1957). In Graf's (1937) work on alcoholic intoxication, I have found no recourse to the Jacksonian idea of "release." In Graf 1930, Graf is sympathetic to Schmiedeberg but nevertheless comes to the same conclusion as Joachimoglu: Alcohol has a stimulating effect as well as an inhibiting effect.

77. Rein 1938, 128.

78. Erich Hippke took a similar position. In a 1943 lecture at the *Haus der Flieger* in Berlin (to celebrate Hermann Goering's fiftieth birthday), Hippke identified tiredness and apathy as symptoms of altitude sickness, rather than research foci in their own right.

79. On Lottig and the Hamburg Institute, see Harsch 2003, chapter 2, sections 3 and 4.

80. Lottig 1936.

81. Livingston 1962, 164–65.

82. Gilbert 1951, 3–19.

83. Lottig 1939, 227.

84. Lottig 1939, 226.

85. Strughold (1950, 39) described Lottig's work as of "a speculative rather than an experimental character."

86. Kraepelin 1919, 176.

87. The quotation is from a translation of Kraepelin's article by Eric J. Engstrom (1992, 262). See also Engstrom 1991.

88. See, e.g., Haffner 1986, chapter 13.

89. Flanagan records that the official directive ordering Lt. Col. Fitts to the European Theatre had already been issued on May 5, 1945.

90. Gimbel 1990.

91. Lovelace 1946. Lovelace's interrogation is dated June 6, 1945.

92. At the end of April 1945, there was still fighting in the area around Munich and Garmisch-Partenkirchen; see Kershaw 2011, 344, and Wilmot 1952, 701.

93. Henschke 1945.

94. The original report, *Eine neue Steueranordnung für Jadgflugzeuge*, October 1944, is now in the Bundesarchiv. An English translation, Henschke and Karlson 1946, was made by the Intelligence (T-2) Air Documents Division.

95. After the war, Hollwich (1909–1991) worked in Munich and was an authority on diseases of the eye (François 1977).

96. Williams 1945, 8. The preference for physiological rather than psychological or psychiatric terms of reference is evident in, e.g., Georg von Knorre's 1943 report *Übermüdungszustände bei Fliegern* (Fatigue States in Pilots). Knorre identifies a group of characteristic symptoms such as irritability, sleep disorders, headaches, and sweating, but he recommends that they be seen as a particular failure of the autonomic nervous system "for which a cause has yet to be found." The recommended treatment is in a medically directed hospital (Knorre 1943, 8).

97. The author of one of the papers included in A. J. Anthony's 1941 survey of work on fatigue was also looking at disturbances to cell permeability, see Parade 1941.

98. Ficker and de Rudder 1942 (second edition 1948).

99. Whittingham 1941a.

100. Whittingham 1941b, 2.

101. Hollwich (1979) developed the theory that light entering the eye had an energetic and metabolic function as well as a visual function.

102. Gerathewohl 1944.

103. Gerathewohl 1953. The lack of attention given to fatigue echoes that of comparable prewar and wartime publications such as Ruff and Strughold 1944, and the literature survey of Schmidt 1938. Gerathewohl's book can be looked on as an updating of the Ruff and Strughold volume. Strughold and Schmidt also worked in the United States in the postwar years.

104. After WWII, Gerathewohl produced a series of reports and papers while based at the School of Aviation Medicine at Randolph Field and the Office of Aviation Medicine of the Federal Aviation Administration, Washington. See, e.g., Gerathewohl 1951, 1952, 1954, 1976.

105. For example, there is a brief mention of *Ermüdungserscheinungen* (fatigue phenomena) in relation to the question of how much pressure should be needed to operate the controls of an aircraft (Gerathewohl 1953, 37–38).

106. Hartcup (2000, 133) says of the Flying Personnel Research Committee that "the British committee enjoyed an unusual degree of independence: its recommendations were quickly approved and trials of new equipment carried out rapidly."

107. Gregory 2000, 41.

108. In a lecture, C. P. Snow described, for the benefit of an American audience, the "apparently casual to-ing and fro-ing by which high English business gets done." Thus, A would lunch with B at the Athenaeum, B would have a cup of tea with C and D—things might then happen "with the smoothness, the lack of friction, and the effortless speed with which things can only happen in England when the Establishment is behind one" (Snow 1961, 28). A less flattering view of the Establishment style is expressed in H. J. Eysenck's remarks about Bartlett's elitism and sense of entitlement (Eysenck 1990, 111).

Chapter Eight

1. For an account of the professional mobilization of American psychologists for the war effort and its complicated entanglement with both academic and domestic politics, see chapter 2 of Capshew's important *Psychologists on the March* (1999).

2. On the overall system of scientific administration in the United States during these years, see Purcell 1979. On the Office of Scientific Research and Development, see Baxter 1968; Bray 1948; Owens 1994.

3. Macmillan, Johnston, and Williams 1946.

4. Macmillan, Johnston, and Williams 1946, 44. Similar negative conclusions can be drawn from Chapanis, Garner, and Morgan 1949 and from Thorndike 1951. Both discuss the Cambridge Cockpit, but neither mentions any American studies that are similar in nature. On the Committee on Selection and Training of Aircraft Pilots, see Capshew 1999, 44.

5. Flanagan 1948.

6. Melton 1947.

7. Fitts 1947b.

8. Bilodeau and Bilodeau 1961, 244–45.

9. Melton 1957, 742.

10. Melton 1947, 1033; Capshew 1999, 165–69.

11. Gagné 1954, 99.

12. Hull 1943. On Hull, see Gondra 2008. Authoritative accounts of the various strands of work that fall under the heading of "behavior theory" and "learning theory" are to be found in Hilgard 1956 and Estes et al. 1954.

13. See pages 379–81 of the third, revised edition of Watson's (1929) classic.

14. Hull (1943, 30) explained his methodological position by reference to P. W. Bridgman's (1938) *The Logic of Modern Physics*. Koch (1954) gives a detailed discussion of Hull's definitional practices.

15. Koch 1954, 106–7.

16. Bilodeau and Bilodeau 1961, 263.

17. Hull 1943, 278.

18. Hull 1943, 208.

19. An asymptote is a limit that is approached closer and closer but never achieved. Zero is the asymptote of the sequence of numbers $(1/n)$ where $n = 1, 2, 3$, etc.

20. Hull 1943, 102.

21. The language of "conditioning" is Pavlovian. Hull cited Pavlov extensively but kept his distance from Pavlov's speculative brain physiology.

22. Hull 1943, 281.

23. Hull 1943, 284.

24. The psychologist's use of the word "reminiscence" differs from common usage. More important is the counterintuitive character of the psychological phenomenon itself. The fact that it is predictable from Hull's theory is an answer to critics of psychology who suggest that psychologists only explain what is obvious anyway. The point is well made in Broadbent 1964a, 123–26.

25. Hull 1943, 282.

26. For Hull's own unease about conditioned inhibition, see the discussion "Problems Connected with the Conditioning of Inhibitory Potential $(_sI_R)$," in Hull 1943, 301–2.

27. Gleitman, Nachmias, and Neisser 1954. The criticisms explored in the paper include important problems presented by experimental results as well as theoretical dilemmas, contradictions, and paradoxes. For example, Hull took for granted that the extinction of a response involved the repeated evocation of the response. The authors challenge this assumption by assembling the evidence that a conditioned response can be extinguished without the repetition of the response, i.e., by what might be called latent extinction. For simplicity, I am selecting for consideration just one strand of the argument in this paper.

28. Gleitman, Nachmias, and Neisser 1954, 30.

29. Gleitman, Nachmias, and Neisser 1954, 30.

30. Hull 1943, 300.

31. Gleitman, Nachmias, and Neisser (1954). In his review of the *Principles*, Sigmund Koch had noticed this implication of Hull's definitions but had failed to exploit its full significance. See Koch 1944, 284–85. McGeoch and Irions (1952) also draw attention to this result in a footnote and describe it as "an unfortunate deduction." They say: "Since $_sH_R$, once it reaches its maximum, is no longer influenced by continuous reinforcement, since I_R does not subtract from $_sH_R$, and since no provision is made for differential increments in I_R under conditions of reinforcement and non-reinforcement, it follows that, as soon as $_sH_R$ reaches its limit, performance should decline as rapidly under conditions of overlearning (continued reinforcement) as under conditions of extinction (non-reinforcement)" (55).

32. Hilgard 1956, 138.

33. Jensen 1961.

34. In addition to Hilgard, Jensen looked at critical suggestions from Iwahara 1957; Jones 1958; Osgood 1953; and Woodworth and Schlosberg 1954. Also relevant are earlier elaborations of Hull's work in Ammons 1947a and 1947b and Kimble 1949.

35. Jensen 1961, 294.

36. Jensen 1961, 278. There is a clear account of Hull's 1943 position, as well as a discussion of the Gleitman, Nachmias, and Neisser objection, in Hilgard and Bower 1975, 159–66. Hilgard seems to dismiss the problem that Hull's postulates imply that the learning curve will decline to zero with continued practice: "These deficiencies . . . could be patched up with closer attention to details of formulation of the inhibition assumptions, or the laws of chaining, etc." (166). Hilgard's stance reminds us of the variability of expert judgements that can surround one and the same anomaly.

37. Bilodeau and Bilodeau 1961, 264.

38. Schmidt 2011; Wickens and Williges 2011.

39. Adams 1956, 189.

40. "It must be concluded therefore, that present-day servo theory stands in an analogous, not a scientific, relationship to man-machine tracking systems" (Adams 1961, 59).

41. Adams 1963.

42. Adams 1964, 193.

43. Adams 1964, 194.

44. Bilodeau and Bilodeau 1961, 267.

45. Adams 1955.

46. Adams 1964, 186.

47. Adams 1955, 390.

48. Adams 1955, 382.

49. Adams 1955, 393.

50. Adams 1967, 18.

51. Adams 1964, 194.

52. Broadbent 1957, 399. Milton's original line was from the poem "On His Blindness." The line became popular in Britain during both World Wars when it was used in reference to those serving on the home front.

53. Broadbent 1957. In a subtle assessment of the literature from animal learning, Broadbent described a number of experiments that demonstrated that performance decrements can be evoked despite the absence of repeated responses—i.e., they can be elicited by perceptual processes.

54. Broadbent says, in *Perception and Communication*, that "the bias of the filter towards novelty . . . is towards channels which have not recently passed information through the filter" (Broadbent 1958, 86). Unfortunately, the misreading was not special to Adams but proved to be endemic.

55. Adams 1964, 181–82.

56. Frankmann and Adams 1962, 269.

57. Melton and Briggs 1960, 72.

58. Taylor 1947. Melton (1957) gives a useful description of the different organizational structures in which the psychologists of the US Army, Air Force, and Navy carried out their work.

59. On the historical role of fire-control research, see Mindell 2002. William Kappauf worked with Birmingham and Taylor on Project N-111, which he describes as having been "organised to investigate psychological problems in the design and operation of new antiaircraft lead computing gun sights and gun directors" (Kappauf 1947, 84–85). Kappauf emphasized how proactive psychologists were in shifting their role away from selection and training and toward equipment design (83–84). In the interwar years, Bartlett was also involved with anti-aircraft fire-control systems, though of an earlier generation; see Bloor 2000. Some of the research questions hinted at by Bartlett in discussing this theme in *Remembering* were later, independently, followed up in the Naval Research Laboratory—e.g., regarding group structures and patterns of interaction. No awareness of this overlap was registered by either party. See the section on "Multiman-Machine Systems" in Melton and Briggs 1960, 88–89.

60. Taylor 1960, 646. There is further discussion in the posthumous Taylor 1963, 866–72.

61. Adams 1961, 56–57.

62. Licklider (1960) provides a technical survey of the cybernetic approach that was influential amongst psychologists.

63. Taylor 1960, 647.

64. Taylor 1957, 253.

65. Ely, Bowen, and Orlansky 1957.

66. Taylor 1960, 647.

67. Occasionally Taylor said a system operator acts "analogously to a complex differential equation solver"; see, e.g., Taylor 1963, 869, 875. This softening of the language may be a concession to readers still in the grip of what Taylor saw as disciplinary conservatism.

68. Birmingham and Taylor 1954a.

69. Birmingham and Taylor 1954b. Henschke of the Garmisch-Partenkirchen group, whose work was discussed in the last chapter, held a position similar to the Birmingham-Taylor design philosophy. See Henschke and Mauch 1950, 90. This design stance is almost certainly connected with their move from Germany to America immediately after the end of WWII. I do not know whether the fact that they held to this stance was a cause or an effect of the move from country to country.

70. Birmingham and Taylor 1954b, 1754.

71. Birmingham and Taylor 1954b, 1752.

72. For a perceptive account of this episode, see Mindell 2008. On a more journalistic level, see Wolfe 1980 (first published 1979).

73. Taylor and Birmingham 1959, 178.

74. Taylor and Birmingham 1959, 178.

75. Taylor and Birmingham 1959, 179.

76. Taylor and Birmingham 1959, 180.

77. Taylor and Birmingham 1959, 180.

78. Taylor and Birmingham 1959, 180.

79. Taylor and Birmingham 1959, 180.

80. Taylor and Birmingham 1959, 181.

81. Taylor and Birmingham 1959, 181.

82. Taylor and Birmingham 1959, 178.

83. Taylor 1963, 859.

84. Mindell 2002. Mindell's study gives examples of work on amplifiers that admirably conveys this inductive technological process and its ramifications: See also Bennett 1963, especially chapter 3, "The Electronic Feedback Amplifier."

85. Taylor 1963, 880.

86. Bartlett 1958a, 165.

87. The classic formulation of Taylor and Birmingham's argument was first articulated by Duhem in 1914 in the field of physics. It took the form of showing that no scientific hypothesis can be tested in isolation. See Duhem 1962.

Chapter Nine

1. Bartlett 1951, 209.

2. Welford 1968. See particularly chapter 8, "Effects of Loading."

3. "Complete systems and schemes of psychological explanation are the biggest stumbling-block to progress in psychology" (Bartlett 1936, 49–50). See also the denunciation of "fixed, final and dogmatic systematizing" in Bartlett 1937, 110.

4. Bartlett 1958a.

5. Bartlett 1958a, 115.

6. Bartlett 1958a, 50.

7. Bartlett 1932, 280; 1958a, chapter 7 and chapter 8.

8. Bartlett 1958a, 11.

9. Bartlett 1958a, 8.

10. Bartlett 1958a, 136.

11. Bartlett 1958a, 12.

12. Gigerenzer 2001, 47.

13. Broadbent 1962.

14. As Broadbent (1971, 479) said: "It is the gradual spread of much-abused reductionism which creates a dialogue between one view and another."

15. Broadbent 1962, 197.

16. Broadbent 1962, 197.

17. Broadbent 1962,197.

18. Broadbent 1962, 199–200.

19. Broadbent 1962, 200.

20. Tanner and Swets 1954; see also Green and Swets 1966.

21. Broadbent 1962, 203.

22. The full details of the experiment were later published in Broadbent and Gregory 1963.

23. Broadbent 1962, 202.

24. Broadbent 1962, 203.

25. Broadbent 1971, 60.

26. Broadbent 1971, 60-63.

27. Jerison and Wallis 1957a, 1957b; Jerison and Wing 1957.

28. Broadbent (1980, 55) mentions the particular role played by the US Navy in furthering these contacts.

29. Broadbent 1963b.

30. Broadbent 1971, 60.

31. Broadbent 1963b.

32. Broadbent 1971, 61. There was also another problem. A button press that failed to generate a light flash could be a signal, or it could be a failure to push the button correctly. Signal recognition was thus rendered ambiguous—a state of affairs that surely contributed to the lack of follow-up using this apparatus.

33. Broadbent 1971, 61.

34. Broadbent 1971, 62.

35. Broadbent 1971, 63.

36. Broadbent 1971, 64. One sees here a good example of the methodology that Broadbent advocated in *Perception and Communication*, namely that of trying to divide the possibilities into broad alternatives and then repeating the process to narrow down the field.

37. Broadbent 1971, 65.

38. Broadbent 1962, 202.

39. Broadbent 1971, 69-70. On page 460, Broadbent notes that "the key achievement of the theory is the disproof of the traditional guessing correction with its linear relation between correct and false responses."

40. Broadbent 1971, 110. See also Wiener 1964 and Broadbent and Gregory 1963, 322. Broadbent and Gregory hinted that two different forms or loci of arousal might be needed, but they did not elaborate. These comments prefigure some the later more speculative discussions of arousal in *Decision and Stress*.

41. In 1928, starting from a discussion of transmission processes of the kind using Morse code, R. V. L. Hartley, of Bell Systems, derived the formula $H = n \log s$, where H is defined as the amount of information, s is the number of symbols available at each selection, and n is the number of selections that comprise a signal. See Hartley 1928. For background, see Cherry 1951; Miller 1953; and Collins 2007.

42. See, e.g., the tidal metaphor in Broadbent 1977a, 181.

43. Broadbent 1971, 4.

44. The modified understanding of the filtering process, which allowed for the "break though of the unattended," did not compromise Broadbent's rereading and criticism of Pavlov. For an amusing but highly informative account of the circumstances that led to the modification of the filter theory, see Broadbent 1964. Much of the important data used against the earlier, over-simple account of filtering came from the work of Treisman (1964) in Oxford. The matter is discussed in Broadbent's 1982 "Task Combination and the Selective Intake of Information," in which he showed that the new data required an early rather than late processing of information in the nervous system. Pavlov was not mentioned by name in this paper, but Pavlov's premises implied a simple form of the "late processing" position—which was the target of Broadbent's original criticism.

45. Broadbent 1982, 259.

46. Hockey 1970a, 1970b, 1970c.

47. Broadbent 1971, 431.

48. Broadbent 1958, 136.

49. Broadbent 1982, 260.

50. In order to help in the identification of friendly and hostile aircraft, service men and women were issued books of silhouettes. These books were also available to the public: see, e.g., the enormously popular *Aircraft Recognition* by Saville-Sneath (1941), which was originally devised for use by the Observer Corps. The psychologist Magda Vernon (who worked with Craik on dark adaptation) studied the perception of aircraft silhouettes for the Flying Personnel Research Committee. See M. D. Vernon 1941a and 1941b.

51. Broadbent 1982, 260.

52. Broadbent 1971, 10.

53. Broadbent 1993, 871.

54. "In general there will be an economical coding for any communication channel such that more probable messages are transmitted more rapidly, and the human nervous system is no exception" (Broadbent 1958, 134). For the later stance toward the doctrine of optimal coding, see Broadbent 1971, 276.

55. Broadbent's information-theoretic derivation of Hick's law is given on pages 274–75, the decision theory derivation on pages 288–90, of *Decision and Stress* (1971). The log N law appears to have first surfaced in the work of German members of the psychotechnic tradition, but only as a purely empirical relation. Hick (perhaps advised by Leonard) attributed the empirical discovery to Georg Blank (1934).

56. Hick 1952. On Hick, see Welford 1975.

57. Broadbent 1971, 280–82.

58. Leonard 1959, 1961.

59. Broadbent 1959, 113.

60. Broadbent 1971, 17.

61. Bartlett 1942c.

62. Duffy 1957; Hebb 1955.

63. Broadbent 1963b, 188; Frankmann and Adams 1962.

64. Malmo and Surwillo 1960.

65. Poulton 1971, 64.

66. Broadbent 1971, 426.

67. Broadbent 1971, 417:

68. Broadbent 1971, 425.

69. Corcoran 1962; Wilkinson 1963.

70. Broadbent 1963b, 190. I am quoting here from an earlier discussion. The argument is the same as in *Decision and Stress*, but the earlier formulation is more explicit.

71. Broadbent 1971, 441.

72. Broadbent 1971, 441.

73. Bartlett 1942c.

74. Hamilton and Copeman 1970. 150.

75. The data comes from Wilkinson and Colquhoun 1968. For the experimental details, see Wilkinson 1961, 263.

76. Broadbent 1971, 447.

Chapter Ten

1. Broadbent 1977a.

2. Broadbent 1977a, 182.

3. Every psychologist will be able to think of examples where hierarchies and levels of some kind were invoked. Two well-known cases are Bryan and Harter 1899 and Hull 1934. In the case of German psychology, a particularly interesting example of a hierarchical and stratified theory of the mind is to be found in the work of G. E. Müller (1850–1934). Müller's case has been studied in detail by Martin Kusch, who demonstrates that a preoccupation with stratification linked Müller's political preferences, his organization of his laboratory, and his model of the mind; see Kusch 1999, chapter 3.

4. Smith 1992.

5. Smith 1992, 162.

6. See, e.g., the discussion of mental energy by Adrian, Head, and Myers 1923.

7. Smith 1992, 13.

8. Broadbent 1977a, 185.

9. Broadbent 1977a, 185.

10. Broadbent 1977a, 191.

11. Broadbent 1977a, 191.

12. Broadbent 1977a, 200.

13. William Shakespeare, *As You Like It*, 2.vii.

14. Broadbent 1977a, 192.

15. On the ideological opposition, see Wapshott 2011. On Keynes, see Skidelsky 1983, 1992, 2000.

16. Broadbent 1977a, 192.

17. Similar conclusions were drawn from an experimental simulation of controlling an economy; see Broadbent and Aston 1978.

18. Broadbent 1977a, 194.

19. Broadbent not only demanded knowledge from his audience, there was also the question of sympathy. Some psychologists, even Cambridge psychologists, will have had reservations about the plausibility of the engineering comparisons. Writing in 1962, in a new appendix to his 1950 *Introduction to Modern Psychology*, Zangwill issued a word of caution. It is difficult to believe, he said, that engineering concepts provide the best guide to the vicissitudes of organic evolution, and he warned against the "seductive" blueprints of the engineer (Zangwill 1962, 232).

20. R. J. G. Pitman's 1966 *Automatic Control Systems Explained* provides an accessible account of the basic ideas used in Broadbent's argument. This book was written for students at Cranfield College of Aeronautics. Cranfield was the location of the important postwar symposium on fatigue whose proceedings were reported in Floyd and Welford 1953.

21. Broadbent 1977a, 195.

22. The analysis of steady-state error is to be found in standard textbooks on control theory—e.g., Pitman 1966 and Bolton 2004. Pitman worked directly with the differential equations; see his appendices 4 and 5. Bolton expresses the results in terms of Laplace transforms; see his section 5.5.

23. Bolton 2004, 294.

24. Broadbent 1977a, 195.

25. Broadbent 1977a, 196.

26. Broadbent 1977a, 197.

27. Broadbent 1977a, 197–98.

28. Broadbent 1977a, 199.

29. Broadbent 1977a, 199.

30. Broadbent 1977a, 199.

31. Broadbent 1977a, 199.

32. Broadbent 1977a, 199.

33. Kuhn 1962, chapters 7 and 8. Kuhn (1962, 83) spoke of a "proliferation of divergent articulations," which come to feel increasingly ad hoc. Kuhn is normally read as implying that there is a single paradigm within some broad, identified field of study. This understanding does not fit the case of learning theory in the United States. Here, the practitioners were divided into competing schools, each of which arguably adhered to and developed its own paradigm. With that qualification, Kuhn's analysis seems remarkably accurate. See *Modern Learning Theory: A Critical Analysis of Five Examples* (Estes et al. 1954).

34. Miller, Galanter, and Pribram 1986; and Neisser 1967. On the "cognitive revolution," see Greenwood 1999. For the view that both the behaviorist and cognitive revolutions were "myths," see Leahey 1992.

35. Hilgard and Bower, in their fourth edition of *Theories of Learning*, said that *Plans* had "become a classic in the current Zeitgeist of cognitive psychology and is required reading for serious students of modern theoretical psychology" (Hilgard and Bower 1975, 148).

36. Miller, Galanter, and Pribram 1986, 32.

37. Miller, Galanter, and Pribram 1986, 34.

38. Broadbent pointed to Hull's valuable use of engineering analogies and Skinner's definition of a "response" as a class of actions that generate a single effect—not a unique pattern of muscular movements. This fact alone, says Broadbent, means that much of Chomsky's criticism of Skinner misses its intended target.

39. Broadbent 1986, xvii.

40. Broadbent 1986, xx.

41. Broadbent also offered detailed technical objections to Chomsky's account of language. In addition, he signaled his dislike of certain aspects of Chomsky's political ideology, particularly the stress on "creative self-expression." Broadbent preferred a form of socialism that put the stress on the removal of material poverty. See Broadbent 1973, 42–43.

42. Broadbent 1986, xxiii.

43. Broadbent 1971, 479.

44. Broadbent 1986, xxiv.

45. Broadbent 1986, xxvii.

46. Broadbent 1986, xix.

47. Broadbent 1986, xxvii.

48. Broadbent 1986, xxviii.

49. Myers 1920a.

50. Broadbent was not alone in remarking on the temporal disparity between the two rather different cognitive revolutions, but as a major participant, he spoke with particular authority. In his *History of Modern Experimental Psychology*, George Mandler also noted that, in his experimental work, Broadbent and his colleagues "anticipated the American revolution" (Mandler 2007, 169). In a footnote on the same page, Mandler indicated the central role of the Cambridge laboratory as an influence, saying: "For the major British influences see Bartlett (1932), Broadbent (1958), and Craik (1943, 1966)." But what is the link that we are invited to discern, *at the level of experimental practice*, between Bartlett's 1932 study of remembering and the work of Craik and Broadbent? Mandler leaves that question unanswered. The analysis given in the present book provides the answer: it lies in the work done with the Cambridge Cockpit.

51. Broadbent's observation about marginality might be illustrated by noting that when Mandler sketched the diverse disciplinary sources that make up cognitive science, he listed, alongside neurophysiology, psychology, artificial intelligence, and linguistics, what he called "some more vague concept like human/machine interaction" (Mandler 2007, 220). Broadbent's position, by contrast, was that the analysis of human-machine interaction was not, in reality, vague, nor should it be seen as peripheral—and in Cambridge (England), it had never been either of these things.

52. Elsewhere, Broadbent makes the connection with Bartlett's work of the 1940s explicit; see, e.g., Broadbent 1986, 159.

53. Broadbent 1977a, 199.

54. Broadbent 1977a, 199.

55. Broadbent cites the work on errors and accidents of James Reason (1975). See also Reason and Mycielska 1982.

56. Broadbent 1977a, 200.

57. Broadbent did not cite the sources for his example or give details. I suspect he was describing the accident involving BAC 1-11, G-ASJJ, on January 14, 1969. This accident was the subject of the civil aircraft accident report CAP no. 347, Board of Trade (1971). The report implicates a range of relevant factors in addition to throttling back a perfectly good engine. These included (i) an engine fault heard as a dull explosion that was real but not incapacitating—a fact that would have emerged if the pilot had followed standard procedure and gained more height before deciding on a course of action; (ii) the inadvertent throttling-back of the suspect-but-still-functioning engine, which remained unnoticed and unaccounted for; (iii) a suggestion by a senior officer on board regarding the nature and origin of the fault that was misunderstood by the pilot as an order—hence the precipitate action; and (iv) the probability that sleep loss played a role in precipitating this sequence of events.

58. Broadbent 1977a, 199.

59. Poulton 1957a, 467; 1974, 150–52.

Summary and Conclusions

1. During WWII, servo-engineers independently developed a methodological principle that was remarkably similar to Bartlett's rule that simple stimuli did not generate simple responses. The point is well expressed by Licklider: "If the transfer characteristic changes when the input parameters change, one cannot determine, through tests with simple signals, the transfer characteristics that describe the behavior set off by complex signals. This was understood by the electrical engineers who, during and after World War II, became interested in the human operators of their servo systems. They were careful to work, therefore, with reasonably realistic test signals" (Licklider 1960, 213).

2. Welford 1968, 252.

3. In the Cockpit experiments, it was the real aircraft in which the trainee-pilot subjects had learned to fly that were, in effect, acting as simulators with respect to the Cambridge Cockpit. Max Hammerton, of the Applied Psychology Unit in Cambridge, has studied the transfer of training based on simulators and found that "first-shot" (i.e., first-attempt) performance, after simulator training, can be as bad as that with no training at all. There is almost zero transfer, though this observation does not apply to later attempts, which invariably show positive transfer. This surprising "first-shot" effect might also help explain the apparently low levels of

performance often observed in the Cockpit. See Hammerton 1963, 1967; and Broadbent 1973, 13–19.

4. Poulton (1974) makes no reference to the Cambridge Cockpit in his monograph *Tracking Skill and Manual Control*, but his discussions are highly relevant. He says: "Tracking becomes more difficult as the degree of pacing is increased. The most difficult part of an airline pilot's task is while he is coming in to land. This is because he has to keep his aircraft on a closely defined glide path. His airspeed has to be held within close limits" (Poulton 1974, 8). See also the important section "Errors Due to Misunderstanding Compensatory Displays" (Poulton 1974, 150–52).

5. Broadbent 1973, ix.

6. Bartlett 1932, 118.

7. Air Ministry 1943.

8. Sherwood 1966, 175.

9. By 1979, Broadbent was even prepared to pose once again Muscio's old Cambridge question: Was a fatigue test possible? Broadbent's answer was "not yet," but he recorded some fascinating developments that took the question of fatigue into the field of social psychology (Broadbent 1979, 1288).

10. For evidence of this confidence, see, e.g., Baddeley and Weiskrantz 1993; Hockey 2013.

11. Kuhn stated the central point when he said: "There is no standard higher than the assent of the relevant scientific community" (1962, 93). In other words, there is no "independent" criterion of success or progress—that is, none that is independent of the collective judgement of the scientific community. Kuhn's sociological insight has many ramifications and (as I have previously indicated in chapter 2, note 19) it casts light on Muscio's stance. Muscio lamented the lack of an "independent criterion" for fatigue when, given the social character of knowledge, there could be no such thing.

12. Broadbent 1971, 436.

13. Broadbent 1973, 63.

14. Chmiel, Totterdell, and Folkard 1995.

15. Chmiel, Totterdell, and Folkard 1995, S39.

16. Chmiel, Totterdell, and Folkard 1995, S52.

17. Chmiel, Totterdell, and Folkard cite, e.g., Rabbitt and Maylor 1991.

18. Taylor and Birmingham 1959.

19. Hockey 2013.

20. On the pioneering character of postwar Cambridge work on fatigue, see Hockey 2013, 94; on the two-level model, see Hockey 2013, 99; on the new perspective, see Hockey 2013, 4. However, Hockey makes no mention of Chmiel, Totterdell, and Folkard 1995.

21. See, e.g., Hockey 2013, 4, 145.

22. Borghini et al. 2014; van Weelden et al. 2022.

23. FPRC meeting no. 29 (February 23, 1943), minute 267 (a), TNA: AIR 57/42.

24. Broadbent 1971, 447.

25. The beta symbol as used in the EEG work must not be confused with the beta symbol conventionally used to define the criterion in signal detection theory.

26. Van Weelden et al. 2022, 6. See also the reports of experimental results on real flight tests by Gorji et al. 2023; and Wu et al. 2019. These papers contain detailed discussions of the use of the machine learning methods necessary to decode the EEG data from the brains of pilots.

27. Van Weelden et al. 2022, 10, table 8.

28. Bartlett 1958b, 510.

29. In Broadbent's later publications, there were signs that his patience was wearing thin. He felt that psychology was in danger of going backwards. Past achievements were being neglected. If psychologists had properly understood the material analyzed in *Perception and Communication*, he declared, then it would be possible to begin to discuss the problems broached in *Decision and Stress*! See Broadbent 1984c, 90. See also Broadbent's valedictory address, "A Word Before Leaving" (Broadbent 1993, 872–73).

30. Rolfe 1996, 71.

31. Broadbent 1973, 89.

References

Abzug, M. J., and E. E. Larrabee. 1997. *Airplane Stability and Control: A History of the Technologies That Made Aviation Possible*. Cambridge University Press.

Adams, J. A. 1955. "A Source of Decrement in Psychomotor Performance." *Journal of Experimental Psychology* 49, no. 6: 390–94.

Adams, J. A. 1956. "Some Implications of Hull's Theory for Human Motor Performance." *Journal of General Psychology* 55: 189–98.

Adams, J. A. 1957. *Some Considerations in the Design and Use of Dynamic -Flight Simulators*. Research report no. TN-57-51. US Air Force Personnel Training Research Center, Lackland Air Force Base, Texas.

Adams, J. A. 1961. "Human Tracking Behavior." *Psychological Bulletin* 58, no. 1: 55–79.

Adams, J. A. 1963. Comments on Feldman's "Reconsideration of the Extinction Hypothesis of Warm Up in Motor Behavior." *Psychological Bulletin* 60, no. 5: 460–63.

Adams, J. A. 1964. "Motor Skills." *Annual Review of Psychology* 15: 181–202.

Adams, J. A. 1967. *Human Memory*. McGraw-Hill.

Addison, P. 1994. *The Road to 1945: British Politics and the Second World War*. Pimlico.

Addison, P., and J. A. Crang, eds. 2000. *The Burning Blue: A New History of the Battle of Britain*. Pimlico.

Adrian, E. D. 1928. *The Basis of Sensation: The Action of the Sense Organs*. Christopher.

Adrian, E. D. 1935. *The Mechanism of Nervous Action: Electrical Studies of the Neurone*. Oxford University Press.

Adrian, E. D. 1949a. *Sensory Integration: The Sherrington Lecture*. University of Liverpool.

Adrian, E. D. 1949b. "Centenary of Pavlov's Birth." *British Medical Journal*, September 16: 553–55.

Adrian, E. D., H. Head, and C. S. Myers. 1923. "The Conception of Nervous and Mental Energy." *British Journal of Psychology: General Section* 14, no. 2: 121–52.

Adrian, E. D., K. J. W Craik, and R. S. Sturdy. 1938. "The Electrical Response of the Auditory Mechanism in Cold-Blooded Vertebrates." *Proceedings of the Royal Society of London* B, no. 175: 435–55.

Air Ministry. 1943. *Notes on the Prevention of Flying Fatigue in Flying Personnel*. FPRC report no. 543, Air Ministry pamphlet no. 154, June. TNA: AIR 57/10.

Air Ministry. 1947. *Psychological Disorders in Flying Personnel of the Royal Air Force Investigated*

During the War 1939–1945. Air Ministry Air Publication no. 3139. His Majesty's Stationery Office.

Albert, D., and H. Gundlach, eds. 1997. *Apparative Psychologie: Geschichtliche Entwicklung und gegenwärtige Bedeutung*. Pabst Science Publishers.

Alexander, L. 1945. *Neuropathology and Neurophysiology, including Encephalography, in Wartime Germany*. Combined Intelligence Objectives Subcommittee, CIOS item no. 24, (medical) file no. XXVII–1, July 20. His Majesty's Stationery Office. British Library Document Supply Services.

Ammons, R. B. 1947a. "Acquisition of Motor Skill I: Quantitative Analysis and Theoretical Formulation." *Psychological Review* 54: 263–81.

Ammons, R. B. 1947b. "Acquisition of Motor Skill II: Rotary Pursuit Performance with Continuous Practice Before and After a Single Rest." *Journal of Experimental Psychology* 37: 393–411.

Anderson, J. R. 1980. *Cognitive Psychology and Its Implications*. W. H. Freeman.

Anon. 2001. "Christopher Poulton: Obituary." *Times*, January 30.

Anthony, A. J. 1940. "Die Wirkung kurzdauernder Sauerstoffatmung auf Herz und Kreislauf des gesunden Menschen." *Zeitschrift für die gesamte experimentelle Medizin* 108: 275–86.

Anthony, A. J. 1941. "Leistungen, Ermüdung, Übermüdung." *Deutsche medizinische Wochenschrift* 67, no. 49: 1327–32.

Anthony, A. J., and S. Atmer. 1936. "Versuche zur medikamentösen Behandlung und Verhütung der Höhenkrankheit." *Luftfahrtmedizin* 1: 185–91.

Anthony, A. J., and G. Schaltenbrand. 1936. "Gibt es eine Abhängigkeit der Muskelspannung von der Sauerstoffkonzentration der Atmungsluft?" *Luftfahrtmedizin* 1: 218–25.

Ash, M. 1980. "Academic Politics in the History of Science: Experimental Psychology in Germany, 1879–1941." *Central European History* 13, no. 3: 255–86.

Ash, M. 1995. *Gestalt Psychology in German Culture, 1890–1967: Holism and the Quest for Objectivity*. Cambridge University Press.

Ash, M., and U. Geuter, eds. 1985. *Geschichte der deutschen Psychologie im 20. Jahrhundert*. Westdeutscher Verlag.

Atkinson, R. L., R. C. Atkinson, E. Smith, and D. J. Bem, eds. 1999. *Introduction to Psychology*. Harcourt Brace.

Atzler, E., ed. 1927. *Körper und Arbeit: Handbuch der Arbeitsphysiologie*. Georg Thieme Verlag.

Baddeley, A. 1972. "Selective Attention and Performance in Dangerous Environments." *British Journal of Psychology* 63, no. 4: 537–46.

Baddeley, A., and L. Weiskrantz, eds. 1993. *Attention: Selection, Awareness and Control: A Tribute to Donald Broadbent*. Clarendon Press.

Barnard, A. 2000. *History and Theory in Anthropology*. Cambridge University Press.

Barnes, B. 1982. *T. S. Kuhn and Social Science*. Macmillan.

Barnes, B. 1983. "Social Life as Bootstrapped Induction." *Sociology* 4: 524–45.

Barnes, B. 1988. *The Nature of Power*. Polity Press.

Bartlett, F. C. 1920a. "Some Experiments on the Reproduction of Folk-Stories." *Folk-Lore* 31: 30–47.

Bartlett, F. C. 1920b. "Psychology in Relation to the Popular Story." *Folk-Lore* 31: 264–93.

Bartlett, F. C. 1923a. *Psychology and Primitive Culture*. Cambridge University Press.

Bartlett, F. C. 1923b. "William Halse Rivers Rivers (1864–1922)." *American Journal of Psychology* 34: 275–77.

Bartlett, F. C. 1924. "Symbolism in Folk Lore." In *Proceedings and Papers of the VIIth Congress of Psychology*, edited by C. S. Myers, 278–89. Cambridge University Press.

Bartlett, F. C. 1926. "Critical Review: Aphasia and Kindred Disorders of Speech." *Brain* 49: 581–87.

Bartlett, F. C. 1927. "Obituary: Prof. Bernard Muscio." *British Journal of Psychology* 17, no. 3: 256.

Bartlett, F. C. 1932. *Remembering: A Study in Experimental and Social Psychology*. Cambridge University Press.

Bartlett, F. C. 1936. "Frederic Charles Bartlett." In *A History of Psychology in Autobiography*, vol. 3, edited by C. Murchison, 39–52. Clark University Press.

Bartlett, F. C. 1937. "Cambridge, England: 1887–1937." *American Journal of Psychology* 50, nos. 1–4: 97–110.

Bartlett, F. C. 1940a. *Notes on a Conference Held at the Air Ministry*. FPRC report no. 165(b), June 3. TNA: AIR 57/2.

Bartlett, F. C. 1940b. *Brief Summary of the Present Position of the Test and Experimental Work at Cambridge*. FPRC report no. 118(a). TNA: AIR 57/2.

Bartlett, F. C. 1940c. *Interim Report in Pre-Selection Tests Developed at the Psychological Laboratory, Cambridge*. FPRC report no. 191.TNA: AIR 57/2.

Bartlett, F. C. 1941a. *Some Notes on R.A.F. Flying Accidents*. FPRC report no. 226. TNA: AIR 57/3.

Bartlett, F. C. 1941b. *Progress Report of All Work Undertaken for the RAF*. FPRC report no. 325, June. TNA: AIR 57/5.

Bartlett, F. C. 1942a. *Pre-Selection Tests in the R.A.F.* FPRC report no. 429, February. TNA: AIR 57/8.

Bartlett, F. C. 1942b. *Some Notes on the Investigation of Flying Accidents*. FPRC report no. 447, April. TNA: AIR 57/9.

Bartlett, F. C. 1942c. *Fatigue in the Air Pilot*. FPRC report no. 488, August. TNA:57/9.

Bartlett, F. C. 1943a. "Anthropology in Reconstruction." *Journal of the Royal Anthropological Institute* 73, nos. 1–2: 9–16.

Bartlett, F. C. 1943b. "Fatigue Following Highly Skilled Work." *Proceedings of the Royal Society B* 131: 247–57.

Bartlett, F. C. 1945. "Some Growing Points in Experimental Psychology." *Endeavour* 4, no. 14: 43–52.

Bartlett, F. C. 1946. "Obituary Notice: Kenneth J. W. Craik, 1914–1945." *British Journal of Psychology* 36, no. 3: 109–16.

Bartlett, F. C. 1947. "Some Problems in 'Display' and 'Control.'" In Universitas Catholica Lovaniensis, *Miscellanea Psychologica Albert Michotte*, 440–52. Institut Supérieur de Philosophie.

Bartlett, F. C. 1948a. "Charles Samuel Myers 1873–1946." *Obituary Notices of Fellows of the Royal Society of London* 5: 767–77.

Bartlett, F. C. 1948b. "The Measurement of Human Skill." *Occupational Psychology* 22: 31–38, 83–91.

Bartlett, F. C. 1948c. "Man, Machines and Productivity." *Occupational Psychology* 22: 190–96.

Bartlett, F. C. 1949. "The Effects of Flying upon Human Performance." *L'Année Psychologique* 50–51: 629–38.

Bartlett, F. C. 1950. "Challenge to Experimental Psychology." In *Proceedings and Papers of the Twelfth International Congress of Psychology*, 23–30. Oliver and Boyd.

Bartlett, F. C. 1951. "The Bearing of Experimental Psychology upon Human Skilled Performance." *British Journal of Industrial Medicine* 8: 209–17.

Bartlett, F. C. 1956. "Changing Scene." *British Journal of Psychology* 47, no. 2: 81–87.

Bartlett, F. C. 1958a. *Thinking: An Experimental and Social Study*. George Allen and Unwin.

Bartlett, F. C. 1958b. "Symposium on Fatigue: (a) Laboratory Work on Fatigue." *Royal Society of Health Journal* 78: 510–13.

Bartlett, F. C., and N. H. Mackworth. 1950. *Planned Seeing: Some Psychological Experiments*. Air Ministry Air Publication no. 3139B. His Majesty's Stationery Office.

Bartlett, F. C., and E. M. Smith. 1920. "On Listening to Sounds of Weak Intensity." *British Journal of Psychology* 10: 101–29, 133–65.

Bauer, R., and G. Ullrich. 1985. "Psychotechnik—Wissenschaft und/oder Ideologie dargestellt an der Zeitschrift für 'Industrielle Psychotechnik.'" *Psychologie und Gesellschaftskritik* 9: 106–27.

Baxter, J. P. 1968 (1946). *Scientists Against Time*. MIT Press.

Beall, A. C., and J. M. Loomis. 1997. "Optic-Flow and Visual Analysis of the Base-to-Final Turn." *International Journal of Aviation Psychology* 7, no. 3: 201–23.

Ben-David, J., and R. Collins. 1966. "Social Factors in the Origin of a New Science: The Case of Psychology." *American Sociological Review* 31: 451–65.

Benary, W., A. Kronfeld, E. Stern, and O. Selz. 1919. *Untersuchungen über die psychische Eignung zum Flugdienst*. Johann Ambrosius.

Bennett, S. 1963. *A History of Control Engineering 1930–1955*. Peter Peregrinus.

Bethe, A., G. von Bergmann, G. Embden, and A. Ellinger, eds. 1930. *Handbuch der normalen und pathologischen Physiologie*. Vol. 15. Springer.

Bills, A. G. 1931. "Blocking: A New Principle in Mental Fatigue." *American Journal of Psychology* 43: 230–45.

Bills, A. G. 1934. *General Experimental Psychology*. Longmans, Green.

Bills, A. G. 1937. "Blocking in Mental Fatigue and Anoxemia Compared." *Journal of Experimental Psychology* 20, no. 5: 437–52.

Bills, A. G., and N. J. Shapin. 1936. "Mental Fatigue Under Automatically Controlled Rates of Work." *Journal of General Psychology* 15: 335–46.

Bilodeau, E. A., and I. M. Bilodeau. 1961. "Motor-Skills Learning." *Annual Review of Psychology* 12: 243–80.

Birmingham, H. P., and F. V. Taylor. 1954a. *A Human Engineering Approach to the Design of Man-Operated Continuous Control Systems*. Interim report no. NRL-4333, April 7. Naval Research Laboratory, Washington, DC.

Birmingham, H. P., and F. V. Taylor. 1954b. "A Design Philosophy for Man-Machine Control Systems." *Proceedings of the I.R.E.* 42: 1748–58.

Blackburn, H. 1975. "Ernst Simonson MD (1898–1974)." *European Journal of Cardiology* 3, no. 1: 77–79.

Blank, G. 1934. "Brauchbarkeit optischer Reaktionsmessungen." *Industrielle Psychotechnik* 11: 140–50.

Bloor, D. 2000. "Whatever Happened to Social Constructiveness?" In *Bartlett, Culture and Cognition*, edited by A. Saito, 194–215. Psychology Press.

Bloor, D. 2019. "Die Psychologie der 1930er Jahre in Cambridge und die Anfänge des Sozialkonstruktivismus." In *Kulturpsychologie in interdisziplinärer Perspektive*, edited by J. Straub, P. Chakklarath, and G. Rebane, 55–86. Psychosozial Verlag.

Board of Trade. 1971. *Civil Aircraft Accident*. Report on the accident to BAC 1-11 G-SJJ, January 14, 1969, CAP no. 347. Her Majesty's Stationery Office.

Boas, F. 1901. "Kathlamet Texts." *Bureau of American Ethnology*, bulletin 26. Smithsonian Institution, Government Printing Office.

Boas, F. 1916. "Tsimshian Mythology." Based on texts recorded by Henry W. Tate. In *Thirty-First Annual Report of the Bureau of American Ethnology*, 29–1037. Government Printing Office.

Bolton, W. 2004. *Instrumentation and Control Systems*. Elsevier.

Boneay, C. A. 1998. "Hermann Ebbinghaus: On the Road to Progress or Down the Garden Path?" In *Portraits of Pioneers in Psychology*, edited by G. A. Kimble and M. Wertheimer, 51–64. American Psychological Association.

Borck, C. 2001. "Electricity as a Medium of Psychic Life: Electrotechnical Adventures in Psychodiagnosis in Weimar Germany." *Science in Context* 14, no. 4: 565–90.

Borck, C. 2008. "Recording the Brain at Work: The Visible, the Readable and the Invisible in Electroencephalography." *Journal of the History of the Neurosciences* 17: 367–79.

Borghini, G., L. Astolfi, G. Vecchiato, D. Mattia, and F. Babiloni. 2014. "Measuring Neurophysiological Signals in Aircraft Pilots and Car Drivers for the Assessment of Mental Workload, Fatigue and Drowsiness." *Neuroscience and Behavioral Reviews* 44: 58–75.

Boring, E. G. 1950. *A History of Experimental Psychology*. 2nd ed. Appleton-Century-Crofts.

Borkenau, F. 1937. *The Spanish Cockpit: An Eye-Witness Account of the Political and Social Conflicts of the Spanish Civil War*. Faber and Faber.

Bracken, H. von. 1956. "Paradoxien der Ermüdung." *Zentralblatt für Arbeitswissenschaft und sociale Betriebspraxis* 10, no. 12: 177–92.

Bradford Hill, A. 1937. *Principles of Medical Statistics*. Lancet.

Bradford Hill, A. 1983. "A Pilot in the First World War." *British Medical Journal* 287: 1947–49.

Bradford Hill, A., and G. O. Williams. 1943. *An Investigation of Landing Accidents in Relation to Fatigue*. FPRC report no. 423, July. TNA: AIR 57/8.

Bramson, A., and N. Birch. 1982. *The Tiger Moth Story*. Airlife.

Bray, C. W. 1948. *Psychology and Military Efficiency: A History of the Applied Psychology Panel of the National Defense Research Committee*. Princeton University Press.

Bridgman, P. W. 1927 [1938]. *The Logic of Modern Physics*. Macmillan.

Bringmann, W. G., H. E. Lück, R. Miller, and C. E. Early, eds. 1997. *A Pictorial History of Psychology*. Quintessence Publishing Co.

Broadbent, D. E. 1953a. "Classical Conditioning and Human Watch Keeping." *Psychological Review* 60: 331–37.

Broadbent, D. E. 1953b. "Neglect of Surroundings in Relation to Fatigue Decrements in Output." In *Symposium on Fatigue*, edited by W. F. Floyd and A. T. Welford, 173–78. H. K. Lewis.

Broadbent, D. E. 1953c. "Noise, Paced Performance and Vigilance Tasks." *British Journal of Psychology* 44, no. 4: 295–303.

Broadbent, D. E. 1954. "Some Effects on Noise on Visual Performance." *Quarterly Journal of Experimental Psychology* 6, no. 1: 1–5.

Broadbent, D. E. 1957. "The Vigilant Man and the Active Man." *Advancement of Science* 13: 399–402.

Broadbent, D. E. 1958. *Perception and Communication*. Pergamon.

Broadbent, D. E. 1959. "Information Theory and Older Approaches to Psychology." *Acta Psychologica* 15: 111–15.

Broadbent, D. E. 1962. "Common Principles in Perception, Reaction and Intellectual Decision." In *Communication Processes*, edited by F. A. Geldard, 197–206. NATO Conference Series, vol. 4. Pergamon Press.

Broadbent, D. E. 1963a. "Human Perception and Animal Learning." In *Current Problems in Animal Behaviour*, edited by W. H. Thorpe and O. L. Zangwill, 248–71. Cambridge University Press.

Broadbent, D. E. 1963b. "Some Recent Research from the Applied Psychology Research Unit, Cambridge." In *Vigilance: A Symposium*, edited by D. N. Buckner and J. J. McGrath, 72–87. McGraw-Hill.

Broadbent, D. E. 1964a. *Behaviour*. Methuen.

Broadbent, D. E. 1964b. "Information Processing in the Nervous System." *Science* 150: 457–62.

Broadbent, D. E. 1970. "Frederic Charles Bartlett (1886–1969)." *Biographical Memoirs of Fellows of the Royal Society* 16: 1–13.

Broadbent, D. E. 1971. *Decision and Stress*. Academic Press.

Broadbent, D. E. 1973. *In Defence of Empirical Psychology*. Methuen.

Broadbent, D. E. 1976. "Noise and the Details of Experiments; A Reply to Poulton." *Applied Ergonomics* 7, no. 4: 231–35.

Broadbent, D. E. 1977a. "Levels, Hierarchies, and the Locus of Control." *Quarterly Journal of Experimental Psychology* 29: 181–201.

Broadbent, D. E. 1977b. "Precautions in Experiments on Noise." *British Journal of Psychology* 68, no. 4: 427–29.

Broadbent, D. E. 1979. "Is a Fatigue Test Now Possible?" *Ergonomics* 22, no. 12: 1277–90.

Broadbent, D. E. 1980. "Donald E. Broadbent." In *History of Psychology in Autobiography*, vol. 7, edited by G. Lindzey, 39–73. Freedman and Co.

Broadbent, D. E. 1982. "Task Combination and Selective Intake of Information." *Acta Psychologica* 50: 253–90.

Broadbent, D. E. 1984a. "Performance and Its Measurement." *British Journal of Clinical Pharmacology* 18: 5S–9S.

Broadbent, D. E. 1984b. "Simulators, Simulation and Driving—Summary." *British Journal of Clinical Pharmacology* 18: 133S–134S.

Broadbent, D. E. 1984c. "The Maltese Cross: A New Simplistic Model for Memory." *Behavioral and Brain Sciences* 7: 55–94.

Broadbent, D. E. 1986. "The Enterprise of Performance." *Quarterly Journal of Experimental Psychology* 38A: 151–62.

Broadbent, D. E. 1993. "A Word Before Leaving." In *Attention and Performance*, vol. 14, edited by D. E. Meyer and S. Kornblum, 863–79. MIT Press.

Broadbent, D. E., and B. Aston. 1978. "Human Control of a Simulated Economic System." *Ergonomics* 21, no. 12: 1035–43.

Broadbent, D. E., and M. Broadbent. 1975. "Some Further Data Concerning the Word Frequency Effect." *Journal of Experimental Psychology: General* 104, no. 4: 297–308.

Broadbent, D. E., and M. Gregory. 1963. "Vigilance Considered as a Statistical Decision." *British Journal of Psychology* 54, no. 4: 309–23.

Broadhurst, P. L. 1959. "The Interaction of Task Difficulty and Motivation: The Yerkes-Dodson Law Revived." *Acta Psychologica* 16: 329–38.

Browne, R. C. 1943–1944. *Day and Night Performance in Teleprinter Switchboard Operators*. FPRC report no. 514, April 1943; 514 (a), August 1943; and 514 (b), February 1944. TNA: AIR 57/10.

Browne, R. C. 1945. *Trial of Two Altitude Indicators*. FPRC report no. 611, February; and 611(a), April. TNA: AIR 57/12.

Browne, R. C. 1949. "Day and Night Performance in Teleprinter Switchboard Operators." *Occupational Psychology* 23: 21–126.

Bryan, W. L., and N. Harter. 1899. "Studies in the Telegraphic Language: The Acquisition of a Hierarchy of Habits." *Psychological Review* 6: 345–75.

Buckner, D. N., and J. J. McGrath, eds. 1963. *Vigilance: A Symposium*. McGraw-Hill.

Burnham, J. C. 2009. *Accident Prone: A History of Technology, Psychology, and Misfits of the Machine Age*. University of Chicago Press.

Bursill, A. E. 1958. "The Restriction of Peripheral Vision During Exposure to Hot and Humid Conditions." *Quarterly Journal of Experimental Psychology* 10, no. 3: 113–29.

Burtscher, M., E. Gnaiger, J. Burtscher, W. Nachbauer, and A. Brugger. 2012. "Arnold Durig (1872–1961): Life, and Work: An Austrian Pioneer in Exercise and High Altitude Physiology." *High Altitude Medicine and Biology* 13, no. 3: 224–31.

Calvert, E. S. 1954. "Visual Judgements in Motion." *Journal of the Institute of Navigation* 7, no. 3: 233–51.

Capshew, J. H. 1999. *Psychologists on the March: Science, Practice, and Professional Identity, 1929–1969*. Cambridge University Press.

Carpenter, A. 1948. "The Rate of Blinking During Long Visual Search." *Journal of Experimental Psychology* 38, no. 5: 587–91.

Chapanis, A., W. Garner, and C. Morgan. 1949. *Applied Experimental Psychology: Human Factors in Engineering Design*. Wiley.

Cherry, C. 1951. "A History of the Theory of Information." *Proceedings of the Institution of Electrical Engineers* 98, no. 3: 383–93.

Cherry, C., ed. 1961. *Communication Processes, NATO Conference Series Fourth London Symposium*. Butterworth.

Chestnut, R. W. 1972. "Psychotechnik: Industrial Psychology in the Weimar Republic 1918–1924." *Proceedings of the 80th Annual Convention, American Psychological Association* 7, no. 2: 781–82.

Chmiel, N., T. Totterdell, and S. Folkard. 1995. "On Adaptive Control, Sleep Loss, and Fatigue." *Applied Cognitive Psychology* 9: S39–S53.

Choyce, C. C., ed. 1923. *A System of Surgery*. Cassell and Son.

Collins, A. F. 2007. "From H = log sⁿ to Conceptual Framework: A Short History of Information." *History of Psychology* 10, no. 1: 44–72.

Collins, A. F. 2008. "Bartlett, Frederic Charles." In *New Dictionary of Scientific Biography*, vol. 1, edited by N. Koertge, 184–89. Thomas Gale.

Collins, A. F. 2012. "An Asymmetric Relationship: The Spirit of Kenneth Craik and the Work of Warren McCulloch." *Interdisciplinary Science Reviews* 37, no. 3: 254–68.

Collins, A. F. 2013. "The Reputation of Kenneth James William Craik." *History of Psychology* 16, no. 2: 93–111.

Collins, H. M. 1992. *Changing Order: Replication and Induction in Scientific Practice*. Sage.

Conrad, R. 1951. "Speed and Load Stress in a Sensori-Motor Skill." *British Journal of Industrial Medicine* 8: 1–7.

Conrad, R. 1954. "Speed Stress." In *Symposium on Human Factors in Equipment Design*, edited by W. F. Floyd and A. T. Welford, 95–102. H. K. Lewis.

Conrad, R. 1955. "Timing." *Occupational Psychology* 29: 73–181.

Conrad, R. and Hille, B. A. 1955. "Comparison of Paced and Unpaced Performance at a Packing Task." *Occupational Psychology* 29: 15–28.

Corcoran, D. W. J. 1962. "Noise and Sleep Loss." *Quarterly Journal of Experimental Psychology* 14: 178–82.

Costall, A. 1991. "Frederic Bartlett and the Rise of Prehistoric Psychology." In A. Still and A. Costall, *Against Cognitivism: Alternative Foundations for Cognitive Psychology*, 39–54. Harvester.

Costall, A. 1992. "Why British Psychology Is Not Social: Frederic Bartlett's Promotion of the New Academic Discipline." *Canadian Psychology* 33, no. 3: 633–39.

Craik, K. J. W. 1938a. "The Effect of Adaptation on Differential Brightness Discrimination." *Journal of Physiology* 92: 406–21.

Craik, K. J. W. 1938b. "The Effect of Adaptation upon Visual Acuity." *British Journal of Psychology* 29: 252–66.

Craik, K. J. W. 1940a. *Fatigue Apparatus*. FPRC report no. 119, March. TNA: AIR 57/2.

Craik, K. J. W. 1940b. "The Effect of Adaptation on Subjective Brightness." *Proceedings of the Royal Society of London* B, no. 128: 232–47.

Craik, K. J. W. 1943a. "Origin of After Images." *Nature* 145: 512.

Craik, K. J. W. 1943b. "Physiology of Colour Vision." *Nature* 151: 727–28.

Craik, K. J. W. 1943c. *The Nature of Explanation*. Cambridge University Press.

Craik, K. J. W. 1945. "The Present Position of Psychological Research in Britain." *British Medical Bulletin* 3: 24–26.

Craik, K. J. W. 1947–1948. "Theory of the Human Operator in Control Systems: I. The Operator as an Engineering System; II. Man as an Element in a Control System." *British Journal of Psychology* 38: 56–61, 142–48.

Craik, K. J. W. 1963. "Psychological and Physiological Aspects of Control Mechanisms with Special Reference to Tank Gunnery." *Ergonomics* 6: part 1, 1–33; part 2, 419–40.

Craik, K. J. W., and M. D. Vernon. 1941. "The Nature of Dark Adaptation." *British Journal of Psychology* 32, no. 1: 62–81.

Craik, K. J. W., and O. L. Zangwill. 1939. "Observations Relating to the Threshold of a Small Figure Within the Contour of a Closed-Line Figure." *British Journal of Psychology* 30: 139–50.

Dale, H. H. 1955. "Edward Mellanby 1884–1955." *Biographical Memoirs of Fellows of the Royal Society* 1: 192–222.

Danziger, K. 1982. "Mid-Nineteenth-Century British Psycho-Physiology: A Neglected Chapter in the History of Psychology." In *The Problematic Science*, edited by W. R. Woodward and M. G. Ash, 119–46. Praeger.

Danziger, K. 2008. *Making the Mind: A History of Memory*. Cambridge University Press.

Daston, L. 1978. "British Responses to Psycho-Physiology 1860–1900." *Isis* 69: 192–208.

Davis, F. B. 1948. "Psychological Research in the AAF Aviation Psychology Program." *Review of Educational Research* 18: 543–74.

Dearnaley, E. J., and P. B. Warr, eds. 1979. *Aircrew Stress in Wartime Operations*. Academic Press.

Deininger, R. L., and P. M. Fitts. 1955. "Stimulus-Response Compatibility, Information Theory, and Perceptual-Motor Performance." In *Information Theory in Psychology*, edited by H. Quastler, 316–41. Free Press.

Doll, R. 1994. "Austin Bradford Hill." *Biographical Memoirs of Fellows of the Royal Society* 40: 128–40.

Dorsch, F. 1963. *Geschichte und Probleme der Angewandten Psychologie*. Verlag Hans Huber.

Douglas, M. 1966. *Purity and Danger: An Analysis of Concepts of Pollution and Taboo*. Routledge and Kegan Paul.

Douglas, M. 1986. *How Institutions Think*. Syracuse University Press.

Douglas, M. 2000. "Memory and Selective Attention: Bartlett and Evans-Pritchard." In *Bartlett, Culture and Cognition*, edited by A. Saito, 179–93. Psychology Press.

Drew, G. C. 1937a. "The Recurrence of Eating in Rats after Apparent Satiation." *Proceedings of the Zoological Society of London* A 107, no. 1: 95–106.

Drew, G. C. 1937b. "The Variation of Sensory Thresholds with the Rate of Application of the Stimulus." *British Journal of Psychology* 27, no. 3: 297–302.

Drew, G. C. 1938. "The Function of Punishment in Learning." *Journal of Genetic Psychology* 52: 257–67.

Drew, G. C. 1939a. "McDougall's Experiments on the Inheritance of Acquired Characteristics." *Nature* 143: 189–91.

Drew, G. C. 1939b. "The Speed of Locomotion Gradient." *Journal of Comparative Psychology* 27: 333–72.

Drew, G. C. 1940. *An Experimental Study of Mental Fatigue.* FPRC report no. 227, December. TNA: AIR 57/3.

Drew, G. C. 1941. *The R.A.F. Pre-Selection Tests.* FPRC report no. 328. TNA: AIR 57/5.

Drew, G. C. 1951. "Variations in Reflex Blink Rate During Visual-Motor Tasks." *Quarterly Journal of Experimental Psychology* 3, no. 2: 73–88.

Duffy, E. 1957. "The Psychological Significance of the Concept of 'Arousal' or 'Activation.'" *Psychological Review* 64: 256–75.

Duhem, P. 1962. *The Aim and Structure of Physical Theory.* Translated by P. P. Wiener. Atheneum.

Duncan, J., L. Phillips, and P. McLeod, eds. 2005. *Measuring the Mind: Speed, Control, and Age.* Oxford University Press.

Dunlap, J. W. 1948. *The Human Factor in the Design of Stick and Rudder Controls for Aircraft.* Division of Bio-Mechanics, Psychological Corporation.

Ebbinghaus, A., and K. Dörner, eds. 2002. *Vernichten und Heilen: Der Nürnberger Ärzteprozeß und seine Folgen.* Aufbau-Verlag.

Ebbinghaus, H. 1885. *Über das Gedächtnis: Untersuchungen zur experimentellen Psychologie.* Duncker und Humblot.

Ebbinghaus, H. 1908. *Abriss der Psychologie.* Verlag von Veit and Comp.

Ebbinghaus, H. 1964. *Memory: A Contribution to Experimental Psychology.* Translated by H. A. Ruger and C. E. Bussenius. Dover.

Eckert, W. 2013. "Instrumental Modernity and the Dictates of Politics—Sponsorship of Medical Research by the Emergency Association/German Research Foundation 1920–1970." In *The German Research Foundation*, edited by M. Walker, K. Orth, U. Herbert, and R. vom Bruch, 201–20. Franz Steiner Verlag.

Eichholtz, F. 1957. "Zum 70. Geburtstag von Professor Dr. Georg Joachimoglu." *Arzneimittel-Forschung* 7: 755–56.

Edgerton, D. 2006. *Warfare State Britain, 1920–1970.* Cambridge University Press.

Edgerton, D. 2012. *Britain's War Machine: Weapons, Resources and Experts in the Second World War.* Penguin Books.

Elliott, T. R. 1941. "Wilfred Batten Lewis Trotter (1872–1939)." *Obituary Notices of Fellows of the Royal Society* 3, no. 9: 325–44.

Ely, J. H., H. M. Bowen, and J. Orlansky. 1957. "Man Machine Dynamics." In *Joint Services Human Engineering Guide to Equipment Design*, USAF WADC technical report 57-582, chapter 5. Wright Air Development Center, Wright-Patterson Air Force Base, Dayton, Ohio.

Engstrom, E. J. 1991. "Emil Kraepelin: Psychiatry and Public Affairs in Wilhelmine Germany." *History of Psychiatry* 2: 11–32.

Engstrom, E. J. 1992. "Emil Kraepelin: Psychiatric Observations on Contemporary Issues." *History of Psychiatry* 3: 253–69.

Erlanger, J., and H. S. Gasser. 1937. *Electrical Signs of Nervous Activity.* University of Pennsylvania Press.

Eschenbach, W. 1938. "Aufgabe und Untersuchung einer Fliegerdrehkammer." PhD diss., Technische Hochschule, Berlin.

Eschenbach, W. 1941. "Eine Fliegerdrehkammer als psychotechnisches Forschungsgerät." *Industrielle Psychotechnik* 18: 24–28.

Estes, W., S. Koch, K. MacCorquodale, P. Meehl, C. Mueller, W. Schoenfeld, and W. Verplanck. 1954. *Modern Learning Theory: A Critical Analysis of Five Examples.* Appleton-Century-Crofts.

Evans, R. 2000. *Gassed: British Chemical Warfare Experiments on Humans at Porton Down.* House of Stratus.

Everling, E. 1926. "Meßgeräte für den Luftverkehr im Nebel." *Verkehrstechnische Woche: Zeitschrift für das gesamte Verkehrswesen* 20, no. 49: 632–37.

Everling, E. 1934a. "Messen von Drehgeschwindigkeiten durch das Gleichgewichtsorgan." *Acta Aerophysiologica* 10: 30–40.

Everling, E. 1934b. "Vereinfachte Fliegerausbildung." *Verkehrstechnische Woche* 28, no. 44: 577–82.

Eysenck, H. J. 1990. *Rebel with a Cause: The Autobiography of H. J. Eysenck.* W. H. Allen.

Farmer, E. 1921. *Time and Motion Study.* Reports of the Industrial Fatigue Research Board, report no. 14. His Majesty's Stationery Office.

Ferris, J. 1999. "Fighter Defence Before Fighter Command: The Rise of Strategic Air Defence in Great Britain, 1917–1934." *Journal of Military History* 63, no. 4: 845–84.

Ficker, H., and B. de Rudder. 1942. *Föhn und Föhnwirkungen: Der gegenwärtige Stand der Frage.* Geest and Portig.

Fitts, P. M. 1946a. "German Applied Psychology During World War II." *American Psychologist* 1: 151–61.

Fitts, P. M. 1946b. "Psychological Requirements in Aviation Equipment Design." *Journal of Aviation Medicine* 17: 270–75.

Fitts, P. M. 1947a. "Psychological Research on Equipment Design in the AAF." *American Psychologist* 2: 93–98.

Fitts, P. M., ed. 1947b. *Report 19, Psychological Research on Equipment Design.* Research report, Army Air Forces Aviation Psychology Program. US Government Printing Office.

Fitts, P. M. 1951. "Engineering Psychology and Equipment Design." In *Handbook of Experimental Psychology*, edited by S. S. Stevens, 1286–340. John Wiley.

Fitts, P. M., and R. E. Jones. 1947a. *Analysis of Factors Contributing to 460 "Pilot-Error" Experiences in Operating Aircraft Controls.* Memorandum Report TSEAA-694-12. Aero Medical Laboratory, Engineering Division, US Air Material Command, Wright-Patterson Air Force Base, Dayton, Ohio.

Fitts, P. M., and R. E. Jones. 1947b. *Psychological Aspects of Instrument Display: Analysis of 270 "Pilot-Error" Experiences in Reading and Interpreting Aircraft Instruments.* Memorandum Report TSEAA-694-12A. Aero Medical Laboratory, Engineering Division, US Air Material Command, Wright-Patterson Air Force Base, Dayton, Ohio.

Fitts, P. M., and C. M. Seeger. 1953. "S-R Compatibility: Spatial Characteristics of Stimulus Response Codes." *Journal of Experimental Psychology* 46: 199–210.

Fitts, P. M., M. Weinstein, M. Rappaport, N. Anderson, and J. A. Leonard. 1956. "Stimulus Correlates of Visual Pattern Recognition: A Probability Approach." *Journal of Experimental Psychology* 51, no. 1: 1–11.

Flanagan, J. C. 1948. *The Aviation Psychology Program in the Army Air Forces.* Report no. 1. Research report, Army Air Forces Aviation Psychology Program. US Government Printing Office.

Flanagan, J. C. 1954. "The Critical Incident Technique." *Psychological Bulletin* 51, no. 4: 327–58.

Floyd, W. F., and A. T. Welford, eds. 1953. *Symposium on Fatigue.* H. K. Lewis.

Floyd, W. F., and A. T. Welford, eds. 1954. *Symposium on Human Factors in Equipment Design.* H. K. Lewis.

Foerster, O. 1936. "The Motor Cortex in Man in the Light of Hughlings Jackson's Doctrines." *Brain* 59, no. 2: 135–59.

Forrester, J. 2004. "Freud in Cambridge." *Critical Quarterly* 46, no. 2: 1–26.

Forrester, J., and L. Cameron. 2017. *Freud in Cambridge.* Cambridge University Press.

François, J. 1977. "Prof. Dr. Fritz Hollwich." *Klinische. Monatsblätter für Augenheilkunde* 171: 809–10.

Frankmann, J. P., and J. A. Adams. 1962. "Theories of Vigilance." *Psychological Bulletin* 59, no. 4: 257–72.

Freud, S. 1940. *Group Psychology and the Analysis of the Ego.* 2nd ed. Hogarth Press.

Fulton, J. F. 1926. *Muscular Contraction and the Reflex Control of Movement.* Baillière, Tindall and Cox.

Gade, H.-G. 1928. *Zur Psychotechnik des Flugzeugführers.* Technische Hochschule, Danzig.

Gagné, R. M. 1954. "Training Devices and Simulators: Some Research Issues." *American Psychologist* 9: 95–107.

Geldard, F. A., ed. 1965. *Communication Processes.* NATO Conference Series, vol. 4. Pergamon Press.

Gerathewohl, S. 1944. "Psychologische Untersuchungen zur Blindflugeignung." *Zeitschrift für angewandte Psychologie* 66: 361–93.

Gerathewohl, S. 1951. *Conspicuity of Flashing and Steady Light Signals, I.* Special report. USAF School of Aviation Medicine, Randolph Field, Texas.

Gerathewohl, S. 1952. "Eye Movements During Radar Operations." *Journal of Aviation Medicine* 23: 597–607.

Gerathewohl, S. 1953. *Die Psychologie des Menschen im Fleugzeug.* Johann Ambrosius.

Gerathewohl, S. 1954. "Conspicuity of Flashing Light Signals of Different Frequency and Duration." *Journal of Experimental Psychology* 48, no. 4: 247–51.

Gerathewohl, S. 1976. "Optimization of Crew Effectiveness in Future Cockpit Design: Biomedical Implications." *Aviation, Space, and Environmental Medicine* 47: 182–187.

Geuter, U. 1984. *Die Professionalisierung der deutschen Psychologie im Nationalsozialismus.* Suhrkamp.

Geuter, U. 1985a. "Polemos panton pater—Militär und Psychologie im Deutschen Reich 1914–1945." In *Geschichte der deutschen Psychologie im 20. Jahrhundert,* edited by M. G. Ash and U. Geuter, 146–71. Westdeutscher Verlag.

Geuter, U. 1985b. "Nationalsozialistische Ideologie und Psychologie." In *Geschichte der deutschen Psychologie im 20. Jahrhundert,* edited by M. G. Ash and U. Geuter, 172–200. Westdeutscher Verlag.

Gibbs, C. B. 1951. "Transfer of Training and Skill Assumptions in Tracking Tasks." *Quarterly Journal of Experimental Psychology* 3, no. 3: 99–110.

Gibbs, C. B. 1954. "The Continuous Regulation of Skilled Response by Kinaesthetic Feedback." *British Journal of Psychology* 45: 24–39.

Gibbs, C. B. 1961. "Controller Design: Interactions of Controlling Limbs, Time-Lags and Gains in Positional and Velocity Systems." *Ergonomics* 5: 385–402.

Gibson, J. J. 1941. "A Critical Review of the Concept of Set in Contemporary Experimental Psychology." *Psychological Bulletin* 38, no. 9: 781–817.

Gibson, J. J. 1955. "The Optical Expansion Pattern in Aerial Locomotion." *American Journal of Psychology* 68, no. 4: 480–84.

Gibson, T. M., and M. H. Harrison. 1984. *Into Thin Air: A History of Aviation Medicine in the RAF.* Robert Hale.

Gigerenzer, G. 2001. "The Adaptive Toolbox." In *Bounded Rationality*, edited by G. Gigerenzer and R. Selten, 37–50. MIT Press.

Gigerenzer, G., and R. Selten, eds. 2001. *Bounded Rationality: The Adaptive Toolbox*. MIT Press.

Gilbert, A. R. 1951. "Recent German Theories of Stratification of Personality." *Journal of Psychology* 31: 3–19.

Gimbel, J. 1990. "The American Exploitation of German Technical Know-How After World War II." *Political Science Quarterly* 105, no. 2: 295–309.

Gleitman, H., J. Nachmias, and U. Neisser. 1954. "The S-R Theory of Extinction." *Psychological Review* 61, no. 1: 23–33.

Goldie, E. A. G. 1943. *On Respiration Rate During a Bombing Sortie*. FPRC report no. 552, October. TNA: AIR 57/11.

Goldie, E. A. G., and J. C. Gilson. 1945. *Report on a Visit to the Physiological Institute, University of Göttingen*. FPRC report no. 632, June 15. TNA: AIR 57/13.

Gondra, J. M. 2008. "Hull, Clark Leonard (1884–1952)." In *New Dictionary of Scientific Biography*, vol. 3, edited by N. Koertge, 405–10. Thomas Gale.

Gorji, H. T., N. Wilson, J. Van Bree, B. Hoffmann, T. Petros, and K. Tavakolian. 2023. "Using Machine Learning Methods and EEG to Discriminate Aircraft Pilot Cognitive Workload During Flight." *Scientific Reports* 13, no. 1: 2507. https://doi.org/10.1038/s41598-023-29647-0.

Graf, O. 1930. "Zur Frage der Wirkung verschiedener alkoholisher Getränke." *Zeitschrift für die gesamte Neurologie und Psychiatrie* 130: 187–218.

Graf, O. 1933. "Über den Zusammenhang zwischen Alkoholblutkonzentration und psychischer Alkoholwirkung." *Arbeitsphysiologie* 6: 169–213.

Graf, O. 1937. "Neue Anschauungen über die Einwirkung des Alkoholgenusses auf die Arbeitsleistung." *Die Alkoholfrage* 33: 223–26.

Graf, O. 1939. "Über den Einfluß von Pervitin auf einige psychische und psychomotorische Funktionen." *Arbeitsphysiologie* 10: 692–705.

Graf, O. 1943. "Eine Methode zur Untersuchung der pharmakologischen Beeinflussung von Koordinationsleistungen." *Arbeitsphysiologie* 12: 449–68.

Graf, O. 1950. "Increase of Efficiency by Means of Pharmaceutics (Stimulants)." In Surgeon General, *German Aviation Medicine*, 1080–103. Department of the Air Force.

Green, D. M., and J. A. Swets. 1966. *Signal Detection Theory and Psychophysics*. Wiley.

Greenwood, J. D. 1999. "Understanding the 'Cognitive Revolution' in Psychology." *Journal of the History of the Behavioral Sciences* 35, no. 1: 1–22.

Gregory, R. 1963. "The Brain as an Engineering Problem." In *Current Problems in Animal Behaviour*, edited by W. H. Thorpe and O. L. Zangwill, 307–30. Cambridge University Press.

Gregory, R. 1966. *Eye and Brain: The Psychology of Seeing*. Weidenfeld and Nicolson.

Gregory, R. 1983. "Forty Years On: Kenneth Craik's *The Nature of Explanation* (1943)." *Perception* 12: 233–38.

Gregory, R. 2000. "Bartlett and Experimental Psychology." In *Bartlett, Culture and Cognition*, edited by A. Saito, 39–45. Psychology Press.

Grether, W. F. 1947a. "Survey of Display Problems in the Design of Aviation Equipment." In *Report 19*, edited by P. M. Fitts, 21–33. US Government Printing Office.

Grether, W. F. 1947b. "Efficiency of Several Types of Control Movements in the Performance of a Simple Compensatory Pursuit Task." In *Report 19*, edited by P. M. Fitts, 227–39. US Government Printing Office.

Grether, W. F., J. T. Cowles, and R. E. Jones. 1947. "The Effect of Anoxia on Visual Illusions." In *Report 19*, edited by P. M. Fitts, 249–55. US Government Printing Office.

Grindley, G. C. 1940. *Notes on Accuracy of Reports Made After Reconnaissance Flights by Coastal Command*. FPRC report no. 120, February. TNA: AIR 57/2.

Guilford, J. P., and J. T. Lacy, eds. 1947. *Printed Classification Tests*. Research report no. 5, Army Air Forces Aviation Psychology Program. US Government Printing Office.

Gundlach, H. 1997. "The Mobile Psychologist: Psychology and the Railroads." In *Pictorial History of Psychology*, edited by W. G. Bringmann, H. E. Lück, R. Miller, and C. E. Early, 506–9. Quintessence Publishing Co.

Gundlach, H. 2004. "Reine Psychologie, Angewandte Psychologie und die Institutionalisierung der Psychologie." *Zeitschrift für Psychologie* 212, no. 4: 183–99.

Gundlach, H. 2005. "Das Psychotechnische Prüflaboratorium der Eisenbahn-Generaldirektion Dresden." In *Illustrierte Geschichte der Psychologie*, edited by H. E. Lück and R. Miller, 257–62. Beltz Verlag.

Haak, R. 1996. "Grundlagen und Entwicklung der Berliner Psychotechnik—Frühe Jahre des Instituts für Industrielle Psychotechnik der TH Charlottenburg/Berlin." In *Untersuchungen zur Geschichte der Psychologie und der Psychotechnik*, edited by H. Gundlach, 165–76. Profil Verlag.

Haffner, S. 1986. *Failure of a Revolution: Germany 1918–1919*. Translated by G. Rapp. Banner Press.

Hagner, M. 2001. "Cultivating the Cortex in German Neuroanatomy." *Science in Context* 14, no. 4: 565–90.

Hallgebauer, ? 1927. "Historisch-kritische Betrachtung zum Fahrzeugfuehrerproblem." *Psychotechnische Zeitung* 2: 15–23.

Hamilton, P., and A. Copeman. 1970. "The Effect of Alcohol and Noise on Components of a Tracking and Monitoring Task." *British Journal of Psychology* 61: 144–56.

Hammerton, M. 1963. "Transfer of Training from a Simulated to a Real Control Situation." *Journal of Experimental Psychology* 66: 450–53.

Hammerton, M. 1967. "Simulators for Training." *Electronics and Power* 13: 8–10.

Hancock, P. A., P. A. Desmond, and G. Mathews. 2012. "Conceptualising and Defining Fatigue." In *Handbook of Operator Fatigue*, edited by G. Mathews, P. A. Hancock, P. A. Desmond, and C. Neubauer, 63–73. Ashgate.

Harsch, V. 2003. *Das Institute für Luftfahrtmedizin in Hamburg-Eppendorf (1927–1945)*. Rethra Verlag.

Harsch, V. 2004. *Leben, Werk und Zeit des Physiologen Hubertus Strughold (1898–1986)*. Rethra Verlag.

Hartcup, G. 2000. *The Effects of Science on the Second World War*. Palgrave.

Hartley, R. V. L. 1928. "Transmission of Information." *Bell Systems Technical Journal* 17: 535–50.

Hartman, B., and P. M. Fitts. 1950. *The Development of Techniques and Procedures for the Study of Alertness in Aviation Personnel*. National Research Council Committee on Aviation Psychology, November.

Hartree, D. R. 1935. "The Differential Analyser." *Nature* 135: 940–43.

Harwood, J. 1997. "German Science and Technology Under National Socialism." *Perspectives on Science* 5, no. 1: 128–51.

Head, H. 1918. *The Sense of Stability and Balance in the Air*. Reports of the Air Medical Investigation Committee, Medical Research Committee, special report no. 28. His Majesty's Stationery Office.

Head, H. 1926. *Aphasia and Kindred Disorders of Speech*. Cambridge University Press.

Head, H., and G. Riddoch. 1917. "The Automatic Bladder, Excessive Sweating and Some Other Reflex Conditions, in Gross Injuries of the Spinal Cord." *Brain* 40, nos. 2–3: 188–263.

Head, H., W. H. R. Rivers, and J. Sherren. 1905. "The Afferent Nervous System from a New Aspect." *Brain* 28, no. 2: 99–115.

Hebb, D. O. 1955. "Drives and the C.N.S. (Conceptual Nervous System)." *Psychological Review* 62: 243–54.

Henschke, U. 1945. *Medizinisches Forschungsinstitut, Garmisch-Partenkirchen.* Air Documents Division, T-2, AMC, Wright Field, microfilm no. R 3984, F 387. Wright Field Technical Documents Library, Washington, DC.

Henschke, U., and P. Karlson. 1946. *A New Control System for Fighter Aircraft.* Air Technical Index no. 7019. Translated by Air Documents Division, Intelligence (T-2), Wright Field, Dayton, Ohio.

Henschke, U. K., and Hans A. Mauch. 1950. "How Man Controls." In Surgeon General, *German Aviation Medicine,* 83–91. Department of the Air Force.

Herle, A., and S. Rouse, eds. 1998. *Cambridge and the Torres Strait: Centenary Essays on the 1898 Anthropological Expedition.* Cambridge University Press.

Herrmann, T. 1966. "Zur Geschichte der Berufeignungsdiagnostik." *Archiv für die gesamte Psychologie* 118: 253–78.

Hick, W. E. 1952. "On the Rate of Gain of Information." *Quarterly Journal of Experimental Psychology* 4: 11–26.

Hick, W. E., and J. A. V. Bates. 1950. *The Human Operator of Control Mechanisms.* Ministry of Supply, Permanent Records of Research and Development, no. 17-204. Ministry of Supply, London.

Hilgard, E. R. 1956. *Theories of Learning.* Appleton-Century-Crofts.

Hilgard, E. R., and G. H. Bower. 1975. *Theories of Learning.* 4th ed. Prentice-Hall.

Hilgard, E. R., and D. G. Marquis. 1940. *Conditioning and Learning.* Appleton-Century.

Hill, A. V. 1936. "A Tribute to Pavlov." *British Medical Journal* 1, no. 3992: 508–9.

Hill, A. V., and H. Upton. 1923. "Muscular Exercise, Lactic Acid, and the Supply and Utilization of Oxygen." *Quarterly Journal of Medicine* 16: 35–171.

Hippke, E. 1943. "Die Flugmedizin im Dienst der Kriegführung." *Luftwissen* 10, no. 1: 3–5.

Hockey, G. R. J. 1970a. "Effect of Loud Noise on Attentional Selectivity." *Quarterly Journal of Experimental Psychology* 22: 28–36.

Hockey, G. R. J. 1970b. "Signal Probability and Spatial Location as Possible Bases for Increased Selectivity in Noise." *Quarterly Journal of Experimental Psychology* 22: 37–42.

Hockey, G. R. J. 1970c. "Changes in Attention Allocation in a Multi-Component Task under Loss of Sleep." *British Journal of Psychology* 61, no. 4: 473–80.

Hockey, G. R. J. 2005. "Operator Functional State: The Prediction of Breakdown in Human Performance." In *Measuring the Mind: Speed, Control, and Age,* edited by J. Duncan, L. Phillips, and P. McLeod, 373–94. Oxford University Press.

Hockey, G. R. J. 2013. *The Psychology of Fatigue: Work, Effort and Control.* Cambridge University Press.

Hodgkin, A. 1979. "Edgar Douglas Adrian 1889–1977." *Biographical Memoirs of Fellows of the Royal Society* 28: 1–73.

Holdstock, D. 2004. "Trotter, Wilfred Batten Lewis (1872–1939)." In *Oxford Dictionary of National Biography,* vol. 55, edited by H. C. G. Mathew and B. Harrison, 430–32. Oxford University Press.

Hollwich, F. 1979. *The Influence of Ocular Light Perception in Man and in Animals.* Springer.

Holmes, F. M., ed. 1990. *Dictionary of Scientific Biography.* Scribner's.

Holmes, M. E. 2013. "The Psychologist and the Bombardier: The Army Air Forces' Aircrew Classification Program in WWII." *Endeavour* 38, no. 1: 43–54.

Hooper, D. 1993. "Professor Derek Russell Davis (1914–1993)." *British Journal of Medical Psychology* 66, no. 3: 209–12.

Howarth, C. I. 1972. "J. Alfred Leonard (1922–1972)." *Bulletin of the British Psychological Society* 25: 227–28.

Hull, C. L. 1934. "The Concept of the Habit-Family Hierarchy and Maze Learning." *Psychological Review* 41: 33–54, 134–52.

Hull, C. L. 1943. *Principles of Behavior: An Introduction to Behavior Theory*. Appleton-Century Company.

Hüttenberger, P.1976. "Nationalsozialistische Polycratie." *Geschichte und Gesellschaft: Zeitschrift für historische Sozialwissenschaft* 2: 417–22.

Iwahara, S. 1957. "Hull's Conception of Inhibition: A Revision." *Psychological Reports* 3: 9–10.

Jackson, J. H. 1884. "Evolution and Dissolution of the Nervous System." *British Medical Journal*, April 5: 591, 660, 703. Reprinted 1932, in *Selected Writings of John Hughlings Jackson*, vol. 2, edited by J. Taylor, 45–75. Hodder and Stoughton.

Jacyna, L. S. 2000. *Lost Words: Narratives of Language and Brain 1825–1926*. Princeton University Press.

Jaeger, S. 1985. "Zur Herausbildung von Praxisfeldern der Psychologie bis 1933." In *Geschichte der deutschen psychologie*, edited by M. Ash and U. Geuter, 83–112. Westdeutscher Verlag.

Jaeger, S., and I. Staeuble. 1981. "Die Psychotechnik und ihre gesellschaftlichen Entwicklungsbedingungen." In *Die Psychologie des 20. Jahrhunderts*, edited by F. Stoll, 53–95. Kindler.

Jane's Fighting Aircraft of World War II. 1989 (1946–1947). Bracken Books.

Jensen, A. R. 1961. "On the Reformulation of Inhibition in Hull's System." *Psychological Bulletin* 58, no. 4: 274–98.

Jerison, H. J., and R. A. Wallis. 1957a. *Experiments on Vigilance. II. One-Clock and Three-Clock Monitoring*. USAF WADC technical report TR-57-206. Wright Air Development Center, Wright-Patterson Air Force Base, Ohio.

Jerison, H. J., and R. A. Wallis. 1957b. *Experiments on Vigilance. III. Performance on a Simple Vigilance Task in Noise and Quiet*. USAF WADC technical report TR-57-318. Wright Air Development Center, Wright-Patterson Air Force Base, Ohio.

Jerison, H. J., and A. Wing. 1957. *Effects of Noise and Fatigue on a Complex Vigilance Task*. USAF WADC technical report TR 47-14. Wright Air Development Center, Wright-Patterson Air Force Base, Ohio.

Jones, H. G. 1958. "The Status of Inhibition in Hull's Theory: A Theoretical Revision." *Psychological Review* 65: 179–82.

Jones, L. V. 2007. "Some Lasting Consequences of US Psychology Programs in World Wars I and II." *Multivariate Behavioral Research* 42, no. 3: 593–608.

Kaminsky, J. 1967. "Art. Spencer, Herbert." In *The Encyclopedia of Philosophy*, edited by P. Edwards, 523–27. Macmillan.

Kappauf, W. 1947. "History of Psychological Studies of the Design and Operation of Equipment." *American Psychologist* 2: 83–86.

Kaufmann, D., ed. 2000. *Geschichte der Kaiser-Wilhelm-Gesellschaft im Nationalsozialismus: Bestandsaufnahme und Perspektiven der Forschung*. Wallstein Verlag.

Kehrt, C. 2010. *Moderne Krieger: Die Technikerfahrungen deutscher Militärpiloten 1910–1945*. Ferdinand Schöningh.

Kennedy, J. L. 1953. "Some Practical Problems of the Alertness Indicator." In *Symposium on Fatigue*, edited by W. F. Floyd and A. T. Welford, 149–53. H. K. Lewis.

Kershaw, I. 2011. *The End: Hitler's Germany, 1944–45*. Allen Lane.

Killen, A. 2007. "Weimar Psychotechnics Between Americanism and Fascism." *Osiris* 22: 48–71.

Kimble, G. A. 1949. "An Experimental Test of a Two-Factor Theory of Inhibition." *Journal of Experimental Psychology* 39, no. 1: 15–23.

Kimble, G. A., and M. Wertheimer, eds. 1998. *Portraits of Pioneers in Psychology*. American Psychological Association.

Klee, E. 2003. *Das Personenlexikon zum Dritten Reich: Wer war was vor und nach 1945?* S. Fischer.

Kluwe, R. H. 1997. "Simulation in der empirisch-psychologischen Forschung." In *Apparative Psychologie*, edited by D. Albert and H. Gundlach, 203–24. Pabst Science Publishers.

Knorre, G. von. 1943. *Übermüdungszustände bei Fliegern*. Forschungsbericht [research report] 18/43. *Mitteilungen aus der Gebiet der Luftfahrtmedizin*, edited by the Inspecteur des Sanitätswesens der Luftwaffe. DMM, R.d.L. Ob.d.L.

Koch, S. 1944. "Review of Hull's *Principles of Behavior*." *Psychological Bulletin* 41: 269–86.

Koch, S. 1954. "Clark L. Hull." In W. Estes, S. Koch, K. MacCorquodale, P. Meehl, C. Mueller, W. Schoenfeld, et al., *Modern Learning Theory*, 1–176. Appleton-Century-Crofts.

Koch, S., ed. 1963. *Psychology: A Study of a Science*. McGraw-Hill.

Koelle, H. H. 1979. "Luft-und Raumfahrt an der technischen Universität Berlin." In *Wissenschaft und Gesselschaft*, vol. 2, edited by R. Rürup, 143–52. Springer.

Koertge, N., ed. 2008. *New Dictionary of Scientific Biography*. Thomas Gale.

Koffka, K. 1936. *Principles of Gestalt Psychology*. Kegan Paul, Trench, Trubner.

Koonce, J. M. 1984. "A Brief History of Aviation Psychology." *Human Factors* 26, no. 5: 499–508.

Kornmüller, A. E. 1944. *Einführung in die Elektrenkephalographie*. J. F. Lehmanns Verlag.

Koschel, E. 1913. "Welche Anforderungen müssen an die Gesundheit der Führer von Luftfahrzeugen gestellt werden?" *Jahrbuch der wissenschaftlichen gesellschaft fuer Flugtechnik* 1: 143–56.

Kosin, R. 1988. *The German Fighter Since 1915*. Translated by K. Thomas. Putnam.

Kraepelin, E. 1902. "Die Arbeitscurve." *Philosophische Studien* 19: 459–507.

Kraepelin, E. 1919. "Psychiatrische Randbemerkungen zur Zeitgeschichte." *Kriegshefte der Süddeutschen Monatshefte April 1919 bis September 1919* 16: 171–83.

Kuhn, T. S. 1961. "The Role of Measurement in Modern Physical Science." *Isis* 52: 161–90.

Kuhn, T. S. 1962. *The Structure of Scientific Revolutions*. University of Chicago Press.

Kusch, M. 1995. *Psychologism: A Case Study in the Sociology of Philosophical Knowledge*. Routledge.

Kusch, M. 1999. *Psychological Knowledge: A Social History and Philosophy*. Routledge.

Leahey, T. H. 1992. "The Mythical Revolutions of American Psychology." *American Psychologist* 47, no. 2: 308–18.

Leaking, T. H. 1987. *A History of Psychology: Main Currents in Psychological Thought*. Prentice-Hall.

Lehmann, G. 1950. "Importance of Oxygen Tension to Efficiency." In Surgeon General, *German Aviation Medicine*, 1104–12. Department of the Air Force.

Lehmann, G. 1963. "Grafs Lebenswerk und seine Bedeutung für die Wissenschaft." *REF-Nachrichten* 16, no. 6: 254–59.

Lehmann, G., and O. Graf. 1942. "Versuche über die Wirkung von Sauerstoffatmung bei normalen Druck auf die Leistungsfähigkeit." *Luftfahrtmedizin* 6: 183–200.

Lehmann, G., H. Straub, and A. Szakáll. 1939. "Pervitin als leistungssteigerndes Mittel." *Arbeitsphysiologie* 10: 680–91.

Leonard, J. A. 1959. "Tactual Choice Reactions I." *Quarterly Journal of Experimental Psychology* 11, no. 2: 78–83.

Leonard, J. A. 1961. "Choice Reaction Time Experiments and Information Theory." In *Communication Processes*, edited by C. Cherry, 137–46. Butterworth.

Lewis, M., and J. M. Haviland-Jones, eds. 2000. *Handbook of Emotions*. 2nd ed. Guilford Press.

Licklider, J. C. R. 1960. "Quasi-Linear Operator Models in the Study of Manual Tracking." In *Developments in Mathematical Psychology*, edited by D. R. Luce, 167–279. Free Press of Glencoe.

Lindzey, G., ed. 1980. *A History of Psychology in Autobiography*. Vol. 7. Freedman and Co.

Livingston, P. 1962. *Fringe of Clouds*. Johnson.

Lottig, H. 1936. "Zur Vereinheitlichung des Schreibversuchs bei der Höhentauglichkeitsprüfung." *Luftfahrtmedizin* 1: 15–19.

Lottig, H. 1939. "Wechselbeziehungen zwischen Psyche und vegitativem Nervensystem beim Flieger." *Luftfahrtmedizinische Abhandlungen* 3: 221–39.

Lovelace, W. R. 1946. *Research in Aviation Medicine for the German Air Force*. Combined Intelligence Objectives Sub-Committee (CIOS) final report no. XXVI-56. His Majesty's Stationery Office. British Library Document Supply Services.

Luce, D. R., ed. 1960. *Developments in Mathematical Psychology: Information, Learning and Tracking*. Free Press of Glencoe.

Lück, H. E., and R. Miller, eds. 2005. *Illustrierte Geschichte der Psychologie*. Beltz Verlag.

Mackworth, N. H. 1944. *Notes on the Clock Test—A New Approach to the Study of Prolonged Perception to Find the Optimum Length of Watch for Radar Operators*. FPRC report no. 586, May. TNA: AIR 57/11.

Mackworth, N. H. 1948–1949. "The Breakdown of Vigilance During Prolonged Visual Search." *Quarterly Journal of Experimental Psychology* 1: 6–21.

Mackworth, N. H. 1950. *Researches on the Measurement of Human Performance*. Medical Research Council, Special Report Series no. 268. His Majesty's Stationery Office.

Macmillan, J. W., R. E. Johnston, and A. C. Williams. 1946. *The Role of Fatigue in Pilot Performance*. National Research Council Committee on Selection and Training of Aircraft Pilots.

Macrakis, K. 2000. "*Surviving the Swastika* Revisited." In *Geschichte der Kaiser-Wilhelm-Gesellschaft*, edited by D. Kaufmann, 586–99. Wallstein Verlag.

Maier, C. S. 1970. "Between Taylorism and Technocracy: European Ideologies and the Vision of Industrial Productivity in the 1920s." *Journal of Contemporary History* 5, no. 2: 27–69.

Malmo, R. B., and W. W. Surwillo. 1960. "Sleep Deprivation: Changes in Performance and Physiological Indicants of Activation." *Psychological Monographs* 74 (whole no. 502).

Mandler, G. 2007. *A History of Modern Experimental Psychology: From James and Wundt to Cognitive Science*. MIT Press.

Mathew, H. C. G., and B. Harrison, eds. 2004. *Oxford Dictionary of National Biography*. Vol. 55. Oxford University Press.

Mathews, G., P. A. Hancock, P. A. Desmond, and C. Neubauer, eds. 2012. *The Handbook of Operator Fatigue*. Ashgate.

McFarland, R. A. 1932. "The Psychological Effects of Oxygen Deprivation (Anoxaemia) on Human Behaviour." *Archives of Psychology* 145.

McFarland, R. A. 1941. "Fatigue in Aircraft Pilots." *New England Journal of Medicine* 225, no. 22: 845–55.

McFarland, R. A. 1953. *Human Factors in Air Transportation: Occupational Health and Safety*. McGraw-Hill.

McGeoch, J., and A. L. Irions. 1952. *The Psychology of Learning*. Longmans, Green.

Melton, A. W., ed. 1947. *Report 4, Apparatus Tests*. Research report, Army Air Forces Aviation Psychology Program. US Government Printing Office.

Melton, A. W. 1957. "Military Psychology in the United States of America." *American Psychologist* 12: 740–46.

Melton, A. W., and G. E. Briggs. 1960. "Engineering Psychology." *Annual Review of Psychology* 11: 71–98.

Metcalf, J. T. 1921. "Cutaneous and Kinesthetic Senses." *Psychological Bulletin* 18, no. 4: 81–202.

Meyer, D. E., and S. Kornblum, eds. 1993. *Attention and Performance*, vol. 14, *Synergies in Experimental Psychology, Artificial Intelligence, and Cognitive Neuroscience*. MIT Press.

Meyer, H. H. 1922. "Oswald Schmiedeberg." *Die Naturwissenschaften* 10, no. 5: 105–7.

Middleton, D., and D. Edwards, eds. 1990. *Collective Remembering*. Sage.

Miller, G. A. 1951. *Language and Communication*. McGraw-Hill.

Miller, G. A. 1953. "What Is Information Measurement?" *American Psychologist* 8: 3–11.

Miller, G. A., E. Galanter, and K. Pribram. 1986. *Plans and the Structure of Behavior*. Adams-Bannister-Cox.

Miller, J. 1972. "The Dog Beneath the Skin." *Listener* 88: 74–76.

Miller, J. 1978. *The Body in Question*. Jonathan Cape.

Mindell, D. A. 2002. *Between Human and Machine: Feedback, Control and Computing Before Cybernetics*. Johns Hopkins University Press.

Mindell, D. A. 2008. *Digital Apollo: Human and Machine in Spaceflight*. MIT Press.

Mitchell, M. J. H., and M. A. Vince. 1951. "The Direction of Movement of Machine Controls." *Quarterly Journal of Experimental Psychology* 3, no. 1: 24–25.

Moede, W. 1930. *Lehrbuch der Psychotechnik, 1 Band*. Verlag von Julius Springer.

Murchison, C., ed. 1936. *A History of Psychology in Autobiography*. Vol. 3. Clark University Press.

Muscio, B. 1922. "Is a Fatigue Test Possible?" *British Journal of Psychology* 12: 31–46.

Myers, C. S. 1911a. *A Text-Book of Experimental Psychology, Part I*. 2nd ed. Cambridge University Press.

Myers, C. S. 1911b. *A Text-Book of Experimental Psychology, Part II: Laboratory Exercises*. 2nd ed. Cambridge University Press.

Myers, C. S. 1920a. *Mind and Work: The Psychological Factors in Industry and Commerce*. University of London Press.

Myers, C. S. 1920b. "The Independence of Psychology." *Discovery* 1: 335–40.

Myers, C. S. 1924a. "The Study of Fatigue." *Journal of Personnel Research* 3, no. 9: 321–34.

Myers, C. S., ed. 1924b. *Proceedings and Papers of the VIIth Congress of Psychology*. Cambridge University Press.

Myers, C. S. 1936. "Charles Samuel Myers." In *A History of Psychology in Autobiography*, vol. 3, edited by C. Murchison, 215–30. Clark University Press.

Myers, C. S. 1940. *Shell Shock in France 1914–18*. Cambridge University Press.

Myers, C. S. 1941. "Obituary Notice: Sir Henry Head, 1861–1940." *British Journal of Psychology* 32: 5–14.

Nature. 1943. "Scientific Research and Development: Research in Wartime." Editorial. *Nature* 151 (February 20): 203–7.

Neisser, U. 1967. *Cognitive Psychology*. Appleton-Century-Crofts.

Neumann, A. 2005. '*Arztum ist immer Kämpftum.' Die Heeressanitätsinspektion und das Ampt, Chef des Wehrmachtsanitätswesens im Zweiten Weltkrieg (1939–1945)*. Droste Verlag.

Neumann, A. 2012. "Das Kaiser-Wilhelm-Institut für Arbeitsphysiologie und der Kampf gegen die Ermüdung." In *Arbeit, Leistung und Ernährung*, edited by T. Plesser and H. Thamer, 171–95. Franz Steiner Verlag.

Ohler, N. 2016. *Blitzed: Drugs in Nazi Germany*. Translated by S. Whiteside. Allen Lane.

Oldfield, R. C. 1937. "Some Recent Experiments Bearing on 'Internal Inhibition.'" *British Journal of Psychology* 28: 28–42.

Oldfield, R. C., and O. L. Zangwill. 1943. "Head's Concept of the Schema and Its Application in Contemporary British Psychology, Part III, Bartlett's Theory of Memory." *British Journal of Psychology* 33: 113–29.

Osgood, C. E. 1953. *Method and Theory in Experimental Psychology*. Oxford University Press.

Ospovat, D. 1978. "Perfect Adaptation and Teleological Explanation: Approaches to the Problem of the History of Life in the Mid-Nineteenth Century." *Studies in History of Biology* 2: 33–66.

Owens, L. 1994. "The Counterproductive Management of Science in the Second World War: Vannevar Bush and the Office of Scientific Research and Development." *Business History Review* 68: 515–76.

Parade, G. W. 1941. "Ermudung." *Deutsche medizinische Wochenschrift* 67, no. 49: 1333–37.

Parker, G. H. 1919. *The Elementary Nervous System*. Lippincott.

Parry, J. B. 1947. "The Selection and Classification of R.A.F. Air Crew." *Occupational Psychology* 21: 158–69.

Parsons, J. H. 1927. *An Introduction to the Theory of Perception*. Cambridge University Press.

Pavlov, I. P. 1927. *Conditioned Reflexes: An Investigation of the Physiological Activity of the Cerebral Cortex*. Translated and edited by G. V. Anrep. Oxford University Press.

Pavlov, I. P. 1928. *Lectures on Conditioned Reflexes*. Translated and edited by W. Horsley Gantt. International Publishers.

Petersen, O. 1939. "Ueber die Sinnfälligkeit von Blindflugmeßgeräten." *Industrielle Psychotechnik* 16, no. 8: 225–39.

Phillips, C. G. 1974. "Francis Martin Rouse Walshe 1885–1973." *Biographical Memoirs of Fellows of the Royal Society* 20: 456–81.

Pitman, R. J. G. 1966. *Automatic Control Systems Explained*. Macmillan.

Platt, B. S. 1956. "Sir Edward Mellanby, G.B.E., K.C.B., M.D., F.R.C.P., F.R.S. (1884–1955): The Man, Research Worker, and Statesman." *Annual Review of Biochemistry* 25: 1–28.

Plesser, T., and H. Thamer, eds. 2012. *Arbeit, Leistung und Ernährung: Vom Kaiser-Wilhelm-Institut für Arbeitsphysiologie in Berlin zum Max-Planck-Institut für molekulare Physiologie und Leibniz Institut für Arbeitsforschung in Dortmund*. Franz Steiner Verlag.

Poulton, E. C. 1950. "Perceptual Anticipation and Reaction Time." *Quarterly Journal of Experimental Psychology* 2: 90–112.

Poulton, E. C. 1952a. "Perceptual Anticipation in Tracking with Two-Pointer and One-Pointer Displays." *British Journal of Psychology* 43: 222–29.

Poulton, E. C. 1952b. "The Basis of Perceptual Anticipation in Tracking." *British Journal of Psychology* 43: 295–302.

Poulton, E. C. 1957a. "On Prediction in Skilled Movements." *Psychological Bulletin* 54: 467–79.

Poulton, E. C. 1957b. "On the Stimulus and Response in Pursuit Tracking." *Journal of Experimental Psychology* 53, no. 3: 89–194.

Poulton, E. C. 1957c. "Learning the Statistical Properties of the Input in Pursuit Tracking." *Journal of Experimental Psychology* 54, no. 1: 28–32.

Poulton, E. C. 1971. "Skilled Performance and Stress." In *Psychology at Work*, edited by P. B. Warr, 55–75. Penguin Books.

Poulton, E. C. 1974. *Tracking Skill and Manual Control*. Academic Press.

Poulton, E. C. 1976. "Continuous Noise Interferes with Work by Masking Auditory Feedback and Inner Speech." *Applied Ergonomics* 7: 79–84.

Poulton, E. C., and R. L. Gregory. 1952. "Blinking During Visual Tracking." *Quarterly Journal of Experimental Psychology* 4: 57–65.

Purcell, C. 1979. "Science Agencies in World War II: The OSRD and Its Challengers." In *The Sciences in the American Context*, edited by N. Reingold, 359–78. Smithsonian Institution Press.

Quastler, H., ed. 1955. *Information Theory in Psychology: Problems and Methods*. Free Press.

Rabbitt, P. M. A, and E. A. Maylor. 1991. "Investigating Models of Human Performance." *British Journal of Psychology* 82: 259–90.

Rabinbach, A. 1990. *The Human Motor: Energy, Fatigue and the Origins of Modernity*. University of California Press.

Rabinbach, A. 2018. *The Eclipse of the Utopias of Labor*. Fordham University Press.

Raehlmann, I. 2005. *Arbeitswissenschaft im Nationalsozialismus: Eine wissenschafts-Soziologische Analyse*. Verlag für Sozialwissenschaften.

Rawdon-Smith, A. F. 1934. "Auditory Fatigue." *British Journal of Psychology* 25: 77–85.

Rawdon-Smith, A. F. 1935. "Experimental Deafness." *Nature* 136 (July 6): 32.

Rawdon-Smith, A. F. 1936. "Experimental Deafness: Further Data upon the Phenomenon of So-Called Auditory Fatigue." *British Journal of Psychology* 26: 233–44.

Reason, J. 1975. "How Did I Come to Do That?" *New Behaviour* 24 (April): 10–13.

Reason, J., and K. Mycielska. 1982. *Absent Minded? The Psychology of Mental Lapses and Everyday Errors*. Prentice-Hall.

Reid, C. 1928. "The Mechanism of Voluntary Muscular Fatigue." *Quarterly Journal of Experimental Physiology* 19: 17–48.

Reid, D. D. 1942. *The Influence of Psychological Disorders on Efficiency in Operational Flying*. FPRC report no. 508, September. TNA: AIR 57/10.

Rein, H. 1938. "Die physiologische Forschung im Dienste neuzeitlicher militärärztlicher Aufgaben inbesondere bei der Luftwaffe." *Der Deutsche Militärärzt* 3, no. 3: 124–29.

Rein, H. 1941. "Blut-Sauerstoff und Blutverteilungs-Regelung." *Zeitschrift für Kreislaufforschung* 33, no. 8: 241–58.

Reingold, N., ed. 1979. *The Sciences in the American Context: New Perspectives*. Smithsonian Institution Press.

Rivers, W. H. R. 1924. *Instinct and the Unconscious: A Contribution to a Biological Theory of the Psycho-Neuroses*. 2nd ed. Cambridge University Press.

Rivers, W. H. R., and H. Head. 1908. "A Human Experiment on Nerve Division." *Brain*, part 2 (November): 323–450.

Rivers, W. H. R., and E. Kraepelin. 1896. "Ueber Ermüdung und Erholung." *Psychologische Arbeiten* 1: 627–78.

Rivers, W. H. R., and T. S. Rippon. 1920. "Mental Aptitude for Aviation." In Medical Research Council, *The Medical Problems of Flying*. His Majesty's Stationery Office.

Rolfe, J. 1995. "The Flight Simulators: Two Cambridge Inventors." *Cambridge: The Magazine of the Cambridge Society* 36: 56–62.

Rolfe, J. 1996. "Craik and the Cambridge Cockpit." *Psychologist* 9, no. 2: 69–71.

Rosenzweig, M. R., and G. Stone. 1948. "Wartime Research in Psycho-Acoustics." *Review of Educational Research* 18, no. 6: 642–45.

Roskill, S. W. 1961. *The War at Sea*, vol. 3, part 2, *1939–1945*. His Majesty's Stationery Office.

Roth, K. H. 2000. "Strukturen, Paradigmen und Mentalität in der luftfahrtmedizinischen Forschung des 'Dritten Reichs' 1933 bis 1941: Der Weg ins Konzentrationslager Dachau." *Zeitschrift für Sozialgeschichte des 20. und 21. Jahrhunderts* 1: 49–77.

Roth, K. H. 2002. "Tödliche Höhen: Die Unterdruckkammer-Experimente im Konzentrationslager Dachau und ihre Bedeutung für die luftfahrtmedizinische Forschung des 'Dritte Reichs.'" In *Vernichten und Heilen*, edited by A. Ebbinghaus and K. Dörner, 110–51. Aufbau-Verlag.

Roughton, F. J. W. 1949. "Joseph Barcroft 1872–1947." *Obituary Notices of Fellows of the Royal Society* 6, no. 18: 315–45.

Ruff, S., and H. Strughold. 1944. *Grundriss der Luftfahrtmedizin*. Johann Ambrosius Barth.

Rürup, R., ed. 1979. *Wissenschaft und Gesselschaft: Beiträge zur Geschichte der Technischen Universität Berlin, 1879–1979*. Springer.

Russell, O. 1993. "Derek Russell Davis." *Psychiatric Bulletin* 17: 314.

Russell Davis, D. 1946a. *A Note on the Causation of Accidents*. FPRC report no. 652. TNA: AIR 57/16.

Russell Davis, D. 1946b. *Analeptics and Fatigue*. Applied Psychology Unit, APU report no. 45. TNA: FD 1/4146.

Russell Davis, D. 1946c. *German Applied Psychology, London, British Intelligence Objectives Sub-Committee (BIOS)*. Final report no. 970. His Majesty's Stationery Office.

Russell Davis, D. 1947. "Post-Mortem on German Applied Psychology." *Occupational Psychology* 21, no. 3: 105–10.

Russell Davis, D. 1948. *Pilot Error: Some Laboratory Experiments*. Air Ministry Air Publication no. 3139A. His Majesty's Stationery Office.

Russell Davis, D. 1951. "The Wonderful Freshness of Memories in Hysteria." *British Journal of Medical Psychology* 24, no. 1: 64–68.

Russell Davis, D. 1952. "Experimental Method in Clinical Psychology." *British Journal of Medical Psychology* 25, no. 1: 21–25.

Russell Davis, D. 1957. *An Introduction to Psychopathology*. Oxford University Press.

Russell Davis, D., and J. H. Cullen. 1958. "Disorganization of Perception in Neurosis and Psychosis." *American Journal of Psychology* 71, no. 1: 229–237.

Ryle, G. 1949. *The Concept of Mind*. Hutchinson.

Saito, A., ed. 2000. *Bartlett, Culture and Cognition*. Psychology Press.

Saville-Sneath, R. A. 1941. *Aircraft Recognition*. Penguin Books.

Schaffer, S. 1994. *From Physics to Anthropology—and Back Again*. Prickly Pear Press.

Schmaltz, F. 2012. "Wehrphysiologische Forschung für die Reichswehr am Kaiser-Wilhelm-Institut für Arbeitsphysiologie." In *Arbeit, Leistung und Ernährung*, edited by T. Plesser and H. Thamer, 197–235. Franz Steiner Verlag.

Schmidt, H. 1938. "Funktions- und Leistungsanalyse des Höhenfliegers nach berufswichtigen Gesichtspunkten." PhD diss., Technische Hochschule, Berlin.

Schmidt, I. 1938. *Bibliographie der Luftfahrtmedizin*. Springer.

Schmidt, R. A. 2011. "Jack Adams, a Giant of Motor Behavior, Has Died." *Journal of Motor Behavior* 43, no. 1: 83–84.

Schmiedeberg, O. 1883. *Grundriss der Arzneimittellehre*. F. C. W. Vogel.

Sharpey-Schafer, E. 1929. "The Effects of De-Nervation of a Cutaneous Area." *Quarterly Journal of Experimental Physiology* 19: 85–107.

Shephard, Ben. 2000. *A War of Nerves*. Jonathan Cape.

Shepherd, M. 1995. "Two Faces of Emil Kraepelin." *British Journal of Psychiatry* 167: 174–83.

Sherwood, S. L., ed. 1966. *The Nature of Psychology: A Selection of Papers, Essays and Other Writings by the Late Kenneth J. W. Craik*. Cambridge University Press.

Siddall, G. J., and D. M. Anderson. 1955. "Fatigue During Long Performance on a Simple Compensatory Tracking Task." *Quarterly Journal of Experimental Psychology* 7: 159–65.

Simonson, E. 1935. "Der heutige Stand der Theorie der Ermüdung." *Der Ergebnisse der Physiologie* 37: 299–365.

Skidelsky, R. 1983, 1992, 2000. *John Maynard Keynes*. 3 vols. Macmillan.

Slobodin, R. 1997. *W. H. R. Rivers: Pioneer Anthropologist, Psychiatrist of "The Ghost Road."* Sutton Publishing Limited.

Smith, R. 1992. *Inhibition: History and Meaning in the Science of Mind and Brain*. Free Association Books.

Snow, C. P. 1961. *Science and Government*. Oxford University Press.

Sperandio, J.-C. 1978. "The Regulation of Working Methods as a Function of Workload Among Air Traffic Controllers." *Ergonomics* 21: 195–202.

Spur, G. 2008. *Industrielle Psychotechnik—Walther Moede: Eine biographische Dokumentation*. Carl Hanser Verlag.

Stevens, S. S., ed. 1951. *Handbook of Experimental Psychology*. John Wiley.

Still, A., and A. Costall, eds. 1991. *Against Cognitivism: Alternative Foundations for Cognitive Psychology*. Harvester.

Stoll, F., ed. 1981. *Die Psychologie des 20. Jahrhunderts, Bd. 13*. Kindler.

Stout, G. F. 1902. *Analytic Psychology*. Swan Sonnenschein.

Straub, J., P. Chakklarath, and G. Rebane, eds. 2019. *Kulturpsychologie in interdisziplinärer Perspektive*. Psychosozial Verlag.

Strughold, H. 1950. "Development, Organisation, and Experience of Aviation Medicine in Germany During World War II." In Surgeon General, *German Aviation Medicine*, 12–51. Department of the Air Force.

Surgeon General, US Air Force. 1950. *German Aviation Medicine: World War II*. Department of the Air Force.

Swets, J. A. 1961. "Is There a Sensory Threshold?" *Science* n.s. 134: 168–77.

Tanner, W. P., and J. A. Swets, 1954. "A Decision-Making Theory of Visual Detection." *Psychological Review* 61, no. 6: 401–9.

Taylor, F. V. 1947. "Psychology at the Naval Research Laboratory." *American Psychologist* 2: 87–92.

Taylor, F. V. 1957. "Psychology and the Design of Machines." *American Psychologist* 12: 249–58.

Taylor, F. V. 1960. "Four Basic Ideas in Engineering Psychology." *American Psychologist* 15, no. 10: 643–49.

Taylor, F. V. 1963. "Human Engineering and Psychology." In *Psychology: A Study of a Science*, vol. 5, edited by S. Koch, 831–907. McGraw-Hill.

Taylor, F. V., and H. P. Birmingham. 1956. "Simplifying the Pilot's Task Through Display Quickening." *Journal of Aviation Medicine* 27: 27–31.

Taylor, F. V., and H. P. Birmingham. 1959. "That Confounded System Performance Measure—A Demonstration." *Psychological Review* 66, no. 3: 178–82.

Taylor, J., ed. 1931–32. *Selected Writings of John Hughlings Jackson*. 2 vols. Hodder and Stoughton.

Terraine, J. 1985. *The Right of the Line: The Royal Air Force in the European War 1939–1945*. Hodder and Stoughton.

Thomson, W., and P. G. Tait. 1896. *Treatise on Natural Philosophy*. Vol. 1. Cambridge University Press.

Thorndike, E. L. 1912. "The Curve of Work." *Psychological Review* 19, no. 3: 165–94.

Thorndike, E. L. 1923. *Mental Work and Fatigue and Individual Differences and the Causes*. New York Teachers College, Columbia University.

Thorndike, E. L., and I. Lorge. 1944. *The Teachers Word Book of 30,000 Words*. Bureau of Publications, New York Teachers College, Columbia University.

Thorndike, R. L. 1951. *The Human Factor in Accidents with Special Reference to Aircraft Accidents*. Project no. 21-30-001, report no. 1. USAF School of Aviation Medicine, Randolph Field, Texas.

Thorpe, W. H., and O. L. Zangwill, eds. 1963. *Current Problems in Animal Behaviour*. Cambridge University Press.

Treisman, A. M. 1964. "Verbal Cues, Language, and Meaning in Selective Attention." *American Journal of Psychology* 7: 206–19.

Trotter, W. 1908. "Herd Instinct and Its Bearing on the Psychology of Civilised Man." *Sociological Review* 1: 227–48.

Trotter, W. 1909. "Sociological Applications of the Psychology of the Herd Instinct." *Sociological Review* 2: 36–54.

Trotter, W. 1916. *Instincts of the Herd in Peace and War*. Fisher Unwin.

Trotter, W. 1923. "The Scalp, Skull, and Brain." In *A System of Surgery*, edited by C. C. Choyce, vol. 3, 436–565. Cassell and Son.

Trotter, W., and H. M. Davies. 1909. "Experimental Studies in the Innervation of the Skin." *Journal of Physiology* 38: 134–246.

Trotter, W., and H. M. Davies. 1913. "The Peculiarities of Sensibility Found in Cutaneous Areas Supplied by Regenerating Nerves." *Journal für Psychologie und Neurologie* 20, no. 2: 102–51.

Turner, F. M. 1974. *Between Science and Religion: The Reaction to Scientific Naturalism in Late Victorian England*. Yale University Press.

Unger, F. 1994. "Der Einsatz von Pervitin im deutschen Heer im 2. Weltkrieg und dessen wissenschaftliche Vorbereitung seit 1937." *Wehrmedizinische Monatsschrift* 38: 374–81.

van Weelden, E., M. Alimardini, T. J. Wiltshire, and M. M. Louwerse. 2022. "Aviation and Neurophysiology: A Systematic Review." *Applied Ergonomics* 105: 103838. https://doi.org/10.1016/j.apergo.2022.103838.

Vernon, H. M. 1916. *Output in Relation to Hours of Work*. Health of Munition Workers Committee, Ministry of Munitions. His Majesty's Stationery Office.

Vernon, M. D. 1941a. *The Ability to Distinguish Dimly Illuminated Aeroplane Silhouettes*. FPRC report no. 247. TNA: AIR 57/4.

Vernon, M. D. 1941b. *Report on Learning to Identify Aeroplanes from Photographs*. FPRC report no. 382. TNA: AIR 57/7.

Vernon, P. E., and J. B. Parry. 1949. *Personnel Selection in the British Forces*. University of London Press.

Vince, M. A. 1944. *Direction of Movement of Machine Controls*. FPRC report no. 637(a). TNA: AIR 57/15.

Vince, M. A. 1946a. *The Psychological Effect of a Non-Linear Relation Between Control and Display*. FPRC report no. 637(b). TNA: AIR 57/15.

Vince, M. A. 1946b. *The Intermittency of Control Movements and the Psychological Refractory Period*. FPRC report no. 637c. TNA: AIR 57/15.

Vince, M. A. 1948. "The Intermittency of Control Movements and the Psychological Refractory Period." *British Journal of Psychology* 38, no. 3: 149–57.

Vince, M. A. 1948–1949. "Corrective Movements in a Pursuit Task." *Quarterly Journal of Experimental Psychology* 1: 85–103.

Vince, M. A. 1949. "Rapid Response Sequences and the Psychological Refractory Period." *British Journal of Psychology* 40, no. 1: 23–40.

Vince, M. A. 1996. "Recollections of Kenneth Craik." *Psychologist* 9, no. 2: 67–68.

Vincenti, W. G. 1990. *What Engineers Know and How They Know It.* Johns Hopkins University Press.

Walker, M., K. Orth, U. Herbert, and R. vom Bruch, eds. 2013. *The German Research Foundation 1920–1970: Funding Poised Between Science and Politics.* Franz Steiner Verlag.

Walshe, F. M. R. 1921. "The Role of Anoxaemia in the Production of Symptoms in Disease and Injury of the Nervous System." *Medical Science Abstracts and Reviews* 4, no. 5: 427–34.

Walshe, F. M. R. 1922. "Review of W. H. R. Rivers, Instinct and the Unconscious, 1922." *Medical Science Abstracts and Reviews* 6, no. 3: 216–22.

Walshe, F. M. R. 1942. "The Anatomy and Physiology of Cutaneous Sensibility: A Critical Review." *Brain* 65: 48–112.

Walshe, F. M. R. 1961. "Contributions of John Hughlings Jackson to Neurology." *Archives of Neurology* 5, no. 2: 119–31.

Wapshott, N. 2011. *Keynes Hayek: The Clash that Defined Modern Economics.* W. W. Norton.

War Office. 1922. *Theory and Use of Anti-Aircraft Sound-Locators.* His Majesty's Stationery Office.

Ward, J. 1920. *Psychological Principles.* 2nd ed. Cambridge University Press.

Warr, P. B., ed. 1971. *Psychology at Work.* Penguin Books.

Watson, J. B. 1929. *Psychology from the Standpoint of a Behaviorist.* Lippincott.

Weber, M. M., W. Burgmair, and E. J. Engstrom. 2006. "Zwischen klinischen Krankheitsbildern und 'psychischer Volkshygiene.'" *Deutsches Arzteblatt* 103, no. 41: A2685–90.

Weddell, G., L. Guttmann, and E. Guttmann. 1941. "The Local Extension of Nerve Fibres into Denervated Areas of Skin." *Journal of Neurology and Psychiatry* 4, no. 2: 206–25.

Weiß, H., ed. 1998. *Biographisches Lexikon zum Dritten Reich.* S. Fischer Verlag.

Welford, A.T. 1970. "Sir Frederic Charles Bartlett (1886–1969)." *Yearbook of the American Philosophical Society*: 109–14.

Welford, A. T. 1968. *Fundamentals of Skill.* Methuen.

Welford, A. T. 1975. "Obituary: William Edward Hick (1912–1974)." *Ergonomics* 18, no. 2: 251–52.

Whittenburg, J. A., S. Ross, and T. G. Andrews. 1956. "Sustained Perceptual Efficiency as Measured by the Mackworth Clock Test." *Perceptual and Motor Skills* 6: 109–16.

Whittingham, H. E. 1939a. Proposed Research at Cambridge. FPRC report no. 58, September. TNA: AIR 57/1.

Whittingham, H. E. 1939b. *A Progress Report on Pre-Selection of Flying Personnel.* FPRC report no. 71, December. TNA: AIR 57/1.

Whittingham, H. E. 1940. *Report to the Secretary of State for Air on the Activities of the FPRC.* FPRC report no. 144, April 27. TNA: AIR 57/2.

Whittingham, H. E. 1941a. *Extracts from German Literature.* FPRC report no. 321 and 321(a), (b), (c), (d), July 1941. TNA: AIR 57/5.

Whittingham, H. E. 1941b. *Medical Services in the German Air Force.* FPRC report no. 384. TNA: AIR 57/7.

Whittle, P. 2000. "W. H. R. Rivers and the Early-History of Psychology at Cambridge." In *Bartlett, Culture and Cognition,* edited by A. Saito, 21–35. Psychology Press.

Wickens, C. D., and R. C. Williges. 2011. "Jack A. Adams (1922–2010)." *American Psychologist* 66, no. 7: 638.

Wiener, E. L. 1964. "Multiple Channel Monitoring." *Ergonomics* 7: 453–60.

Wilkinson, R. T. 1961. "Interaction of Sleep Loss with Knowledge of Results, Repeated Testing and Individual Differences." *Journal of Experimental Psychology* 62, no. 3: 263–71.

Wilkinson, R. T. 1963. "Interaction of Noise with Knowledge of Results and Sleep Deprivation." *Journal of Experimental Psychology* 66: 332–37.

Wilkinson, R. T. 1964. "Effects of up to 60 Hours' Sleep Deprivation on Different Types of Work." *Ergonomics* 7, no. 2: 175–86.

Wilkinson, R. T., and W. P. Colquhoun. 1968. "Interaction of Alcohol with Incentives and Sleep Deprivation." *Journal of Experimental Psychology* 76, no. 4: 523–629.

Williams, D. 1945. *Neuropsychiatric Organisation in the German Air Force.* Combined Intelligence Objectives Sub-Committee (CIOS) report no. XXVI-81. His Majesty's Stationery Office. British Library Document Supply Services.

Williams, G. O. 1942. *Fatigue and Flying Accidents.* FPRC report no. 492, September. TNA: AIR 579.

Wilmot, C. 1952. *The Struggle for Europe.* Collins.

Wolfe, T. 1980. *The Right Stuff.* Bantam Books.

Wood, D., and D. Dempster. 1969. *The Narrow Margin: The Battle of Britain and the Rise of Air Power 1930–1940.* Arrow Books.

Woodward, W. R. 1990. "Stevens, Stanley Smith." In *Dictionary of Scientific Biography,* vol. 18, no. 2, edited by F. M. Holmes, 869–75. Scribner's.

Woodward, W. R., and M. G. Ash, eds. 1982. *The Problematic Science: Psychology in Nineteenth-Century Thought.* Praeger.

Woodworth, R. S. 1938. *Experimental Psychology.* Holt.

Woodworth, R. S., and H. Schlosberg. 1954. *Experimental Psychology.* Holt.

Wu, E. Q., X. Y. Peng, C. Z. Zhang, J. X. Lin, and R. S. F. Sheng. 2019. "Pilot's Fatigue Status Recognition Using Deep Contractive Autoencoder Network." *IEEE Transactions on Instrumentation and Measurement* 68, no. 10: 3907–19.

Yerkes, R. M., and J. D. Dodson. 1908. "The Relation of Strength of Stimulus to Rapidity of Habit-Formulation." *Journal of Comparative Neurology and Psychology* 18, no. 5: 458–82.

Young, Allan. 1995. *The Harmony of Illusions: Inventing Post-Traumatic Stress Disorder.* Princeton University Press.

Young, R. M. 1970. *Mind, Brain and Adaptation in the Nineteenth Century: Cerebral Localization and Its Biological Context from Gall to Ferrier.* Clarendon Press.

Zangwill, O. L. 1950. *An Introduction to Modern Psychology.* 2nd ed. 1962. Methuen.

Zangwill, O. L. 1980. "Kenneth Craik: The Man and His Work." *British Journal of Psychology* 11, no. 1: 1–16.

Zülch, K. Z. 1973. "Otfrid Foerster: 1873–1941." *Zeitschrift für Neurologie* 205: 177–84.

Zusne, L., ed. 1984. *Biographical Dictionary of Psychology.* Greenwood Press.

Index

www.ingramcontent.com/pod-product-compliance
Lightning Source LLC
Chambersburg PA
CBHW022137020426
42334CB00015B/933